Signal Digitization and Reconstruction in Digital Radios

For a listing of recent titles in the
Artech House *Signal Processing Library*,
turn to the back of this book.

Signal Digitization and Reconstruction in Digital Radios

Yefim S. Poberezhskiy
Gennady Y. Poberezhskiy

ARTECH
HOUSE

BOSTON | LONDON
artechhouse.com

Library of Congress Cataloging-in-Publication Data
A catalog record for this book is available from the U.S. Library of Congress

British Library Cataloguing in Publication Data
A catalog record for this book is available from the British Library.

ISBN 13: 978-1-63081-380-2

Cover design by John Gomes

© 2019 Artech House
685 Canton Street
Norwood, MA

All rights reserved. Printed and bound in the United States of America. No part of this book may be reproduced or utilized in any form or by any means, electronic or mechanical, including photocopying, recording, or by any information storage and retrieval system, without permission in writing from the publisher.

All terms mentioned in this book that are known to be trademarks or service marks have been appropriately capitalized. Artech House cannot attest to the accuracy of this information. Use of a term in this book should not be regarded as affecting the validity of any trademark or service mark.

10 9 8 7 6 5 4 3 2 1

In memory of S. I. Poberezhskiy and E. D. Kravets

To Galina Poberezhskiy

To Katherine, Matthew, and Simone

Contents

Preface   xiii

CHAPTER 1
Signals and Waveforms   1

1.1   Overview   1
1.2   Signals and Their Processing   2
      1.2.1   Analog, Discrete-Time, and Digital Signals   2
      1.2.2   Deterministic and Stochastic Signals   5
      1.2.3   Basic Operations on Signals   13
1.3   Expansions of Signals   16
      1.3.1   Orthogonal Expansions   16
      1.3.2   Trigonometric and Exponential Fourier Series   18
      1.3.3   Fourier Transform and Its Properties   21
      1.3.4   Spectral Distribution of Signal Energy and Power   25
      1.3.5   Transmission of Signals Through LTI Systems   28
1.4   Baseband and Bandpass Signals   29
      1.4.1   Baseband Signals and Modulation   29
      1.4.2   Bandpass Signals and Their Complex-Valued Equivalents   31
      1.4.3   Bandwidths of Signals and Circuits   35
1.5   Summary   39
      References   40

CHAPTER 2
Radio Systems   43

2.1   Overview   43
2.2   Radio Systems and Radio Spectrum   44
      2.2.1   Diversity of Radio Systems   44
      2.2.2   RF Spectrum and Its Utilization   45
2.3   Radio Communication Systems   48
      2.3.1   General   48
      2.3.2   Communication Txs and Rxs   51
      2.3.3   Channel Coding, Modulation, and Spreading   55
2.4   Other Radio Systems   59
      2.4.1   Broadcasting Systems   59
      2.4.2   Radio Navigation and Positioning Systems   61
      2.4.3   Radio Methods in Positioning and Geolocation   65

		2.4.4 Radar and EW Systems	68
2.5	Summary		70
	References		72

CHAPTER 3
Digital Transmitters 75

3.1	Overview		75
3.2	Digital Tx Basics		76
	3.2.1	Txs of Different Categories of Digital Radios	76
	3.2.2	Architecture of a Digital Tx	77
	3.2.3	Direct Digital Synthesis	80
3.3	D&R in a Digital Tx		86
	3.3.1	Digitization of TDP Input Signals	86
	3.3.2	Reconstruction of TDP Output Signals	89
	3.3.3	Comparison of Reconstruction Techniques and Conversion Block Architectures	96
3.4	Power Utilization Improvement in Txs		100
	3.4.1	Power Utilization in Txs with Energy-Efficient Modulation	100
	3.4.2	AQ-DBPSK Modulation	102
	3.4.3	Power Utilization in Txs with Bandwidth-Efficient Modulation	107
3.5	Summary		108
	References		109

CHAPTER 4
Digital Receivers 113

4.1	Overview		113
4.2	Digital Rx Basics		114
	4.2.1	First Steps of Digital Radio Development	114
	4.2.2	Main Characteristics of Rxs	117
	4.2.3	Digital Rxs and Txs	122
4.3	Dynamic Range of a Digital Rx		126
	4.3.1	Factors Limiting Rx Dynamic Range	126
	4.3.2	Intermodulation	128
	4.3.3	Required Dynamic Range of an HF Rx	136
4.4	Digitization in a Digital Rx		140
	4.4.1	Baseband Digitization	140
	4.4.2	Bandpass Digitization	142
	4.4.3	Comparison of Digitization Techniques and Architectures of AMFs	145
4.5	Demodulation of Energy-Efficient Signals		148
	4.5.1	Demodulation of Differential Quadrature Phase-Shift Keying Signals with DS Spreading	148
	4.5.2	Demodulation of AQ-DBPSK Signals	151
4.6	Summary		157
	References		159

CHAPTER 5
Sampling Theory Fundamentals   163

5.1 Overview   163
5.2 S&I from a Historical Perspective   164
    5.2.1 Need for S&I at the Dawn of Electrical Communications   164
    5.2.2 Discovery of Classical Sampling Theorem   165
    5.2.3 Sampling Theory After Shannon   167
5.3 Uniform Sampling Theorem for Baseband Signals   169
    5.3.1 Sampling Theorem and Its Constructive Nature   169
    5.3.2 Interpretations of Sampling Theorem   177
    5.3.3 Baseband S&I Corresponding to Indirect Interpretation   184
5.4 Uniform Sampling Theorem for Bandpass Signals   187
    5.4.1 Baseband S&I of Bandpass Signals   187
    5.4.2 Bandpass S&I of Bandpass Signals   189
    5.4.3 Comparison of Baseband and Bandpass S&I of Bandpass Signals   193
5.5 Summary   194
    References   195

CHAPTER 6
Realization of S&I in Digital Radios   199

6.1 Overview   199
6.2 S&I Based on the Sampling Theorem's Indirect Interpretation   200
    6.2.1 Sampling Based on the Indirect Interpretation   200
    6.2.2 Interpolation Based on the Indirect Interpretation   205
6.3 S&I Based on the Sampling Theorem's Hybrid Interpretation   207
    6.3.1 Sampling Based on the Hybrid Interpretation   207
    6.3.2 Interpolation Based on the Hybrid Interpretation   212
6.4 S&I Based on the Sampling Theorem's Direct Interpretation   215
    6.4.1 Sampling Based on the Direct Interpretation   215
    6.4.2 Interpolation Based on the Direct Interpretation   219
6.5 Channel Mismatch Mitigation   222
    6.5.1 Approaches to the Problem   222
    6.5.2 Separation of Signal and Error Spectra   222
    6.5.3 Channel Mismatch Compensation   224
6.6 Selection and Implementation of Weight Functions   226
    6.6.1 Theoretical Basis   226
    6.6.2 B-Spline-Based Weight Functions   229
    6.6.3 Additional Remarks on Weight Function Implementation   233
6.7 Need for Hybrid and Direct Interpretations   234
    6.7.1 Evaluation of Hybrid and Direct Interpretations' Advantages   234
    6.7.2 Two-Stage Spatial Suppression of ISs   236
    6.7.3 Virtual-Antenna-Motion-Based Spatial Suppression of ISs   238
6.8 Summary   244
    References   246

CHAPTER 7
Improving Resolution of Quantization — 249

7.1	Overview	249
7.2	Conventional Quantization	249
	7.2.1 Quantization of Rx Input Signals	249
	7.2.2 Quantization of Tx Input Signals	254
7.3	Joint Quantization of Samples	256
	7.3.1 Principles of Joint Quantization	256
	7.3.2 Design Considerations	262
7.4	Compressive Quantization of Images	264
	7.4.1 Basic Principles	264
	7.4.2 Design Considerations	267
	7.4.3 Assessment of Benefits	270
7.5	Summary	275
	References	276

APPENDIX A
Functions Used in the Book — 279

A.1	Rectangular and Related Functions	279
A.2	Delta Function	281
A.3	B-Splines	285

APPENDIX B
Sampling Rate Conversion in Digital Radios — 287

B.1	Downsampling by an Integer Factor	287
B.2	Upsampling by an Integer Factor	287
B.3	Sampling Rate Conversion by a Noninteger Factor	288
B.4	Optimization of Sampling Rate Conversion	290
B.5	Generalization	293

APPENDIX C
On the Use of Central Limit Theorem — 295

C.1	Paradox Statement	295
C.2	Paradox Resolution	296
C.3	Discussion	297

APPENDIX D
Sampling Theorem for Bandlimited Signals — 299

D.1	Sampling Theorem for Baseband Signals	299
	D.1.1 Theorem	299
	D.1.2 Proof	300
	D.1.3 Discussion	300
D.2	Sampling Theorem for Bandpass Signals	300
	D.2.1 Sampling of Bandpass Signals Represented by $I(t)$ and $Q(t)$	301

   D.2.2 Sampling of Bandpass Signals Represented by $U(t)$ and $\theta(t)$ 302
   D.2.3 Sampling of Bandpass Signals' Instantaneous Values   303

List of Acronyms   305
About the Authors   311
Index   313

Preface

At present, most signal processing is performed in the digital domain. Digitization and reconstruction (D&R) circuits form the interfaces between digital signal processing units and the analog world. These interfaces significantly influence the overall quality, effectiveness, and efficiency of processing. The D&R circuits in digital radios convert the largest variety of signals that can be baseband and/or bandpass, real-valued and/or complex-valued, and narrowband and/or wideband. The signals can correspond to voice, music, images, results of measurements, sensible transmissions, and emissions of natural sources. Therefore, digital radios are a perfect case study for investigating D&R of analog signals in general.

This book provides information on D&R in digital radios, analyzes D&R techniques in detail, and outlines other signal processing operations performed in the radios, demonstrating their interdependence with D&R. However, the book's main objective is introducing new theoretically sound concepts and approaches that enable radical improvement of D&R circuits' characteristics. Despite the focus on digital radios, many results presented in the book are also applicable to general-purpose analog-to-digital converters (A/Ds) and digital-to-analog converters (D/As) as well as to D&R circuits of other applications.

Chapter 1 and the Appendixes are refreshers on the theoretical information repeatedly used in the book (readers' familiarity with signal processing approximately at the undergraduate electrical engineering level is presumed). They introduce notions and functions important for other chapters (orthogonal basis, complex-valued equivalents of bandpass signals, delta functions, and B-splines), contain some explanations, and provide initial information on modulation, frequency conversion, filtering, upsampling and downsampling, and several other operations. This material can also be used as a concise reference source on the topics covered there. Some clarifications presented in Chapter 1 and the Appendixes may be new even for very knowledgeable readers (e.g., the paradox related to the central limit theorem in Appendix C and the description of typical and atypical random sequences in Section 1.2.2).

The main objectives of Chapter 2 are to demonstrate the similarity of requirements for D&R in different radio systems with comparable bandwidths and similar radio frequency (RF) environments and to substantiate focusing on the D&R techniques used in digital communication radios. Simultaneously, the chapter outlines the division of the RF spectrum into frequency bands, radio wave propagation modes in these bands, and the spectrum utilization by various RF systems. It also provides concise information on communication, broadcasting, navigation, radar, and some other systems. For communication systems, Chapter 2 outlines the principles of channel coding, modulation, and spreading, describes high-level structures of digital receivers and transmitters, as well as introduces the notions of energy-efficient and bandwidth-efficient modulations.

In Chapters 3 and 4, the studies of digital transmitters and receivers, respectively, differ from those in other publications by the focus on D&R procedures. Conventional realization of D&R in digital radios and factors that determine their complexity are described in detail, whereas other operations are examined mostly from the standpoint of their connection to D&R. The relation between power utilization and complexity of reconstruction in transmitters is discussed. Several traditional energy-efficient low crest factor modulation and spreading techniques as well as approaches to effective power utilization and signal reconstruction simplification in transmitters with bandwidth-efficient modulations are only outlined. In contrast, alternating quadratures differential binary phase-shift keying (AQ-DBPSK) modulation, which is not described in other books, is thoroughly explained. The early history of digital radios, presented in Section 4.2.1, shows that the key decisions made by the late 1980s still influence the current D&R techniques. It is noted that although most of them were correct, the choice between sample-and hold amplifiers (SHAs) and track-and-hold amplifiers (THAs) in favor of THAs was erroneous. Description and comparison of baseband and bandpass D&R of bandpass signals, evaluation of different architectures of transmitters and receivers, as well as receiver dynamic range analysis are among the primary topics presented in Chapters 3 and 4.

Chapters 2 to 4 introduce some unconventional viewpoints and describe several advanced techniques. The examples of unconventional viewpoints are clarification of the reasons for dividing generalized modulation into three distinct stages: channel encoding, modulation, and spreading (see Section 2.3.3); explanation why several receiver performance characteristics are used instead of one universal characteristic that actually exists (see Section 4.2.2); determining the ultimate boundaries of receiver dynamic range (see Section 4.3.1); and demonstration of the possibility to analytically calculate the required receiver dynamic range when statistical characteristics of multiple interfering signals are known (see Sections 4.3.2 and 4.3.3). Several advanced techniques allowing effective power utilization and signal reconstruction simplification in transmitters with bandwidth-efficient modulations are outlined in Section 3.4.3. In addition, original AQ-DBPSK techniques, which not only improve power utilization and simplify reconstruction in transmitters but also provide higher overall energy efficiency than DBPSK in additive white Gaussian noise (AWGN) channels and allow frequency-invariant demodulation, are analyzed in Sections 3.4.2 and 4.5.2.

Despite presenting such viewpoints and techniques, the bulk of material in Chapters 2 to 4 provides information on the existing radio systems and conventional technology used in digital radios and other equipment. In contrast with those chapters, Chapters 5 to 7 are specifically intended to introduce original concepts and innovative approaches to D&R. Therefore, most of the information presented there cannot be found in any other book.

The brief history of the sampling theory in Chapter 5 exhibits not only its continuing development but also insufficient attention currently paid by theorists to sampling and interpolation (S&I) of bandlimited signals, although such S&I are most widely used in practice and the research potential in this field is not exhausted. This chapter explains the constructive nature of the sampling theorem and presents new concepts related to S&I. The concepts are based on three fundamental facts. First, the classical sampling theorem for bandlimited signals allows several interpretations that correspond to different forms of the theorem's equations. Second, these interpretations are equally optimal in the least-squares sense if ideally realized. Third, neither interpretation can be ideally realized, and the optimality of a nonideal realization cannot be determined within the theorem's scope. Therefore, optimization of feasible S&I algorithms and circuits requires including in their theoretical basis, besides the sampling theory, the theories of linear and nonlinear circuits, and optimal filtering.

Chapter 6 shows intrinsic drawbacks of currently used S&I circuits based on the sampling theorem's indirect interpretation. It also describes and analyzes novel S&I techniques based on the theorem's hybrid and direct interpretations. While these techniques' description is focused on their conceptual design, the analysis emphasizes the key advantages they provide to digital radios, that is, improvement in dynamic range, attainable bandwidth, scale of integration, flexibility, and power consumption. In addition, two original methods of spatial interference rejection are presented in Chapter 6 as examples of potential applications of the novel S&I techniques.

Chapter 7 outlines several currently used effective quantization techniques and shows that, despite the significant increase in speed, accuracy, sensitivity, and resolution of quantizers over the last three decades, new concepts and approaches in this field can still be suggested. Two innovative techniques, one of which is based on joint processing of several samples and another based on combining predictive quantization with instantaneous adjustment of resolution, are described and concisely analyzed in that chapter.

To simplify the understanding of the book's material by wide and diverse readership, the physical and technical substances of the theoretical concepts and approaches introduced there are explained and often clarified at an intuitive level, mathematically intensive proofs are maximally avoided or simplified, and all signal transformations are illustrated by block, timing, and/or spectral diagrams. A historical approach is widely used to explain the reasons for choosing one or another technical and/or technological solution, delays in implementation of useful innovations, and to identify the development trends.

This book is intended for engineers, scientists, and graduate students involved in the research, development, and design of digital radios, general-purpose A/Ds and D/As, as well as sonar, lidar, measurement and instrumentation, biomedical, control, surveillance, and other digital equipment. It will be useful to engineering managers responsible for the aforementioned research, development, and design. Engineers and scientists working in other fields of electrical and computer engineering may also find it informative. Readers can use the book's material for various purposes, such as improving their knowledge on the subject, further development of the new concepts and techniques presented there, practical implementation of these techniques, and forecasting technological trends. We hope this book will be a good addition to the readers' libraries.

CHAPTER 1
Signals and Waveforms

1.1 Overview

This chapter is a concise refresher on the signal theory aspects repeatedly used in the book. The most essential concepts are clarified and their physical meanings are explained. Simultaneously, some rigorous definitions, most proofs, and mathematical subtleties unimportant for the book are omitted. It is expected that readers are familiar with the signal theory approximately at the undergraduate electrical engineering level. Therefore, the material is presented in the order that allows its most concise explanation, but may not be optimal for initial study. For example, bandlimited signals are mentioned prior to the Fourier transform explanation, and the notion of stochastic (random) processes is introduced before discussing random events and variables.

Section 1.2 provides initial information on signals, their processing, and the signal theory methodology. Analog, discrete-time, and digital signals are compared and advantages of digital signals and processing are summarized. Digitization and reconstruction (D&R) of analog signals are explained. Roles of deterministic and stochastic signals are discussed. In connection with stochastic signals, probabilistic characteristics of random events, variables, and processes are described. Basic operations on signals are outlined.

Expansions of signals with respect to orthogonal bases are considered in Section 1.3. The generalized Fourier series expansion is explained and the classical sampling theorem is mentioned as its special case. Since the trigonometric and complex exponential Fourier series and Fourier transform are widely used in the book, their properties are examined. Relations between the energy and power spectra of signals and their correlation functions as well as transmission of signals through linear time-invariant (LTI) circuits are discussed.

Section 1.4 shows that bandpass signals, used for transmission of information over radio channels, are usually represented by their baseband complex-valued equivalents in the digital portions of transmitters (Txs) and receivers (Rxs). Digitization of bandpass signals in Rxs and their reconstruction in Txs can be baseband or bandpass. Relations between the bandpass signals and their equivalents as well as some transformations of the equivalents are explained. Various definitions of signal and circuit bandwidths are discussed.

1.2 Signals and Their Processing

1.2.1 Analog, Discrete-Time, and Digital Signals

Any physical phenomenon changing with time, space, and/or other independent variables can be considered a signal if it reflects or may reflect a state of an object or system. Electrical, optical, acoustic, mechanical, chemical, and other signals are used in applications. Independently of their nature, all signals are finally converted into electrical ones for convenience of processing. Signals presented as functions of time are called waveforms. The terms "signal" and "waveform" are often used interchangeably. The information carried by signals usually determines their importance. This information can be stored, transmitted, received, and/or processed. During these procedures, the signals undergo intentional transformations that preserve the carried information. Yet they are also affected by undesired phenomena (e.g., noise, interference, distortions, and equipment failures) that can damage the information. Transmission, reception, and processing of information consume energy, and its transmission consumes electromagnetic spectrum that is also a limited resource. It is desirable to transmit, receive, process, and store signals with maximum accuracy, reliability, and speed as well as minimum energy and bandwidth consumption, using equipment of minimum weight, size, and cost. These objectives require a strong theoretical foundation. Any theory operates not with real objects (whose mathematical representation is too complex or impossible) but with their simplified mathematical models. The models are considered adequate if the results of calculations and simulations based on them are consistent with experimental results. For instance, a cosine signal

$$u(t) = U_0 \cos[\psi(t)] = U_0 \cos\left(2\pi f_0 t + \varphi_0\right) \tag{1.1}$$

where U_0, $\psi(t)$, f_0, and φ_0 are the amplitude, current phase, frequency, and initial phase of $u(t)$, respectively, is a mathematical model because such an endless signal, which emerged infinitely long time ago and has invariable U_0 and f_0, cannot exist. Still, it is an adequate model of actual cosine signals generated over long time intervals with sufficient accuracy and stability. The use of mathematical models requires caution because their properties differ from those of real objects, but the required attention is justified by the advantages these models provide.

There are various ways to classify signals. For this book, the differences among analog, discrete-time (or sampled), and digital signals are most important. Analog signals are continuous in time and value. The values of discrete-time signals are specified only at certain instants. They are sequences of analog samples. Digital signals are discrete in time and value, that is, they are sequences of numbers typically represented by sets of binary digits (bits). Human speech, voltage at a power outlet, speed of a mechanical object, and most other physical processes considered at the macro level are analog. Their periodic analog measurements produce discrete-time signals, and digital measurements produce digital signals. Figure 1.1 provides examples of analog $u(t)$, discrete-time $u(nT_s)$, and digital $u_q(nT_s)$ signals. Here, $u(nT_s)$ is a result of sampling $u(t)$, and $u_q(nT_s)$ is a result of quantizing $u(nT_s)$ (subscript q means quantized). Signals of the fourth type, which are discrete in value and continuous in time, are not shown due to their lesser importance to this book. Still, they are used, for example, in nonuniform digitization. Although all the signals in Figure 1.1 are functions of time, they can be functions of other scalar or vector variables: for example, functions of time and frequency are widely used in electrical engineering, TV images are functions of the pixels' coordinates and time, and signals at antennas are functions of time, frequency, and three-dimensional spatial coordinates. Signals that depend on several variables may look different along different axes. For instance, signals at antenna arrays are continuous in time and frequency but discrete in space, while analog periodic signals, presented as functions of time and frequency, are discrete along the frequency axis and continuous along the time axis.

The analog signal designation $u(t)$ in Figure 1.1(a) may correspond to its voltage $v(t)$ or current $i(t)$. The instantaneous power $p_u(t)$ of $u(t)$ across resistor R is

$$p_u(t) = \frac{v^2(t)}{R} = i^2(t)R \tag{1.2}$$

For signal reception, the signal-to-noise power ratio (SNR) across R is more important than the absolute values. To simplify calculations, the power is often normalized by assuming that $R = 1\Omega$. This allows rewriting (1.2) as

$$p_u(t) = u^2(t) \tag{1.3}$$

When the instantaneous power $p_u(t)$ of $u(t)$ is expressed by (1.3), the energy and average power of $u(t)$ during the time interval $[-0.5T, 0.5T]$ are, respectively,

$$E_{u.T} = \int_{-0.5T}^{0.5T} u^2(t)dt \text{ and } P_{u.T} = \frac{1}{T}\int_{-0.5T}^{0.5T} u^2(t)dt \tag{1.4}$$

For complex-valued signals, $u^2(t)$ in (1.3) and (1.4) should be replaced with $|u(t)|^2 = u(t)u^*(t)$ where $|u(t)|$ is the magnitude (absolute value) of $u(t)$, and $u^*(t)$ is its complex conjugate. Two models of signals are used most often: power signals and energy signals. Power signals have finite average power P_u, and, according to (1.4), their energy over infinite time $E_u = \infty$. Energy signals have finite energy E_u, and, therefore, zero average power over infinite time. Although no actual signal can exist infinitely long or have infinite energy, power signals are adequate models

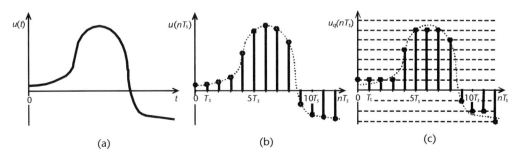

Figure 1.1 Types of signals: (a) analog, (b) discrete-time, and (c) digital.

of many periodic signals and random processes, while energy signals are adequate models of pulses, groups of pulses, and short messages. The use of models requires tackling the theoretical constructs that are neither energy or power signals. For instance, the unit ramp function, which results from integrating the unit step function (see Section A.1), has infinite energy and infinite average power, whereas the delta function (see Section A.2) has infinite energy and undetermined average power over infinite time.

Figure 1.2 shows analog, discrete-time, and digital signals processed by the corresponding circuits. Here, sampling circuits convert analog signals into discrete-time ones, and interpolating circuits perform the inverse operation. An analog-to-digital converter (A/D) quantizes discrete-time signals, transforming them into digital ones, whereas a digital-to-analog converter (D/A) performs analog decoding of digital signals, transforming them into discrete-time ones. A cascade structure of a sampling circuit and an A/D digitizes analog signals, whereas a cascade structure of a D/A and an interpolating circuit reconstructs them. In Figure 1.2, analog signals are processed by an analog processor, discrete-time signals by a discrete-time processor, and digital signals by a digital signal processor (DSP). The actual situation is more complex because analog circuits can, in principle, process not only analog but also discrete-time and digital signals. For example, analog amplifiers can amplify signals of all types. At the same time, analog signals can trigger digital circuits.

In radio systems, sampling is usually uniform (i.e., the sampling rate is constant) and based on the classical sampling theorem, applicable only to bandlimited signals. Therefore, it consists of two operations: antialiasing filtering that limits the signal bandwidth and sample generation carried out by samplers that are typically track-and-hold amplifiers (THAs). Currently, these operations are performed by separate circuits at the A/D input, as shown in Figure 1.3(a). Similar to sampling, analog interpolation also comprises two operations: pulse shaping and interpolating

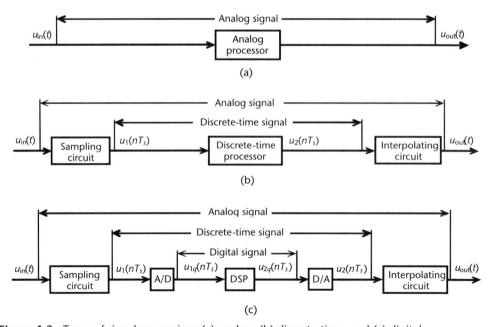

Figure 1.2 Types of signal processing: (a) analog, (b) discrete-time, and (c) digital.

filtering, currently performed separately at the D/A output (see Figure 1.3(b)). As explained in Chapters 3 and 4, the DSP input stage also executes digitization-related operations, and the DSP output stage performs reconstruction-related operations. Comparison of digitization and reconstruction procedures shows that, whereas A/Ds and D/As perform opposite operations, the operations carried out by samplers and pulse shapers as well as by analog antialiasing and interpolating filters are similar but usually have different requirements. As shown in Chapter 5, the sampling theorem's direct interpretation allows combining antialiasing filtering and sample generation as well as pulse shaping and interpolating filtering.

Digitization is an approximation of analog signals by their digital equivalents. It cannot be exact for two reasons. First, quantization maps a continuum of sample values into a finite set of discrete values, producing a quantization error (quantization noise). Second, ideal sampling according to the sampling theorem is physically unrealizable, and practical one is approximate. Signal reconstruction is also inexact. The acceptable information loss during D&R depends on the purpose of processing. Despite the D&R losses and complexity, the use of digital signals and processing quickly expands due to their advantages over analog ones: (1) possibility of regenerating information from signals distorted during transmission, storage, or processing, (2) much higher and independent of destabilizing factors accuracy of processing, (3) superior versatility and flexibility, (4) much larger scale of integration, and (5) reduced cost of equipment development and production. At the dawn of electrical communications, only digital data could be transmitted (telegraphy). Technology development enabled transmission of analog information like voice and images (telephony and television). For a long time, digital signals were transmitted, received, and processed by digital and/or analog devices, whereas analog signals were transmitted, received, and processed only by analog devices. Now technological progress has made digital processing of both digital and analog signals most accurate, reliable, and efficient. At present, samplers and A/Ds are usually placed in the same packages, and the term A/D is often applied to the whole package. To avoid ambiguity, A/Ds are called quantizers in most of this book.

1.2.2 Deterministic and Stochastic Signals

All signals contain information in their parameters. When these parameters are known to users, the signals are deterministic. Deterministic signals are used as test, pilot, synchronization, and carrier signals. Information that requires transmission is

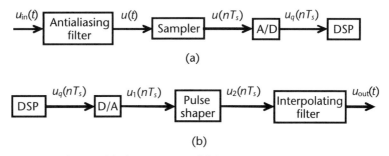

Figure 1.3 Conventional D&R: (a) digitization and (b) reconstruction.

contained only in the parameters unknown to its recipient(s) in advance. The latter signals are modeled by stochastic functions that are characterized by probabilities of their realizations. A complete set of possible realizations with their probability distribution forms an ensemble. Both deterministic and stochastic signals can be analog, discrete-time, or digital. In digital communications, for example, the Rx knows the signals and their a priori probabilities p_m, but is uncertain about which of them is transmitted. The uncertainty is reflected by the entropy that for an ensemble of M signals is

$$H = -\sum_{m=1}^{M} p_m \log_2 p_m \text{ where } \sum_{m=1}^{M} p_m = 1 \quad (1.5)$$

If signals were deterministic, their a priori probabilities would be $p_k = 1$ and $p_m = 0$ for $m \neq k$ when the kth signal is transmitted, and the a priori entropy would be $H_1 = 0$ according to (1.5). This reflects the absence of any uncertainty even before the signal transmission (that would be unneeded due to its known-in-advance outcome). Since transmitted signals are random, their a priori probabilities $0 < p_m < 1$, and, consequently, $H_1 > 0$. As follows from (1.5), H_1 reaches its maximum $H_{1\max} = \log_2 M$ when $p_m = 1/M$ for all m. The signal reception makes the entropy equal to H_2. For ideal reception, $H_2 = 0$. Since the presence of noise does not allow ideal reception, $0 < H_2 < H_1$, and the amount of received information is $I = H_1 - H_2$. Thus, only stochastic signals can be used for information transmission. Interfering signals are also stochastic. Otherwise, they, in principle, could be compensated.

Since thermal noise is the most widely known example of stochastic signals, and engineers often see its realizations on oscilloscope screens, it is worthwhile to emphasize that realizations of other stochastic signals may look much less stochastic. For example, a cosine signal (1.1) is deterministic only if all its parameters (i.e., U_0, f_0, and φ_0) are not random. If at least one of them is random, this signal is stochastic, although its realizations do not look that way. Another example is a sequence of K random binary numbers, each of which can be 0 or 1 with equal probabilities $P(0) = P(1) = 0.5$ and independently of other numbers. When $K = 2$, all the realizations (00, 01, 10, and 11) look "regular." An increase in K changes the situation. Table 1.1 shows 12 selected realizations for $K = 16$. The first six realizations look "regular," whereas the next six look more stochastic. Actually, the fraction of regular realizations is very small when $K = 16$ and an increase in K monotonically reduces it, although their absolute number grows. When $K \to \infty$, their fraction tends to zero. Stochastic fields are stochastic functions of several variables. Stochastic processes are stochastic functions of time. Most stochastic signals discussed in this book are processes. Prior to their analysis, basic information on random events and random variables is provided.

The examples of random events are: appearance of zero or one in a certain position of the sequence described above, arrival of desired and/or interfering signals, equipment failure, false alarm, miss of a target, and loss of synchronization. Random events are characterized by their probabilities that reflect statistically stable relative frequencies of their occurrences. Probabilities $P(A)$ and $P(B)$ of events A and B meet conditions $0 \leq P(A) \leq 1$ and $0 \leq P(B) \leq 1$. Event A is impossible if $P(A) = 0$ and certain if $P(A) = 1$. If random mutually exclusive events A and B constitute all

Table 1.1 Realizations of a Stochastic Binary Sequence

No.	Realizations
1	1 1 1 1 1 1 1 1 1 1 1 1 1 1 1
2	0 0 0 0 0 0 0 0 0 0 0 0 0 0 0
3	1 0 1 0 1 0 1 0 1 0 1 0 1 0 1 0
4	0 1 0 1 0 1 0 1 0 1 0 1 0 1 0 1
5	1 1 0 0 1 1 0 0 1 1 0 0 1 1 0 0
6	0 0 1 1 0 0 1 1 0 0 1 1 0 0 1 1
7	1 0 0 1 1 1 0 0 0 0 1 0 1 1 0 1
8	0 1 0 1 1 1 0 0 1 0 0 0 1 0 1 1
9	0 1 0 0 0 0 1 0 1 1 1 0 1 0 1 1
10	1 0 1 1 0 1 0 0 1 1 1 0 1 0 0 0
11	1 1 1 0 0 1 0 0 0 0 1 0 1 0 1 1
12	1 0 1 1 0 1 0 0 0 0 1 0 1 1 0 1

possible outcomes of a trial, then $P(A \cup B) = P(A) + P(A) = 1$. If binary symbols of a message are transmitted with probabilities $P(0)$ and $P(1)$, the transmission of one of them is a certain event whose probability is one, whereas the transmission of none of them is an impossible event whose probability is zero. Mutually nonexclusive events can be statistically dependent or independent. $P(A|B)$ is a conditional probability of A if B has occurred. Let us assume that the unconditional probability of event A, which is coming across the word "me" in a sensible text, is $P(A)$. It is clear that the occurrence of event B, which is the words "Please call" prior to A, increases its probability, and $P(A|B) > P(A)$. The joint probability $P(AB)$ of A and B is $P(AB) = P(A|B)P(B) = P(B|A)P(A)$. If these events are statistically independent, $P(A|B) = P(A)$, $P(B|A) = P(B)$, and $P(AB) = P(A)P(B)$. An example of independent events is an arrival of a signal and a simultaneous spike of noise.

The values of a random variable obtained as a result of measurements or experiments cannot be predicted with certainty. A number of correctly demodulated symbols in a message, noise level at an amplifier output, result of an ambient temperature measurement, and time before an equipment failure are examples of random variables. The first variable is intrinsically discrete, whereas others are intrinsically continuous. However, even intrinsically continuous variables become discrete if they are measured by digital devices. Note that the words "continuous" and "discrete" in the probability theory correspond, respectively, to the words "analog" and "digital" in digital signal processing (DSP). Random variables are characterized by their probability distributions. The probability mass function (PMF) $p(x_k) = P(X = x_k)$ and the cumulative distribution function (CDF) $F(x) = P(X \leq x)$ are applicable to a discrete random variable X. As follows from these definitions,

$$\sum_{k=1}^{K} p(x_k) = 1 \qquad (1.6)$$

$$F(-\infty) = P(X \leq -\infty) = 0, \quad F(\infty) = P(X \leq \infty) = 1,$$
$$\text{and } P(a < X \leq b) = F(b) - F(a) \tag{1.7}$$

Figure 1.4 demonstrates the PMF and CDF for the number of successes in the series of 5 independent trials when the probability of success in each trial is $p_s = 0.5$. This random variable is distributed according to the binomial law that, in general, describes the probability of m successes in a series of n independent trials:

$$p(m) = P(X = m) = \binom{n}{m} p_s^m (1 - p_s)^{n-m} = \frac{n!}{m!(n-m)!} p_s^m (1 - p_s)^{n-m} \tag{1.8}$$

While the CDF is also applicable to continuous random variables, the PMF is not because the probability of being equal to any particular value is zero for these variables. The probability density function (PDF) $f(x) = F'(x)$ describes the distribution of continuous random variables in a way similar to that provided by the PMF for discrete random variables. As follows from (1.7) and the PDF definition,

$$F(x) = \int_{-\infty}^{x} f(u) du \quad \text{and} \quad P(a < X \leq b) = \int_{a}^{b} f(x) dx \tag{1.9}$$

Figure 1.5 illustrates the PDF and CDF of a continuous random variable distributed according to the Gaussian (normal) law, widely used in applications for two reasons. The first one is that most variables become random due to the impact of a very large number of additive, comparable, and independent random phenomena, so their actual distributions are close to Gaussian according to the central limit theorem. The second reason is that this distribution allows analytical solutions of many problems and, therefore, is often employed even if the Gaussian model is not completely adequate. The Gaussian PDF is

$$f(x) = \frac{1}{\sqrt{2\pi}\sigma} \exp\left[-\frac{(x-m)^2}{2\sigma^2}\right] \tag{1.10}$$

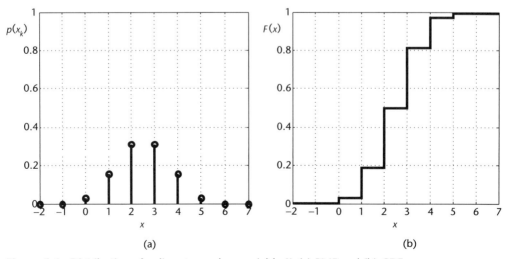

Figure 1.4 Distribution of a discrete random variable X: (a) PMF and (b) CDF.

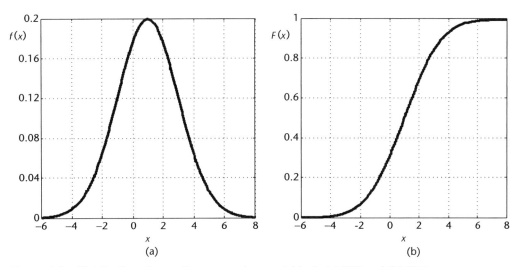

Figure 1.5 Distribution of a continuous random variable X: (a) PDF and (b) CDF.

In multidimensional cases, the probability distributions are described by joint functions. For two dimensions, for example, the joint CDF $F(x, y) = P(X \leq x, Y \leq y)$ and the joint PDF $f(x, y)$ are related as follows:

$$f(x,y) = \frac{\partial^2 F(x,y)}{\partial x \, \partial y} \quad \text{and} \quad F(x,y) = \int_{-\infty}^{x} \int_{-\infty}^{y} f(x_1, y_1) \, dx_1 dy_1 \qquad (1.11)$$

Two-dimensional and one-dimensional distributions are connected:

$$F_1(x) = \int_{-\infty}^{x} \int_{-\infty}^{\infty} f(x_1, y_1) \, dx_1 dy_1 \quad \text{and} \quad F_2(y) = \int_{-\infty}^{\infty} \int_{-\infty}^{y} f(x_1, y_1) \, dx_1 dy_1 \qquad (1.12)$$

$$f_1(x) = \int_{-\infty}^{\infty} f(x,y) \, dy \quad \text{and} \quad f_2(y) = \int_{-\infty}^{\infty} f(x,y) \, dx \qquad (1.13)$$

$$\begin{aligned} F(x,y) &= F_1(x) F_2(y \mid X \leq x) = F_2(y) F_1(x \mid Y \leq y) \quad \text{and} \\ f(x,y) &= f_1(x) f_2(y \mid x) = f_2(y) f_1(x \mid y) \end{aligned} \qquad (1.14)$$

where $F_1(x \mid Y \leq y)$, $F_2(y \mid X \leq x)$, $f_1(x \mid y)$, and $f_2(y \mid x)$ are conditional CDFs and PDFs. When X and Y are statistically independent, $F_1(x \mid Y \leq y) = F_1(x)$, $F_2(y \mid X \leq x) = F_2(y)$, $f_1(x \mid y) = f_1(x)$, and $f_2(y \mid x) = f_2(y)$; so

$$F(x,y) = F_1(x) F_2(y) \quad \text{and} \quad f(x,y) = f_1(x) f_2(y) \qquad (1.15)$$

The probability distributions are exhaustive characteristics of random variables. Their moments contain less detailed information but still allow solving many probabilistic problems. The nth moment α_n of a random variable X is the statistical average of the nth power of X, that is, $\alpha_n(X) = E(X^n)$ where E denotes statistical averaging. For discrete and continuous random variables, respectively, the moments are

$$\alpha_n(X) = E(X^n) = \sum_{k=1}^{K} x_k^n p(x_k) \quad \text{and} \quad \alpha_n(X) = E(X^n) = \int_{-\infty}^{\infty} x^n f(x) dx \quad (1.16)$$

The first moment $\alpha_1(X) = E(X) = m_x$ is the mean value (or statistical average) of X. The central moments of X are defined as $\mu_n(X) = E[(X - m_x)^n]$. According to the definition, the first central moment $\mu_1(X) = 0$, and the second one $\mu_2(X)$ is called variance, denoted as D_x or σ_x^2, and calculated as

$$\mu_2(X) = D_x = \sigma_x^2 = \sum_{k=1}^{K} (x_k - m_x)^2 p(x_k) \quad \text{and}$$

$$\mu_2(X) = D_x = \sigma_x^2 = \int_{-\infty}^{\infty} (x - m_x)^2 f(x) dx \quad (1.17)$$

for discrete and continuous random variables respectively. The positive square root σ_x of D_x is called the standard deviation. While m_x characterizes the position of X on the x-axis, σ_x characterizes its spread. The higher moments characterize other properties of X. If a probability distribution of X is symmetric, its $\mu_n(X) = 0$ for odd n. Therefore, the skewness of a probability distribution is reflected by the ratio $\mu_3(X)/\sigma_x^3$. Its tailedness relative to that of the Gaussian distribution is reflected by the excess kurtosis $\mu_4(X)/\sigma_x^4 - 3$. In the Gaussian PDF (1.10), its parameters m and σ correspond to the mean and standard deviation of X, respectively. Since this PDF is unimodal and symmetric, m is also its median and mode. Thus, m unambiguously characterizes the X position.

Multidimensional random variables are concisely characterized by their joint moments. For a two-dimensional random variable, the $(n + l)$th joint moment $\alpha_{nl}(X, Y) = E(X^n Y^l)$ is calculated as

$$\alpha_{nl}(X, Y) = E(X^n Y^l) = \sum_{k=1}^{K} \sum_{s=1}^{S} x_k^n y_s^l p(x_k, y_s) \quad \text{and}$$

$$\alpha_{nl}(X, Y) = E(X^n Y^l) = \int_{-\infty}^{\infty} \int_{-\infty}^{\infty} x^n y^l f(x, y) \, dx \, dy \quad (1.18)$$

and its central joint moment $\mu_{nl}(X, Y) = E[(X - m_x)^n (Y - m_y)^l]$ as

$$\mu_{nl}(X, Y) = \sum_{k=1}^{K} \sum_{s=1}^{S} (x_k - m_x)^n (y_s - m_y)^l p(x_k, y_s) \quad \text{and}$$

$$\mu_{nl}(X, Y) = \int_{-\infty}^{\infty} \int_{-\infty}^{\infty} (x - m_x)^n (y - m_y)^l f(x, y) \, dx \, dy \quad (1.19)$$

The widely used second central joint moment $\mu_{11}(X, Y) = E[(X - m_x)(Y - m_y)]$ is called the covariance of X and Y. The normalized covariance (or correlation coefficient) is

$$\rho_{xy} = \frac{\mu_{11}(X,Y)}{\sigma_x \sigma_y} = \frac{E\left[(X-m_x)(Y-m_y)\right]}{\sigma_x \sigma_y} \qquad (1.20)$$

This coefficient meets condition $-1 \leq \rho_{xy} \leq 1$ and characterizes a linear statistical dependence between X and Y. When $\rho_{xy} = 0$, X, and Y are uncorrelated (i.e., linearly independent). In general, the absence of correlation between X and Y does not mean their statistical independence, which is guaranteed only by (1.15). However, if the joint distribution of X and Y is Gaussian, the statistical dependence between them can be only linear, and $\rho_{xy} = 0$ means their statistical independence. A linear statistical dependence between X and Y becomes linear functional dependence when $\rho_{xy} = -1$ or $\rho_{xy} = 1$.

Random variables can be considered samples of stochastic processes. Four realizations $n_1(t)$, $n_2(t)$, $n_3(t)$, and $n_4(t)$ of Gaussian noise $N(t)$ generated by a baseband amplifier in a steady-state mode are shown in Figure 1.6(a). The samples of these realizations, taken at the instants t_1, t_2, and t_3, represent values of the random variables $N(t_1)$, $N(t_2)$, and $N(t_3)$. The joint distribution of $N(t_i)$, taken sufficiently often, characterizes the statistical properties of $N(t)$. Since the number of the samples has to be unlimited in general case, analysis of stochastic processes is difficult. Fortunately, many important classes of these processes satisfy constraints that simplify their analysis. Stationary stochastic processes form the most notable class of such processes. A stochastic process $X(t)$ is strictly stationary if its joint probability distribution is time-shift invariant, that is, the joint CDF $F_X(x_{t1}, x_{t2}, ..., x_{tk}, ...)$ of any set of its samples $\{X(t_k)\}$ is equal to $F_X(x_{t1+\tau}, x_{t2+\tau}, ..., x_{tk+\tau}, ...)$ for all τ, k, and t_1, t_2, ..., t_k. In reality, there are no processes with never-changing properties, but stationary stochastic processes are adequate mathematical models of most processes generated in a steady-state mode.

While the joint probability distributions exhaustively characterize the stochastic processes, their moment functions concisely describe them. At each instant t, the moment functions $\alpha_n[X(t)]$ and central moment functions $\mu_n[X(t)]$ are equal,

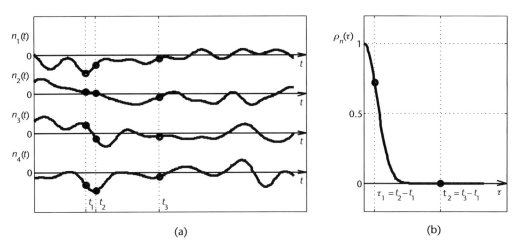

Figure 1.6 Realizations $n_k(t)$ and normalized autocovariance function $\rho_n(t)$ of $N(t)$: (a) $n_k(t)$ and (b) $\rho_n(t)$.

respectively, to the moments and central moments of the corresponding random variables and can be calculated using (1.16) and (1.17). Thus, the position of a stochastic process $X(t)$ at instant t_1 is reflected by $m_x(t_1)$ and its spread by $\sigma_x(t_1)$. Similarly, the joint moment functions $\alpha_{nl}[X(t_1), X(t_2)]$ and central joint moment functions $\mu_{nl}[X(t_1), X(t_2)]$ are defined according to (1.18) and (1.19), respectively. The second joint moment function and the second joint central moment function are, respectively, called autocorrelation (or correlation) and autocovariance (or covariance) functions and calculated as

$$R_x(t_1, t_2) = E[X(t_1)X(t_2)] \text{ and}$$
$$C_x(t_1, t_2) = E\{[X(t_1) - m_x(t_1)][X(t_2) - m_x(t_2)]\} \quad (1.21)$$

Since $R_x(t_1, t_2) = \alpha_2[X(t_1)]$ and $C_x(t_1, t_2) = \mu_2[X(t_1)] = \sigma_x^2(t_1)$ for $t_1 = t_2$, there is no need to use variance function as a separate characteristic. The linear statistical dependence between $X(t_1)$ and $X(t_2)$ is characterized by normalized autocovariance (or normalized covariance) function $\rho_x(t_1, t_2) = C_x(t_1, t_2)/[\sigma_x(t_1)\sigma_x(t_2)]$. Consequently, $-1 \leq \rho_x(t_1, t_2) \leq 1$, and $\rho_x(t_1, t_2) = 0$ if $X(t_1)$ and $X(t_2)$ are linearly independent. The linear statistical independence means the complete statistical independence only for Gaussian processes. To assess the linear statistical dependence between two stochastic processes $X(t)$ and $Y(t)$, cross-correlation function $R_{xy}(t_1, t_2)$ that can also be normalized is widely used. For real-valued processes, $R_{xy}(t_1, t_2) = E[X(t_1)Y(t_2)]$. When stochastic processes are stationary, their moment and central moment functions are time-invariant and their joint moment functions depend just on the difference $\tau = t_2 - t_1$. Thus, $m_x(t) = m_x$, $\sigma_x(t) = \sigma_x$, $R_x(t_1, t_2) = R_x(\tau)$, $R_{xy}(t_1, t_2) = R_{xy}(\tau)$, $C_x(t_1, t_2) = C_x(\tau)$, $\rho_x(t_1, t_2) = \rho_x(\tau) = C_x(\tau)/\sigma_x^2$, and $C_x(\tau) = R_x(\tau) - m_x^2$ for stationary processes. Correlation functions of such processes are even. Figure 1.6(b) presents $\rho(\tau)$ of the Gaussian noise $N(t)$ whose realizations are shown in Figure 1.6(a). Since the mean of this noise $m_n = 0$, its $C_n(\tau) = R_n(\tau)$. $N(t_1)$ and $N(t_2)$ are statistically dependent because $\rho_n(\tau_1) = \rho_n(t_2 - t_1) \neq 0$, whereas $N(t_1)$ and $N(t_3)$ as well as $N(t_2)$ and $N(t_3)$ are statistically independent because, being Gaussian, they are uncorrelated. Many problems concerning stationary $X(t)$ can be solved using just m_x and $R_x(\tau)$. Therefore, the notion of wide-sense stationarity has been introduced: a stochastic process $X(t)$ is wide-sense stationary if $m_x(t) = m_x$ and $R_x(t_1, t_2) = R_x(\tau)$. Any strictly stationary stochastic process is also wide-sense stationary. The inverse statement is correct only for Gaussian processes, where all high-order moment functions are determined by their first and second moment functions.

Ergodic processes constitute other important class of stochastic processes. For an ergodic stochastic process, the results of statistical and time averaging are identical, that is, each its realization reflects all the process properties. In principle, ergodicity does not presume stationarity. In most applications, however, ergodic stochastic processes can be considered stationary. In the example shown in Table 1.1, the first six realizations, which look regular, are atypical and cannot represent this sequence, whereas the next six realizations, which look stochastic, are typical and reflect the statistical properties of the entire random sequence. Although the sequence is relatively short ($K = 16$), the significant majority of the realizations are typical. When K is of the order of hundreds or thousands, the probability of atypical realizations is negligible and the sequence can be considered ergodic. Other examples of stationary and ergodic stochastic processes are the output noise $N(t)$ of an amplifier in a steady-state mode and a cosine signal with a random initial phase φ_0 uniformly distributed within the interval $[-\pi, \pi]$, whereas a cosine signal with a random amplitude is neither stationary nor ergodic, and a cosine signal with a random amplitude and a random initial phase, uniformly distributed within the interval $[-\pi, \pi]$ is stationary but not ergodic. The following equations illustrate two ways of averaging acceptable for ergodic continuous stochastic processes:

$$\alpha_1(X) = m_x = \int_{-\infty}^{\infty} x f(x) dx = \lim_{T \to \infty} \frac{1}{T} \int_{-0.5T}^{0.5T} x(t) dt \quad (1.22)$$

$$\alpha_2(X) = \int_{-\infty}^{\infty} x^2 f(x) dx = \lim_{T \to \infty} \frac{1}{T} \int_{-0.5T}^{0.5T} x^2(t) dt \quad (1.23)$$

$$\mu_2(X) = \sigma_x^2 = \int_{-\infty}^{\infty} (x - m_x)^2 f(x) dx = \lim_{T \to \infty} \frac{1}{T} \int_{-0.5T}^{0.5T} [x(t) - m_x]^2 dt \quad (1.24)$$

$$R_x(\tau) = \int_{-\infty}^{\infty} \int_{-\infty}^{\infty} x_1 x_2 f(x_1, x_2; \tau) dx_1 dx_2 = \lim_{T \to \infty} \frac{1}{T} \int_{-0.5T}^{0.5T} x(t) x(t + \tau) dt \quad (1.25)$$

Equations (1.22) through (1.24) demonstrate that if an ergodic process $X(t)$ represents an electrical signal, m_x is its direct current (dc), $\alpha_2(X)$ is its average power, σ_x^2 is the average power of its alternating current (ac) part, and σ_x is the effective or root-mean-square (rms) value of the ac part. Many stochastic signals do not contain dc. For them, $\alpha_2(X) = \sigma_x^2$ and $C_x(\tau) = R_x(\tau)$. If $X(t)$ is a complex-valued stochastic process, its m_x and $R_x(\tau)$ are also complex-valued, and $R_x(\tau)$ is defined as $R_x(\tau) = E[X(t)X^*(t + \tau)]$ where $X^*(t)$ is the complex conjugate of $X(t)$. Therefore, $R_x(\tau)$ of a complex-valued $X(t)$ is a Hermitian function, that is, $R_x(-\tau) = R_x^*(\tau)$, and $R_x(0) = \alpha_2(X)$ is the average power of $X(t)$. Two cross-correlation functions are defined for complex-valued stochastic processes $X(t)$ and $Y(t)$: $R_{xy}(t_1, t_2) = E[X(t_1)Y^*(t_2)]$ and $R_{yx}(t_1, t_2) = E[Y(t_1)X^*(t_2)]$. The decision about the ergodicity of a stochastic process is usually based on physical analysis of its nature. However, there is a formal sign that a process $X(t)$ is nonergodic: its $C_x(\tau)$ does not tend to zero when $\tau \to \infty$. Most signals that undergo D&R in digital radios can be considered locally stationary and ergodic stochastic processes. More detailed information on the probability theory and stochastic processes can be found, for instance, in [1–13].

1.2.3 Basic Operations on Signals

This section describes the simplest operations, most widely used in signal processing. Scaling changes the level of a signal $u_1(t)$ without changing its position:

$$u_2(t) = c u_1(t) \quad (1.26)$$

where c is a real-valued scaling factor. As shown in Figure 1.7, the $u_1(t)$ magnitude is increased if $|c| > 1$, decreased if $|c| < 1$, and its sign is inverted if $c < 0$. Technically, magnitudes of analog and discrete-time signals are increased by amplifiers and decreased by attenuators. Their signs are changed by inverters. Digital signals are scaled by digital multipliers.

Time shifting changes the signal time position without altering its level or shape (see Figure 1.8). If $u_2(t)$ is a copy of $u_1(t)$ delayed by t_0, two equivalent statements are true: (1) the $u_1(t)$ value at instant t is equal to the $u_2(t)$ value t_0 seconds later, and (2) the $u_2(t)$ value at instant t is equal to the $u_1(t)$ value t_0 seconds earlier:

$$u_1(t) = u_2(t + t_0) \quad \text{and} \quad u_2(t) = u_1(t - t_0) \tag{1.27}$$

If $u_3(t)$ is an advanced copy of $u_1(t)$, the statements are: (1) the $u_1(t)$ value at instant t is equal to the $u_3(t)$ value t_0 seconds earlier, and (2) the $u_3(t)$ value at instant t is equal to the $u_1(t)$ value t_0 seconds later:

$$u_1(t) = u_3(t - t_0) \quad \text{and} \quad u_3(t) = u_1(t + t_0) \tag{1.28}$$

While signals can be delayed using delay lines or by playing back earlier records, their advancing is physically impossible. Fortunately, only relative signal positions are important in most applications. This allows delaying $u_1(t)$ relative to $u_3(t)$ instead of advancing $u_3(t)$ relative to $u_1(t)$.

Time scaling of signals compresses or expands them, keeping their amplitudes unchanged, as shown in Figure 1.9. When scaling factor $k > 1$, $u_1(t)$ is compressed, and when $0 < k < 1$, $u_1(t)$ is expanded. The value of time-scaled signal $u_s(t)$ at instant t is equal to the $u_1(t)$ value at instant kt:

$$u_s(t) = u_1(kt) \tag{1.29}$$

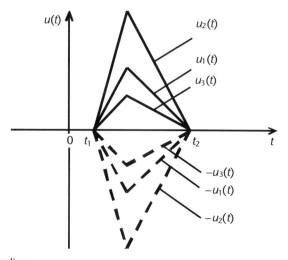

Figure 1.7 Signal scaling.

1.2 Signals and Their Processing

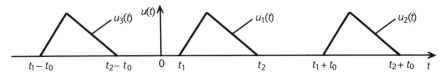

Figure 1.8 Time shifting.

Time scaling can be performed by recording signals and playing them back at different speeds. There are more sophisticated methods of time scaling (see, for instance, [14–18]). Note that the Doppler effect is also time scaling.

Time reversal (or time inversion) of signals changes the direction of the time axis without changing the signals' levels (see Figure 1.10). It is a special case of time scaling with $k = -1$. Since the value of $u_1(t)$ at instant t is equal to the value of $u_2(t)$ at instant $-t$, $u_2(t)$ is a mirror image of $u_1(t)$ about the ordinate axis:

$$u_1(t) = u_2(-t) \text{ and } u_2(t) = u_1(-t) \tag{1.30}$$

Time reversal of signals is often combined with their time shifting. In Figure 1.11, $u_1(t)$ is an original signal, $u_1(-t)$ is its mirror image about the ordinate axis,

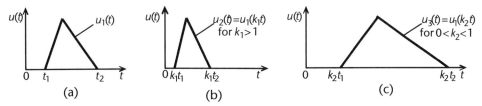

Figure 1.9 Time scaling: (a) original signal $u_1(t)$, (b) compressed signal $u_2(t)$, and (c) expanded signal $u_3(t)$.

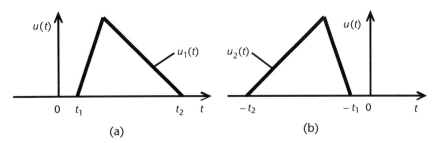

Figure 1.10 Time reversal: (a) original signal $u_1(t)$ and (b) time-reversed signal $u_2(t)$.

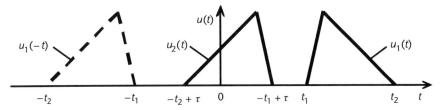

Figure 1.11 Combined time reversal and shifting.

and $u_2(t)$ is the result of time advance by τ and time reversal. Thus, the value of $u_1(t)$ at instant t is equal to the value of $u_2(t)$ at instant $\tau - t$. This operation can be described by any of the equations:

$$u_1(t) = u_2(\tau - t) \text{ and } u_2(t) = u_1(\tau - t) \tag{1.31}$$

1.3 Expansions of Signals

1.3.1 Orthogonal Expansions

A signal $u(t)$ is square-integrable on the time interval $t_1 \leq t \leq t_2$ if

$$\int_{t_1}^{t_2} |u(t)|^2 \, dt = \|u(t)\|^2 < \infty \tag{1.32}$$

where $\|u(t)\|$ and $\|u(t)\|^2$ are, respectively, the signal norm and energy on the interval $t_1 \leq t \leq t_2$. To uniquely represent $u(t)$ by a linear combination of other continuous or piecewise-continuous square-integrable signals $\varphi_n(t)$ where $n = 0, 1, 2, \ldots$, set $\{\varphi_n(t)\}$ should form a basis in a signal space containing $u(t)$:

$$u(t) = c_0 \varphi_0(t) + c_1 \varphi_1(t) + c_2 \varphi_2(t) + \ldots + c_n \varphi_n(t) + \ldots \tag{1.33}$$

where c_n are coefficients. To this end, $\{\varphi_n(t)\}$ should span the space of all possible $u(t)$, and $\varphi_n(t)$ should be linearly independent, that is, the equality $c_0 \varphi_0(t) + c_1 \varphi_1(t) + c_2 \varphi_2(t) \ldots = 0$ should be true if and only if $c_0 + c_1 + c_2 \ldots = 0$. All real-world analog signals are continuous or piecewise-continuous and square-integrable. However, precautions should be taken since the theory is dealing not with actual signals but with their models. Representation of $u(t)$ by $\{\varphi_n(t)\}$ simplifies its analysis and/or processing, especially when $\{\varphi_n(t)\}$ is orthogonal on the interval $t_1 \leq t \leq t_2$, that is, for every pair of $\varphi_n(t)$ and $\varphi_m(t)$ with $n \neq m$,

$$\int_{t_1}^{t_2} \varphi_n(t) \varphi_m^*(t) \, dt = 0 \tag{1.34}$$

After multiplying both sides of (1.33) by $\varphi_n^*(t)$ and integrating the products, we obtain

$$\int_{t_1}^{t_2} u(t) \varphi_n^*(t) \, dt = c_n \|\varphi_n(t)\|^2 \tag{1.35}$$

because the orthogonality of $\{\varphi_n(t)\}$ makes all the terms with $n \neq m$ on the right side of (1.35) equal to zero, and terms with $m = n$

$$\int_{t_1}^{t_2} c_n \varphi_n(t) \varphi_n^*(t) \, dt = c_n \int_{t_1}^{t_2} |\varphi_n(t)|^2 \, dt = c_n \|\varphi_n(t)\|^2 \tag{1.36}$$

From (1.35),

$$c_n = \frac{1}{\|\varphi_n(t)\|^2} \int_{t_1}^{t_2} u(t)\varphi_n^*(t)\,dt \qquad (1.37)$$

Series (1.33) with coefficients (1.37) is a generalized Fourier series with respect to $\{\varphi_n(t)\}$. It minimizes the rms error in the $u(t)$ approximation for a given number N of the series terms, and this error tends to zero when $N \to \infty$. The energy E_u and average power P_u of $u(t)$, represented by such a series on the interval $t_1 \le t \le t_2$, are:

$$E_u = \|u\|^2 = \sum_{n=0}^{\infty} |c_n|^2 \|\varphi_n\|^2 \quad \text{and} \quad P_u = \frac{\|u\|^2}{t_2 - t_1} = \frac{1}{t_2 - t_1} \sum_{n=0}^{\infty} |c_n|^2 \|\varphi_n\|^2 \qquad (1.38)$$

If, in addition to (1.34), $\|\varphi_n(t)\| = 1$ for any n, $\{\varphi_n(t)\}$ is orthonormal. Orthonormal basis simplifies (1.38):

$$E_u = \|u\|^2 = \sum_{n=0}^{\infty} |c_n|^2 \quad \text{and} \quad P_u = \frac{\|u\|^2}{t_2 - t_1} = \frac{1}{t_2 - t_1} \sum_{n=0}^{\infty} |c_n|^2 \qquad (1.39)$$

According to (1.38) and (1.39), known as Parseval's identity, the energies and average powers of signals are equal to the respective sums of the energies and average powers of their orthogonal components. The decomposition of signals into their orthogonal components (analysis) according to (1.37) is illustrated by Figure 1.12(a), whereas their reconstruction (synthesis) according to (1.33) is illustrated by Figure 1.12(b).

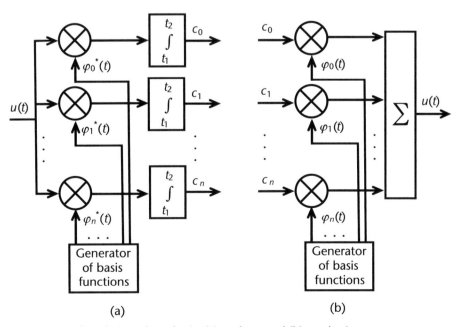

Figure 1.12 Signal analysis and synthesis: (a) analyzer and (b) synthesizer.

Major requirements for $\{\varphi_n(t)\}$ are clear characterization of $u(t)$ in the domain of interest, fast convergence to $u(t)$, and simplicity of $\varphi_n(t)$ generation as well as decomposition and reconstruction of $u(t)$. As shown in Chapter 5, the classical sampling theorem represents signals by generalized Fourier series with respect to the set of sampling functions. Scientific and technological progress increases the number of practically used $\{\varphi_n(t)\}$ and changes their relative importance. The importance of signal representation by trigonometric and exponential Fourier series for this book is explained below.

1.3.2 Trigonometric and Exponential Fourier Series

The response of an LTI system to an input cosine (or sine) signal is also a cosine (or sine) signal of the same frequency with the amplitude and phase determined by the system parameters. Complex exponential signals have this property too, as follows from the Euler's formula $\exp(j\psi) = \cos\psi + j\sin\psi$ where $j = (-1)^{0.5}$. Since the change of the amplitude and phase of a complex exponential signal is equivalent to multiplying it by a complex-valued constant, these signals are eigenfunctions of LTI systems. Cosine signal (1.1) can be presented as

$$u(t) = \text{Re}\left\{U_0 \exp\left[j\left(2\pi f_0 t + \varphi_0\right)\right]\right\} \text{ or}$$
$$u(t) = 0.5U_0 \left\{\exp\left[j\left(2\pi f_0 t + \varphi_0\right)\right] + \exp\left[-j\left(2\pi f_0 t + \varphi_0\right)\right]\right\}$$
(1.40)

As follows from the first equation in (1.40) and shown in Figure 1.13(a), the cosine signal is the real part of phasor $U_0\exp[j(2\pi f_0 t + \varphi_0)]$ with amplitude U_0 and initial phase φ_0, which rotates counterclockwise at angular rate $\omega_0 = 2\pi f_0$. The second equation in (1.40) proves and Figure 1.13(b) illustrates that this signal can also be presented as the sum of two counter-rotating phasors $0.5U_0\exp[j(2\pi f_0 t + \varphi_0)]$ and $0.5U_0\exp[-j(2\pi f_0 t + \varphi_0)]$ that have the same magnitude $0.5U_0$ but opposite

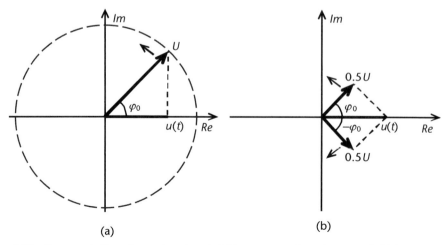

Figure 1.13 Representation of a cosine signal: (a) by one rotating phasor and (b) by two counter-rotating phasors.

initial phases (φ_0 and $-\varphi_0$) and angular rates ($\omega_0 = 2\pi f_0$ and $-\omega_0 = -2\pi f_0$). The negative frequency is meaningless for trigonometric functions, but the frequency sign indicates the phasor rotation direction for complex exponential signals. Since LTI systems do not change the shape of cosine, sine, and complex exponential signals, trigonometric and complex exponential Fourier series are widely used in the analysis and synthesis of analog and, to a certain extent, discrete-time and digital signals and systems. The possibility to use the same expansions for all types of signals and systems has made the Fourier series and transform convenient for D&R investigation. Therefore, they are discussed below.

In the orthogonal basis $\{1, \cos(\omega_0 t), \sin(\omega_0 t), \cos(2\omega_0 t), \sin(2\omega_0 t), \cos(3\omega_0 t), \sin(3\omega_0 t), \ldots, \cos(n\omega_0 t), \sin(n\omega_0 t), \ldots\}$, the cosine and sine signals with angular frequency $\omega_0 = 2\pi f_0 = 2\pi/T_0$ are the fundamentals, whereas the signals with frequencies $n\omega_0$ are their nth harmonics. The 0th harmonic in this basis is represented by 1 because $\cos(0\omega_0 t) = 1$ and $\sin(0\omega_0 t) = 0$. As follows from (1.32), the norms $\|\cos(n\omega_0 t)\| = \|\sin(n\omega_0 t)\| = (0.5T_0)^{0.5}$ for $n > 0$ and $\|\cos(n\omega_0 t)\| = T_0^{0.5}$ for $n = 0$ over an interval T_0. If a periodic signal $u(t)$ with period T_0 is represented by trigonometric Fourier series

$$u(t) = a_0 + \sum_{n=1}^{\infty} \left[a_n \cos(n\omega_0 t) + b_n \sin(n\omega_0 t) \right] = c_0 + \sum_{n=1}^{\infty} c_n \cos(n\omega_0 t + \theta_n), \quad (1.41)$$

the series coefficients, calculated according to (1.37), are

$$a_0 = \frac{1}{T_0} \int_{-0.5T_0}^{0.5T_0} u(t)\,dt, \quad a_n = \frac{2}{T_0} \int_{-0.5T_0}^{0.5T_0} u(t)\cos(n\omega_0 t)\,dt, \text{ and}$$

$$b_n = \frac{2}{T_0} \int_{-0.5T_0}^{0.5T_0} u(t)\sin(n\omega_0 t)\,dt, \quad (1.42)$$

$$c_0 = a_0, \quad c_n = \left(a_n^2 + b_n^2\right)^{0.5}, \text{ and } \theta_n = -\text{atan2}(b_n, a_n) \quad (1.43)$$

where atan2(b_n, a_n) is the four-quadrant arctangent. Coefficients c_0 and c_n plotted versus the frequency axis (ω or f) form the $u(t)$ amplitude spectrum, while phases θ_n plotted versus this axis form its phase spectrum.

The second equation (1.40), applied to the right side of (1.41), converts the trigonometric Fourier series into the complex exponential one:

$$u(t) = c_0 + \sum_{n=-\infty, n\neq 0}^{\infty} 0.5 c_n \exp\left[j(n\omega_0 t + \theta_n)\right] = \sum_{n=-\infty}^{\infty} D_n \exp(jn\omega_0 t) \quad (1.44)$$

where $D_0 = c_0$, while $D_n = 0.5 c_{|n|} \exp[j\,\text{sgn}(n)\theta_n]$ for $n \neq 0$. Thus, each harmonic of series (1.41) is a sum of two phasors of the corresponding series (1.44), and the magnitude of each phasor is half the harmonic's amplitude (see Figure 1.13(b)). In (1.41) to (1.44) $a_0 = c_0 = D_0$ represent the signal dc, whereas the sums of the other spectral components represent its ac part. As follows from (1.43) and (1.44), the amplitude (or magnitude) spectrum is even and the phase spectrum is odd for real-valued signals. Figure 1.14 shows square wave $u_{sq}(t)$ and its trigonometric and complex exponential Fourier series:

$$u_{sq}(t) = 0.5 + \sum_{n=1}^{\infty} (-1)^{0.5(n-1)} \frac{2}{n\pi} [1 - \cos(n\pi)] \cos(n\omega_0 t) \tag{1.45}$$

$$u_{sq}(t) = 0.5 + \sum_{n=-\infty, n \neq 0}^{\infty} (-1)^{0.5(|n|-1)} \frac{1}{n\pi} [1 - \cos(n\pi)] \exp(jn\omega_0 t) \tag{1.46}$$

When series (1.45) and (1.46) are infinite, they converge to $u(t)$ everywhere except at discontinuities where they produce overshoots (the Gibbs phenomenon noticed by J. W. Gibbs (United States) in 1899, but first discovered by H. Wilbraham (United Kingdom) in 1848, as was found later). Otherwise, (1.45) and (1.46) are approximate. The distances between the neighboring spectral components of $u_{sq}(t)$ are equal to $f_0 = 1/T_0$, and all the components with even $n \neq 0$ are equal to zero. Since $u_{sq}(t)$ in Figure 1.14(a) is real-valued and even, all θ_n are multiples of π. This allows displaying the $u_{sq}(t)$ spectrum in one two-dimensional plot. In general case, θ_n are not necessarily multiples of π, and separate plots for the amplitude and phase spectra are required. Although the $u_{sq}(t)$ spectral components have constant amplitudes, their phase relations cause them to add constructively within the pulses and destructively within the pauses. Due to the common nature of the Fourier series and transform, most of their properties are similar.

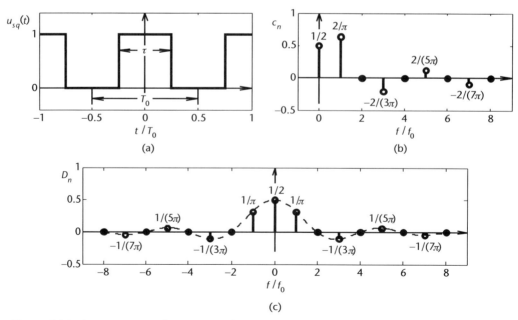

Figure 1.14 Square wave and its spectra: (a) square wave $u_{sq}(t)$, (b) cosine spectrum, and (c) exponential spectrum.

1.3.3 Fourier Transform and Its Properties

The trigonometric and exponential Fourier series reflect the amplitudes and phases of periodic signals' spectral components. A similar characterization of aperiodic signals requires transition from discrete spectral components to spectral densities (i.e., from the Fourier series to the Fourier transform). According to (1.41) through (1.44), an increase in T_0 without changing the pulse amplitude and length reduces both spectral components' amplitudes $|D_n|$ and distances f_0 between them. When $T_0 \to \infty$, any periodic signal tends to an aperiodic one, while $|D_n|$ and f_0 tend to zero at an equal rate, so the limit of D_n/f_0 remains finite as $T_0 \to \infty$. This limit is the spectral density $S_u(f)$ or $S_u(\omega)$ of the resulting $u(t)$, related to it by the Fourier transform pair:

$$S_u(f) = \int_{-\infty}^{\infty} u(t)\exp(-j2\pi ft)\,dt \quad \text{or} \quad S_u(\omega) = \int_{-\infty}^{\infty} u(t)\exp(-j\omega t)\,dt \quad (1.47)$$

$$u(t) = \int_{-\infty}^{\infty} S_u(f)\exp(j2\pi ft)\,df = \frac{1}{2\pi}\int_{-\infty}^{\infty} S_u(\omega)\exp(j\omega t)\,d\omega \quad (1.48)$$

The direct and inverse Fourier transforms (1.47) and (1.48) are often denoted as $S_u(\omega) = F[u(t)]$ and $u(t) = F^{-1}[S_u(\omega)]$, respectively. Argument ω is preferable for analysis, whereas f is more convenient for assessing the bandwidth and frequency distributions of the signal energy and/or power. When $u(t)$ is voltage, $u(t)$, c_n, and D_n have units V, while $S_u(f)$ has unit V/Hz = V · s. Generally, $S_u(f)$ of a real-valued $u(t)$ is complex-valued:

$$S_u(f) = \operatorname{Re}[S_u(f)] + j\operatorname{Im}[S_u(f)] = |S_u(f)|\exp[j\theta_u(f)] \quad (1.49)$$

where $\operatorname{Re}[S_u(f)]$ and $\operatorname{Im}[S_u(f)]$ are, respectively, the real and imaginary parts of $S_u(f)$, while $|S_u(f)|$ and $\theta_u(f)$ are, respectively, the amplitude and phase spectra of $u(t)$. As follows from (1.47), $S_u(-f) = S_u^*(f)$ when $u(t)$ is real-valued, that is, $|S_u(-f)| = |S_u(f)|$ and $\theta_u(-f) = -\theta_u(f)$. This is the Fourier transform conjugate symmetry property. If $u(t)$ is real-valued and even, $S_u(f)$ is also real-valued and even. If $u(t)$ is real-valued and odd, $S_u(f)$ is imaginary-valued and odd. When $u(t)$ consists of even $u_{\text{even}}(t)$ and odd $u_{\text{odd}}(t)$ parts, $\operatorname{Re}[S_u(f)]$ is the spectrum of $u_{\text{even}}(t)$, whereas $j\operatorname{Im}[S_u(f)]$ is the spectrum of $u_{\text{odd}}(t)$. When aperiodic signals contain dc and periodic components, their spectral densities include delta functions (see Section A.2).

Figure 1.15 shows rectangular pulse $u(t)$ and its spectral density $S_u(f)$, calculated according to (1.47),

$$S_u(f) = \frac{U\tau \sin(\pi f\tau)}{\pi f\tau} = U\tau \operatorname{sinc}(\pi f\tau) \quad (1.50)$$

As follows from (1.50) and Figure 1.15(b), the first zero crossings by $S_u(f)$ happen at $f = \pm 1/\tau$ and $S_u(0) = U\tau$. Since $u(t)$ is real-valued and even, $S_u(f)$ is also real-valued and even. This allows displaying $S_u(f)$ in one two-dimensional plot. In general case, displaying $S_u(f)$ requires separate plots for $|S_u(f)|$ and $\theta_u(f)$.

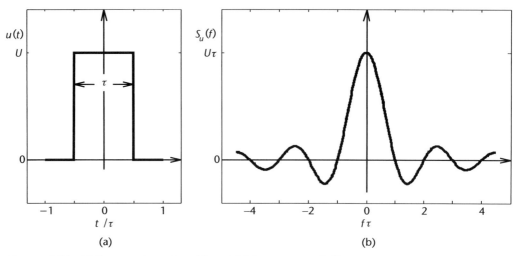

Figure 1.15 (a) Rectangular pulse $u(t)$ and (b) its spectrum $S_u(f)$.

Similar to the conjugate symmetry property, all other properties of the Fourier transforms follow from (1.47) and (1.48). The time-frequency duality property is probably the most obvious one. It states that if $S_u(\omega) = F[u(t)]$, then $F[S_u(t)] = 2\pi u(-\omega)$. In particular, this property proves that since the spectral density of a rectangular signal is the sinc function, a sinc signal has a rectangular spectral density (see Section A.1).

The linearity property can be expressed as

$$F\left[a_1 u_1(t) \pm a_2 u_2(t)\right] = a_1 F\left[u_1(t)\right] \pm a_2 F\left[u_2(t)\right] \text{ and}$$
$$F^{-1}\left[a_1 S_{u1}(\omega) \pm a_2 S_{u2}(\omega)\right] = a_1 F^{-1}\left[S_{u1}(\omega)\right] \pm a_2 F^{-1}\left[S_{u2}(\omega)\right] \quad (1.51)$$

Equations (1.51) can be extended to any finite number of terms.

The time-shifting property states that a signal delay by t_0 (see Figure 1.8) does not change its amplitude spectrum but shifts the frequency components' phases proportionally to their frequencies because the delay is the same for all of them. In mathematical terms, if

$$S_u(\omega) = F[u(t)], \text{ then } F\left[u(t - t_0)\right] = S_u(\omega)\exp(-j\omega t_0) \quad (1.52)$$

Note that time shifts regularly occur in signal processing.

The time-frequency scaling property states that time compression of signals expands their spectra and reduces the spectral densities of the original frequency components, whereas their time expansion causes the opposite effect. Thus, if

$$S_u(\omega) = F[u(t)], \text{ then } F[u(kt)] = \frac{1}{|k|} S_u\left(\frac{\omega}{k}\right) \quad (1.53)$$

Figure 1.9 allows the following simplified explanation of this property. Time compression reduces the pulse area and accelerates its variation. The pulse area reduction decreases the spectral density at zero frequency and around it. Simultaneously, the higher variation rate increases the frequencies of its spectral components. Time expansion has the opposite consequences. Note that the Doppler effect is the time-frequency scaling of signals, not simply their frequency shift as it is often presented.

The time-reversal property is a special case of the previous one with $k = -1$ (see Section 1.2.3). From (1.53), if

$$S_u(\omega) = F[u(t)], \text{ then } F[u(-t)] = S_u(-\omega) \qquad (1.54)$$

Time reversal affects only the phase spectra of real-valued signals because their amplitude spectra are even.

The frequency-shifting property states that multiplication of $u(t)$ by $\exp(j\omega_0 t)$ shifts $S_u(\omega)$ by ω_0. Thus, if

$$S_u(\omega) = F[u(t)], \text{ then } F[u(t)\exp(j\omega_0 t)] = S_u(\omega - \omega_0) \qquad (1.55)$$

This property (dual to the time-shifting property) is a basis of modulation, demodulation, and frequency conversion. It is applicable to real-valued and complex-valued signals and, using (1.40), can be rewritten: if

$$S_u(\omega) = F[u(t)], \text{ then } F[u(t)\cos(\omega_0 t)] = 0.5[S_u(\omega - \omega_0) + S_u(\omega + \omega_0)] \qquad (1.56)$$

If $u(t)$ is baseband, (1.55) and (1.56) reflect amplitude modulation. If $u(t)$ is bandpass, they reflect the first stage of frequency conversion or coherent demodulation. Figure 1.16 illustrates the amplitude modulation (AM) of carrier $\cos(2\pi f_0 t)$ by $u(t)$. Here, LSB and USB stand, respectively, for the lower and upper sidebands of modulated signal $u_{AM}(t) = u(t)\cos(2\pi f_0 t)$. The amplitude spectra $|S_u(f)|$ and $|S_{uAM}(f)|$ of $u(t)$ and $u_{AM}(t)$, respectively, are triangular for illustrative purposes. In Figure 1.16, $u(t)$ does not contain a dc component U_{dc} that otherwise would be shown as $U_{dc}\delta(f)$ in $S_u(f)$, and the modulator performs double-sideband suppressed-carrier (DSB-SC) AM. If U_{dc} existed and were larger than the $u(t)$ ac part, the modulator would perform double-sideband full-carrier (DSB-FC) AM, and $S_{uAM}(f)$ would contain $0.5U_{dc}\delta(f - f_0)$ and $0.5U_{dc}\delta(f + f_0)$ components (see Section A.2). Both modulation techniques not only shift $|S_u(f)|$ but also double its one-sided bandwidth.

The time and frequency convolution properties are dual. The first one states for real-valued $u_1(t)$ and $u_2(t)$ that if

$$S_{u1}(\omega) = F[u_1(t)], \; S_{u2}(\omega) = F[u_2(t)],$$

$$\text{and } u_1(t) * u_2(t) = \int_{-\infty}^{\infty} u_1(\tau) u_2(t - \tau) \, d\tau, \qquad (1.57)$$

$$\text{then } F[u_1(t) * u_2(t)] = S_{u1}(\omega) \cdot S_{u2}(\omega) = S_{u1}(f) \cdot S_{u2}(f) \qquad (1.58)$$

According to the frequency convolution property,

$$F\left[u_1(t) \cdot u_2(t)\right] = \frac{1}{2\pi} S_{u1}(\omega) * S_{u2}(\omega) = S_{u1}(f) * S_{u2}(f) \quad (1.59)$$

The time-differentiation property states that if

$$S_u(\omega) = F[u(t)], \text{ then } F\left[\frac{du(t)}{dt}\right] = j\omega S_u(\omega) = j2\pi f S_u(f) \quad (1.60)$$

Differentiation eliminates the signal's dc part and scales its other spectral components proportionally to their frequencies because the derivative reflects the signal variation rate. The generalized property is:

$$F\left[\frac{du^n(t)}{dt^n}\right] = (j\omega)^n S_u(\omega) = (j2\pi f)^n S_u(f) \quad (1.61)$$

The time-integration property states that if

$$S_u(\omega) = F[u(t)], \text{ then}$$

$$F\left[\int u(t)\,dt\right] = \frac{1}{j\omega} S_u(\omega) + \pi S_u(0)\delta(\omega) = \frac{1}{j2\pi f} S_u(f) + 0.5 S_u(0)\delta(f) \quad (1.62)$$

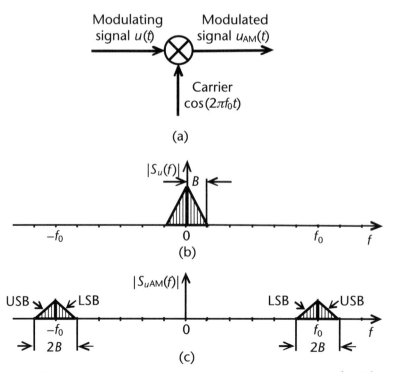

Figure 1.16 Amplitude modulation: (a) modulator, (b) amplitude spectrum $|S_u(f)|$ of $u(t)$, and (c) amplitude spectrum $|S_{uAM}(f)|$ of $u_{AM}(t)$.

To understand this property, recall that integration is inverse to differentiation. The Fourier transform properties described above for real-valued signals can be extended to complex-valued ones.

1.3.4 Spectral Distribution of Signal Energy and Power

The time-domain calculations of signal energy as well as instantaneous and average power are described in Section 1.2.1. Equation (1.38) shows that the energy and average power of signals, represented by their orthogonal components, are equal to the respective sums of the energies and average powers of these components. The Fourier series are special cases of the power signals' orthogonal representation, and the second equation in (1.38), applied to the trigonometric Fourier series (1.41) to (1.43), gives the average power P_u of a periodic signal $u(t)$:

$$P_u = c_0^2 + \sum_{n=1}^{\infty} 0.5 c_n^2 \tag{1.63}$$

Since c_n for $n > 0$ are the amplitudes of the cosine components, the right side of (1.63) is the sum of their average powers, in accordance with Parseval's identity.

For aperiodic energy signals, the Plancherel's theorem, which is a version of Parseval's identity, establishes the connection between the time and spectral distributions of the signal energy:

$$E_{u.e} = \int_{-\infty}^{\infty} |u_e(t)|^2 dt = \frac{1}{2\pi} \int_{-\infty}^{\infty} |S_{u.e}(\omega)|^2 d\omega = \int_{-\infty}^{\infty} |S_{u.e}(f)|^2 df \tag{1.64}$$

In (1.64), which is also called Rayleigh's identity, both $|S_{u.e}(\omega)|^2$ and $|S_{u.e}(f)|^2$ represent the energy spectral density (ESD), determined by the signal amplitude spectrum and independent of its phase spectrum. ESD, presented as a function of f, makes the formulas for energy calculations in the time and frequency domains symmetric. If $E_{u.e}$ is expressed in joules, the $|S_{u.e}(f)|^2$ unit is J/Hz = J · s. As follows from (1.64), the energy $E_{u.e1.2}$ of a real-valued $u_e(t)$ within the interval $[f_1, f_2]$ is

$$E_{u.e1.2} = \int_{-f_2}^{-f_1} |S_{u.e}(f)|^2 df + \int_{f_1}^{f_2} |S_{u.e}(f)|^2 df \tag{1.65}$$

Correlation and covariance functions, described for stochastic signals in Section 1.2.2, are also used for deterministic signals. While these functions of deterministic signals are based on time averaging, those of stochastic signals are based on statistical averaging, which can be replaced by time averaging only for ergodic stochastic signals. The correlation (or autocorrelation) function $R_{u.e}(\tau)$ of an energy signal $u_e(t)$ is

$$R_{u.e}(\tau) = \int_{-\infty}^{\infty} u_e(t) u_e^*(t + \tau) dt \tag{1.66}$$

Equation (1.66) shows that $R_{u.e}(\tau)$ reflects the similarity between $u_e(t)$ and its time-shifted copy $u_e(t - \tau)$, and the units of $R_{u.e}(\tau)$ are joules. The maximum similarity and, consequently, the maximum value of $R_{u.e}(\tau)$ is achieved when $\tau = 0$, and $R_{u.e}(0) = E_{u.e}$ where $E_{u.e}$ is the signal energy. When $u_e(t)$ is real-valued, $R_{u.e}(\tau)$ is even, that is, $R_{u.e}(\tau) = R_{u.e}(-\tau)$. It has been proven that $R_{u.e}(\tau)$ and $|S_{u.e}(f)|^2$ constitute a Fourier transform pair:

$$|S_{u.e}(f)|^2 = F[R_{u.e}(\tau)] \quad \text{and} \quad R_{u.e}(\tau) = F^{-1}[|S_{u.e}(f)|^2] \tag{1.67}$$

Hence, (1.64) is a special case of the second equation in (1.67).

ESD is not defined for power signals due to their infinite energy. Such signals are characterized by the power spectral density (PSD):

$$G_u(\omega) = \lim_{T \to \infty} \frac{|S_{u.T}(\omega)|^2}{T} \quad \text{or} \quad G_u(f) = \lim_{T \to \infty} \frac{|S_{u.T}(f)|^2}{T} \tag{1.68}$$

where $|S_{u.T}(\omega)|^2$ and $|S_{u.T}(f)|^2$ denote the ESD of a real-valued power signal $u_{p.T}(t)$ on the time interval T. If the signal power $P_{u.p}$ is expressed in watts, the $G_u(f)$ unit is W/Hz = W · s = J, that is, the units of the signal PSD and its energy are identical. The infinite energy of power signals also makes the correlation function definition (1.66) unacceptable. The most general definition for them

$$R_{u.p}(\tau) = \lim_{T \to \infty} \frac{1}{T} \int_{-0.5T}^{0.5T} u_p(t) u_p^*(t + \tau) dt \tag{1.69}$$

is identical to that used for the time averaging of ergodic stochastic signals in (1.25). Periodic signals are a special case of power signals. The correlation function of a periodic signal $u_{T0}(t)$ with period T_0 is

$$R_{u.T0}(\tau) = \frac{1}{T_0} \int_{-0.5T_0}^{0.5T_0} u_{T0}(t) u_{T0}^*(t + \tau) dt \tag{1.70}$$

Correlation functions (1.69) and (1.70) are expressed in watts and are even for real-valued signals. Yet, while $R_{u.p}(\tau)$ for nonperiodic signals has only one global maximum $R_{u.p}(0) = P_{u.p}$ at $\tau = 0$, $R_{u.T0}(\tau)$ is a periodic function of τ whose maxima equal to $P_{u.T0}$ appear every period. $R_{u.p}(\tau)$ and $R_{u.T0}(\tau)$ constitute Fourier transform pairs, respectively, with $G_{u.p}(f)$ and $G_{u.T0}(f)$:

$$G_{u.p}(f) = F[R_{u.p}(\tau)] \quad \text{and} \quad R_{u.p}(\tau) = F^{-1}[G_{u.p}(f)] \tag{1.71}$$

$$G_{u.T0}(f) = F[R_{u.T0}(\tau)] \quad \text{and} \quad R_{u.T0}(\tau) = F^{-1}[G_{u.T0}(f)] \tag{1.72}$$

Consequently, the average powers of $u_p(t)$ in general, and $u_{T0}(t)$ specifically, are, respectively,

$$P_{u.p} = \frac{1}{2\pi}\int_{-\infty}^{\infty} G_{u.p}(\omega)d\omega = \int_{-\infty}^{\infty} G_{u.p}(f)df \text{ and}$$

$$P_{u.T0} = \frac{1}{2\pi}\int_{-\infty}^{\infty} G_{u.T0}(\omega)d\omega = \int_{-\infty}^{\infty} G_{u.T0}(f)df \qquad (1.73)$$

For power signals $u_p(t)$ with a dc component u_{dc}, the covariance function $C_{u.p}(\tau) = R_{u.p}(\tau) - u_{dc}^2$ can be introduced. Since $R_{u.e}(\tau)$ and $|S_{u.e}(f)|^2$ as well as $R_{u.p}(\tau)$ and $G_{u.p}(f)$ constitute Fourier transform pairs, it can be stated that the wider the signal's ESD or PSD, the shorter its correlation interval, and vice versa.

The similarity between two different deterministic signals is characterized by their cross-correlation function. For real-valued energy signals $u_{e1}(t)$ and $u_{e2}(t)$, this function is

$$R_{u1.u2.e}(\tau) = \int_{-\infty}^{\infty} u_{e1}(t)u_{e2}(t+\tau)dt \qquad (1.74)$$

Although $R_{u1.u2.e}(\tau)$ turns into $R_{u.e}(\tau)$ when $u_{e1}(t) = u_{e2}(t)$, their properties differ in general case. For instance, $R_{u1.u2.e}(\tau)$ is not necessarily an even function of τ, and its maximum may not correspond to $\tau = 0$. The Fourier transform of $R_{u1.u2.e}(\tau)$ is called the cross-spectral density function, which is complex-valued in the general case because $R_{u1.u2.e}(\tau)$ is not necessarily even.

Stationary stochastic signals are power signals. The PSD $G_{xi}(f)$ of a realization $x_i(t)$ of a stationary stochastic signal $X(t)$ can be determined according to (1.68):

$$G_{xi}(f) = \lim_{T \to \infty} \frac{|S_{xi.T}(f)|^2}{T} \qquad (1.75)$$

However, $G_{xi}(f)$ does not characterize the PSD of $X(t)$. Determining the PSD $G_x(f)$ of $X(t)$ requires statistical averaging of $G_{xi}(f)$ over the ensemble of all $X(t)$ realizations:

$$G_x(f) = E[G_{xi}(f)] \qquad (1.76)$$

When a stationary stochastic signal is also ergodic, statistical averaging can be replaced by time averaging. The correlation function $R_x(\tau)$ and $G_x(f)$ of a stationary stochastic process $X(t)$ constitute a Fourier transform pair:

$$G_x(f) = F[R_x(\tau)] \text{ and } R_x(\tau) = F^{-1}[G_x(f)] \qquad (1.77)$$

This result, known as Wiener-Khinchin theorem, should actually be called Einstein-Wiener-Khinchin theorem because, being proven for deterministic functions by N. Wiener in 1930 and for stationary stochastic processes by A. Khinchin in 1934, it was first derived by A. Einstein no later than 1914. As follows from (1.77), the average power P_x of $X(t)$ is

$$P_x = \frac{1}{2\pi} \int_{-\infty}^{\infty} G_x(\omega) d\omega = \int_{-\infty}^{\infty} G_x(f) df \qquad (1.78)$$

Thus, relations between correlation functions and PSDs are similar for deterministic and stochastic signals. Spectral and correlation analyses are widely used for both types of signals.

1.3.5 Transmission of Signals Through LTI Systems

The input $u_{in}(t)$ and output $u_{out}(t)$ signals of an LTI system are connected in the time domain as

$$u_{out}(t) = u_{in}(t) * h(t) = \int_{-\infty}^{\infty} u_{in}(\tau) h(t - \tau) d\tau \qquad (1.79)$$

where $h(t)$ is the system impulse response. The Fourier transform of $h(t)$ is the system transfer function $H(f)$. The Fourier transform time convolution property applied to (1.79) allows determining the relation between the spectra $S_{u.in}(f)$ and $S_{u.out}(f)$ of the system input and output signals

$$S_{u.out}(f) = S_{u.in}(f) H(f) \quad \text{or} \quad S_{u.out}(\omega) = S_{u.in}(\omega) H(\omega) \qquad (1.80)$$

To separate the distortions of $u_{in}(t)$ introduced by the system's amplitude-frequency response (AFR) $|H(f)|$ and phase-frequency response (PFR) $\theta_h(f)$, (1.80) should be rewritten as

$$|S_{u.out}(f)| \exp[j\theta_{u.out}(f)] = |S_{u.in}(f)| \cdot |H(f)| \exp\{j[\theta_{u.in}(f) + \theta_h(f)]\} \qquad (1.81)$$

When $H(f) = H_0 \exp(-j2\pi f t_0)$, that is, the AFR is uniform and the PFR is linear at least within the signal bandwidth, the LTI system does not distort input signals because it multiplies all their frequency components by the same factor H_0 and delays them by the same time t_0. When $h(t)$ is even about its midpoint, its PFR is linear, and the input signals can be distorted only by its AFR. In general case, the distortions can be caused by both PFR and AFR.

Since correlation and convolution are integrals of the products of two shifted functions, let us determine the relationship between them. The cross-correlation function of two complex-valued signals $u_1(t)$ and $u_2(t)$ is

$$R_{u1.u2}(\tau) = \int_{-\infty}^{\infty} u_1(t) u_2^*(t + \tau) dt \qquad (1.82)$$

Rewriting (1.82) for a new integration variable $t' = -t$ yields

$$R_{u1.u2}(\tau) = -\int_{\infty}^{-\infty} u_1(-t') u_2^*(-t' + \tau) dt' = \int_{-\infty}^{\infty} u_1(-t') u_2^*(\tau - t') dt' = u_1(-\tau) * u_2^*(\tau) \qquad (1.83)$$

Similarly, the correlation function

$$R_u(\tau) = \int_{-\infty}^{\infty} u(t)u^*(t+\tau)dt = u(-\tau) * u^*(\tau) \qquad (1.84)$$

For real-valued signals, (1.83) and (1.84) become, respectively, $R_{u1.u2}(\tau) = u_1(-\tau) * u_2(\tau)$ and $R_u(\tau) = u(-\tau) * u(\tau)$. More information on the topics discussed in this section can be found, for instance, in [19–30].

1.4 Baseband and Bandpass Signals

1.4.1 Baseband Signals and Modulation

Spectra of the Txs' input signals and Rxs' output signals are located close to zero frequency and may include dc. The bandwidths of these signals can be as narrow as a fraction of a hertz or as wide as several gigahertz, but they occupy the base of the spectrum and, therefore, are baseband. Baseband signals may be analog, discrete-time, or digital; may carry different types of information: voice, music, video, text, results of analog and digital measurements and/or processing; and may represent single-source or multisource signals. They cannot be sent over radio channels due to the impossibility of their effective transmission by reasonable-size antennas. Although they can be directly transmitted over a pair of wires or coaxial cable, even then it is more efficient to preprocess them. The preprocessing typically includes amplification, filtering, and spectrum shifting for frequency-division multiplexing (FDM). For discrete-time signals, it may include time-division multiplexing (TDM). Currently, analog and discrete-time signals are usually digitized prior to baseband transmissions. Digital signals may undergo formatting and multiplexing.

Modulation of carriers, which varies their parameters according to the information contained in baseband signals, allows transmitting this information over radio channels. Cosine waves or groups of cosine waves with equidistantly spaced frequencies are most frequently used carriers. When a cosine signal (1.1) is a carrier, its amplitude, phase, and/or frequency can be varied proportionally to a baseband signal or a function of it. Varying only one parameter produces amplitude, phase, or frequency modulation (respectively, AM, PM, or FM). These modulation techniques have several versions. For instance, DSB-SC AM, DSB-FC AM (see Section 1.3.3), and double-sideband reduced-carrier (DSB-RC) AM are versions of AM. Since PM varies the carrier phase proportionally to a modulating signal and FM proportionally to the integral of this signal, these techniques are versions of a more general angle modulation that has many other versions.

While some techniques vary only amplitude or angle of a carrier, others vary both. For instance, single-sideband (SSB) modulation, which is often considered a version of AM, actually combines amplitude and angle modulations because AM-only spectra are symmetric about their center frequencies. The asymmetry of SSB spectrum indicates the presence of angle modulation. Therefore, all SSB versions with full, reduced, and suppressed carriers as well as vestigial sideband (VSB) modulation are actually combinations of amplitude and angle modulations. Yet not all combinations of amplitude and angle modulations have asymmetric spectra. For example, quadrature amplitude modulated (QAM) signals, which are sums of two amplitude-modulated sinusoids of the same frequency phase-shifted by 90°, have symmetric spectra. Modulation by digital signals is often called keying (from a key in the Morse telegraph). Depending on the varied carrier parameter, the basic keying techniques are amplitude-shift keying (ASK), frequency-shift keying (FSK), and phase-shift keying (PSK). They have several versions. Binary ASK is sometimes called on-off keying (OOK). Yet the term "keying" is not always used: digital QAM, for example, is called modulation. In an analog Tx, modulation is usually performed on its intermediate frequency (IF), producing a bandpass information-carrying signal emitted by the antenna after its translation to RF. Examples of bandpass signals with analog and digital (binary) modulations are shown in Figures 1.17 and 1.18, respectively.

In digital Txs (see Figure 1.19(a)), modulation is performed in the Txs' digital portions (TDPs). Prior to it, the Txs' input signals typically undergo source encoding, encryption, and channel encoding. Analog input signals are digitized before entering the TDPs. Modulated signals can be frequency and/or time spread and multiplexed with other signals. All these operations are executed in TDPs mostly using the baseband complex-valued signals that are complex envelopes or equivalents of bandpass real-valued signals intended for transmission. The analog bandpass real-valued signals are reconstructed from these equivalents and prepared for transmission in the Txs' analog and mixed-signal back-ends (AMBs). The reconstruction can be baseband or bandpass.

In digital Rxs (see Figure 1.19(b)), received signals, which are mixtures of desired signals, noise, and interference, are initially processed in their analog and mixed-signal front-ends (AMFs) and then, after digitization, in the Rxs' digital portions (RDPs). The digitization can also be bandpass or baseband. In any case, the received analog bandpass real-valued signals are typically converted into their digital baseband complex-valued equivalents. The RDP processing includes demultiplexing (for multiplexed signals), despreading (for spread signals), demodulation, channel decoding, decryption (for encrypted signals), and source decoding. The

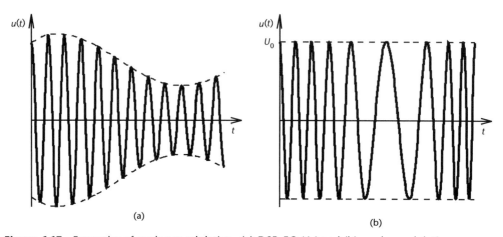

Figure 1.17 Examples of analog modulation: (a) DSB-FC AM and (b) angle modulation.

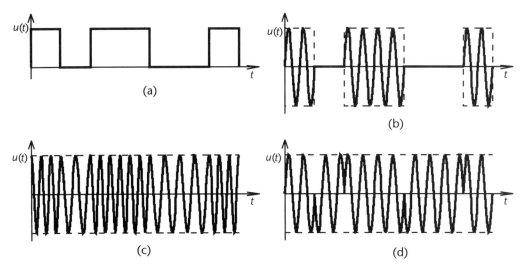

Figure 1.18 Examples of digital modulation: (a) modulating binary sequence, (b) ASK, (c) FSK, and (d) PSK.

reconstruction circuits (RCs) restore digitally-encoded analog signals. More detailed information on digital Txs and Rxs is provided in Chapters 2 to 4.

1.4.2 Bandpass Signals and Their Complex-Valued Equivalents

Real-valued bandpass signals can be represented in TDPs and RDPs by the digital samples of their instantaneous values, amplitudes and phases of their baseband complex-valued equivalents, or I and Q components of these equivalents. As shown, for instance, in [31], each of these representations has certain advantages, but the last one usually utilizes TDPs' and RDPs' computational power most efficiently, especially in multipurpose, multistandard radios. The relation between bandpass signals and their baseband complex-valued equivalents is discussed below. As follows from (1.40), cosine signal (1.1) can be represented as the real part of phasor

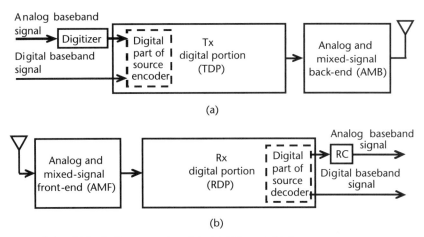

Figure 1.19 High-level block diagrams of a digital (a) Tx and (b) Rx.

$U_0\exp[j(2\pi f_0 t + \varphi_0)]$, which rotates counterclockwise, or as the sum of two counter-rotating phasors $0.5U_0\exp[j(2\pi f_0 t + \varphi_0)]$ and $0.5U_0\exp[-j(2\pi f_0 t + \varphi_0)]$. In both cases, the phasors have constant magnitudes and rotation rates. The exponential Fourier series of the second representation has two spectral components with frequencies f_0 and $-f_0$ (see Figure 1.20(a)). This signal requires single-channel processing. In Figure 1.20(b), $u_c(t)$ is represented by the I and Q components of its phasor that contains only a positive-frequency component but requires two-channel processing.

This concept is extended to bandpass real-valued signals. In such a signal $u(t) = U(t)\cos[2\pi f_0 t + \theta(t)]$, information is carried by its envelope $U(t)$ and/or phase $\theta(t)$. Its center frequency f_0 is usually known and therefore noninformative. This allows excluding f_0 from the most complex signal processing operations in Txs and Rxs. For that reason, the baseband complex-valued equivalents $Z(t)$ of $u(t)$ are processed in the TDPs and RDPs. The relations among $u(t)$, $Z(t)$, and $Z(t)$ components are as follows:

$$u(t) = U(t)\cos\left[2\pi f_0 t + \theta(t)\right] = I(t)\cos(2\pi f_0 t) - Q(t)\sin(2\pi f_0 t) \qquad (1.85)$$

$$Z(t) = U(t)\exp[\theta(t)] = I(t) + jQ(t) \qquad (1.86)$$

where $I(t)$ and $Q(t)$ are, respectively, I and Q components of $Z(t)$. Since

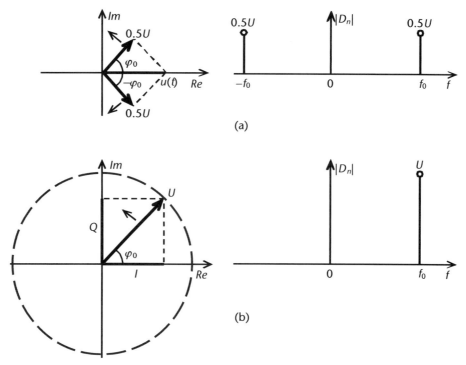

Figure 1.20 Phasor and spectral diagrams of a cosine signal: (a) cosine signal and (b) its complex-valued equivalent.

$$u(t) = \text{Re}\left[Z(t)\exp(j2\pi f_0 t)\right] = \text{Re}\left\{U(t)\exp\left\{j\left[2\pi f_0 t + \theta(t)\right]\right\}\right\}, \quad (1.87)$$

$Z(t)$ is also called the complex envelope of $u(t)$. Any of the pairs $\{I(t), Q(t)\}$ and $\{U(t), \theta(t)\}$ contains all the information carried by $u(t)$. The relations between these pairs follow from (1.86) and (1.87):

$$I(t) = U(t)\cos[\theta(t)] \quad \text{and} \quad Q(t) = U(t)\sin[\theta(t)] \quad (1.88)$$

$$U(t) = \left[I^2(t) + Q^2(t)\right]^{0.5} \quad \text{and} \quad \theta(t) = \text{atan2}[Q(t), I(t)] \quad (1.89)$$

where atan2$[Q(t), I(t)]$ is the four-quadrant arctangent.

Although the transition from $u(t)$ to $Z(t)$ necessitates processing two real-valued signals, $I(t)$ and $Q(t)$ or $U(t)$ and $\theta(t)$, instead of one, the resultant sampling rate does not increase because the rate for each $Z(t)$ component does not exceed a half of that required for $u(t)$. As mentioned above, the representation of $Z(t)$ by $I(t)$ and $Q(t)$ is usually most efficient. According to the conjugate symmetry property of the Fourier transform (see Section 1.3.3), the amplitude spectrum $|S_u(f)|$ of $u(t)$ is symmetric about zero frequency (the amplitude spectra of $I(t)$ and $Q(t)$ have the same property), although each of its sides may be asymmetric about f_0. In contrast with $|S_u(f)|$ and the amplitude spectra of $I(t)$ and $Q(t)$, the amplitude spectrum $|S_Z(f)|$ of $Z(t)$ is, in the general case, asymmetric about zero frequency and can be located on either side of the frequency axis or on both sides simultaneously. Analytic signals, whose imaginary parts are the Hilbert transforms of their real parts, are a special case of complex-valued signals. Since this transform introduces quarter-cycle delay of each spectral component, analytic signals' spectra are located only at positive frequencies.

Frequency conversions are common operations in TDPs and RDPs. Therefore, they are discussed below for digital baseband complex-valued equivalents $Z_q(nT_s)$ and their I and Q components $I_q(nT_s)$ and $Q_q(nT_s)$ where T_s is a sampling period. According to (1.55), the spectrum $S_{Z1}(f)$ of $Z_{q1}(nT_s)$ can be shifted by f_1 as a result of multiplying $Z_{q1}(nT_s) = I_{q1}(nT_s) + jQ_{q1}(nT_s)$ by $\exp_q(j2\pi f_1 nT_s) = \cos_q(2\pi f_1 nT_s) + j\sin_q(2\pi f_1 nT_s)$. Thus, the spectrum $S_{Z2}(f)$ of the obtained complex-valued signal

$$Z_{q2}(nT_s) = \left[I_{q1}(nT_s)\cos_q(2\pi f_1 nT_s) - Q_{q1}(nT_s)\sin_q(2\pi f_1 nT_s)\right] \\ + j\left[I_{q1}(nT_s)\sin_q(2\pi f_1 nT_s) + Q_{q1}(nT_s)\cos_q(2\pi f_1 nT_s)\right] \quad (1.90)$$

is a frequency-shifted copy of $S_{Z1}(f)$: $S_{Z2}(f) = S_{Z1}(f - f_1)$. The frequency converter, realizing (1.90), and positions of $S_{Z1}(f)$, $S_{\exp}(f)$, and $S_{Z2}(f)$ are shown in Figure 1.21. Note that all the spectra are located within $[-0.5f_s, 0.5f_s[$ where $f_s = 1/T_s$ is a sampling rate, and $S_{\exp}(f) = \delta(f - f_1)$ according to (A.20).

In a digital Tx with bandpass reconstruction, a bandpass real-valued signal $u_q(nT_s)$ is generated from its baseband complex-valued equivalent $Z_q(nT_s)$ in the TDP. This generation is a special case of the frequency conversion depicted in Figure 1.21. Indeed, $Z_q(nT_s)$ should be translated to the TDP output frequency f_0 first,

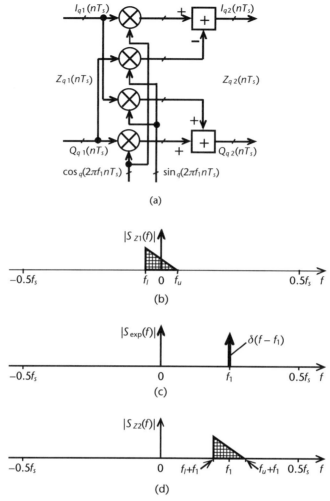

Figure 1.21 Digital frequency conversion of complex-valued signals and their amplitude spectra: (a) converter block diagram, (b) $S_{Z1}(f)$, (c) $S_{exp}(f)$, and (d) $S_{Z2}(f)$.

and then the obtained complex-valued signal $Z_{q1}(nT_s)$ should be transformed into real-valued $u_q(nT_s)$. Since $Z_{q1}(nT_s)$ does not have spectral components at negative frequencies, it is an analytic signal whose transformation into $u_q(nT_s)$ requires only discarding its imaginary part. Consequently, this part just should not be calculated, as shown in Figure 1.22 where $|S_Z(f)|$ and $|S_u(f)|$ are the amplitude spectra of $Z_q(nT_s)$ and $u_q(nT_s)$, respectively, and f_0 is the center frequency of $u_q(nT_s)$. In contrast with $Z_q(nT_s)$ and $\exp_q(j2\pi f_0 nT_s)$, $u_q(nT_s)$ is real-valued, and, therefore, $|S_u(f)|$ is symmetric about zero frequency.

The frequency conversions in Figures 1.21 and 1.22 do not produce undesired spectral images and, therefore, do not require filters because their input signals are complex-valued. The situation changes when at least one of the input signals is real-valued. In a digital Rx with bandpass digitization, a digital bandpass real-valued signal $u_q(nT_s)$ enters the RDP where its baseband complex-valued equivalent $Z_q(nT_s)$ is generated as illustrated in Figure 1.23. Here, two identical digital lowpass filters

1.4 Baseband and Bandpass Signals

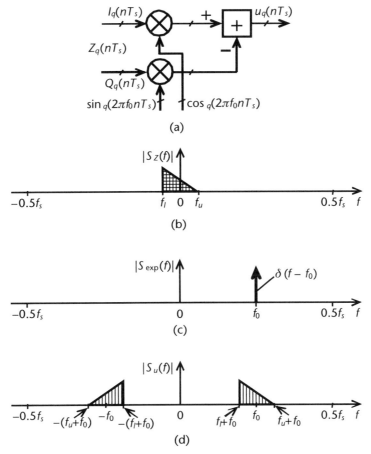

Figure 1.22 Generator of $u_q(nT_s)$ and amplitude spectra of its input and output signals: (a) generator block diagram, (b) $|S_Z(f)|$, (c) $|S_{\exp}(f)|$, and (d) $|S_u(f)|$.

(LPFs) with AFR $|H(f)|$, depicted with a dashed line in Figure 1.23(d), are placed at the multipliers' outputs to suppress the undesired sum-frequency products shown with a dotted line.

1.4.3 Bandwidths of Signals and Circuits

The amounts of information carried by signals and transmitted over circuits depend on their bandwidths. However, it is difficult to uniquely and unambiguously define "bandwidth" because spectral densities of time-limited functions are nonzero at any finite frequency interval, and convenience of definitions depends on applications. Since the bandwidths of pulses are connected to their durations, bandwidths of stochastic processes to their correlation intervals, and bandwidths of circuits to the durations of their impulse responses, the approaches to defining both notions should be identical in each pair. For a baseband real-valued energy signal $u(t)$, the most theoretically consistent definitions are its rms bandwidth and duration (B_{rms} and τ_{rms}). The one-sided rms bandwidth and duration are equal to the positive square roots of the normalized second moments of $S_u(f)$ and $u(t)$, respectively:

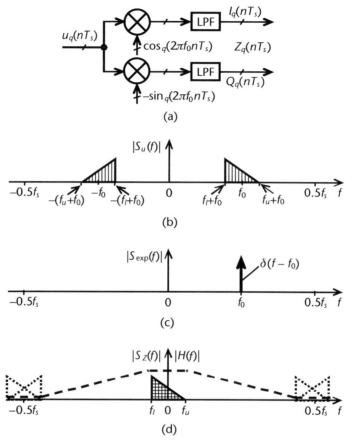

Figure 1.23 Generator of $Z_q(nT_s)$ and amplitude spectra of its input and output signals: (a) generator block diagram, (b) $|S_u(f)|$, (c) $|S_{exp}(f)|$, and (d) $|S_Z(f)|$.

$$B_{rms1} = \left[\frac{\int_{-\infty}^{\infty} f^2 |S_u(f)|^2 df}{\int_{-\infty}^{\infty} |S_u(f)|^2 df} \right]^{0.5} \quad (1.91)$$

$$\tau_{rms1} = \left[\frac{\int_{-\infty}^{\infty} t^2 |u(t)|^2 dt}{\int_{-\infty}^{\infty} |u(t)|^2 dt} \right]^{0.5} \quad (1.92)$$

where t is time relative to t_0 of $u(t)$, calculated as:

$$t_0 = \frac{\int_{-\infty}^{\infty} t |u(t)|^2 dt}{\int_{-\infty}^{\infty} |u(t)|^2 dt} \quad (1.93)$$

1.4 Baseband and Bandpass Signals

The two-sided rms bandwidth and duration are, respectively, $B_{rms2} = 2B_{rms1}$ and $\tau_{rms2} = 2\tau_{rms1}$. Calculating the rms bandwidths of bandpass energy signals requires determining f in (1.91) relative to their center frequencies. Equations (1.91) and (1.92) allow proving the duration-bandwidth uncertainty relation for signals and circuits:

$$\tau_{rms1} B_{rms1} \geq \frac{1}{4\pi} \qquad (1.94)$$

This fundamental relation limits simultaneous reduction of the durations and bandwidths of pulses, correlation intervals and bandwidths of stochastic processes, as well as durations of impulse responses and bandwidths of filters (recall the Fourier transform time-frequency scaling property). Functions with a small duration-bandwidth product are of great interest in radar, sonar, communications, and so forth. They are also important for sampling and interpolating (S&I) circuits. The uncertainty relation does not limit the accuracy of simultaneous time-frequency signal analysis because an analyzed signal (in contrast with an elementary particle in quantum mechanics) can be simultaneously sent to two channels: one with a high temporal resolution and another with a high frequency resolution.

The rms bandwidth and duration, being convenient for theoretical analysis, are not applicable to the functions that do not have finite second moments. Even if these moments exist, they may insufficiently characterize the signal energy distribution and filter's attenuation. The most widely used alternative definition of the $u(t)$ bandwidth B is the difference between the highest f_h and the lowest f_l positive frequencies, for which $|S_u(f)| \geq \alpha |S_u(f)|_{max}$ where $\alpha < 1$. Typically for signals $\alpha = 0.1$, that is,

$$\frac{|S_u(f)|_{max}}{|S_u(f_l)|} = \frac{|S_u(f)|_{max}}{|S_u(f_h)|} = 10 = 20 \text{ dB} \qquad (1.95)$$

When this definition is applied to filters, $\alpha = 1/(2^{0.5}) \approx 0.707$ is often used, and the corresponding bandwidth is called 3-dB or half-power bandwidth. To stricter limit in-band ripple, $0.9 \leq \alpha < 1$ can be selected. Filter stopband is a frequency band where the attenuation exceeds a predetermined level $1/\beta$ with $0 < \beta << \alpha$. The frequency bands between the passbands and stopbands are called transition bands or skirts. In LPFs, $f_l = 0$, and f_h is specified. In bandpass filters (BPFs), f_l and f_h are specified. In highpass filters (HPFs), f_l is specified, and f_h is infinite. The AFRs of LPFs and BPFs are illustrated respectively, by Figure 1.24(a, b).

Antialiasing and interpolating filters used for D&R may have multiple stopbands. In antialiasing filters, they suppress noise and interference within the frequency intervals where replicas of the analog signal spectrum appear after sampling. In interpolating filters, they suppress unwanted replicas of the signal spectrum. The gaps between the stopbands are designated as "don't care" bands in Figure 1.24(c, d) where the AFRs of multi-stopband LPF and BPF are depicted.

Prior to the transmission of a signal over a communication channel of any nature, its spectrum is usually limited to the bandwidth that is called essential, and the spectral components outside this bandwidth are suppressed to minimize

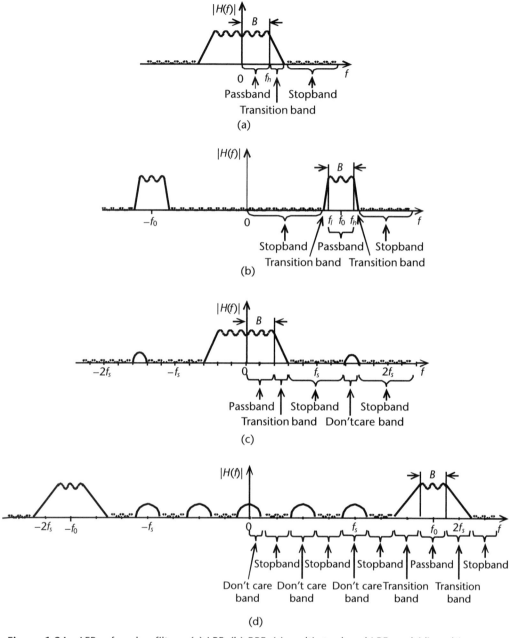

Figure 1.24 AFRs of analog filters: (a) LPF, (b) BPF, (c) multi-stopband LPF, and (d) multi-stopband BPF.

interchannel interference. Most of the signal energy should be located within the essential bandwidth, but its particular fraction depends on the application. In communications, for instance, it is necessary to take into account, besides the signal-to-noise ratio, intersymbol interference (ISI) and synchronization accuracy. In radar, excessive restriction of the signal bandwith reduces range resolution. The out-of-band roll-off rate of the $u(t)$ spectrum $S_u(f)$ is $1/f^{n+1}$ where n is the lowest order of the $u(t)$ derivative with a jump discontinuity. Consequently, the spectral roll-off of a Gaussian pulse is fastest because it has derivatives of all orders. Since the spectral density of a Gaussian pulse is also Gaussian, the duration-bandwidth product of this signal is minimal.

1.5 Summary

A strong theoretical foundation is needed to transmit, receive, and/or process information accurately, reliably, and quickly, with minimum energy and bandwidth consumption, using equipment of minimum weight, size, and cost. Any theory operates not with real objects (whose mathematical representation is too complex or impossible) but with their simplified models that are considered adequate if the results of calculations and simulations based on them are consistent with experimental results. The use of models, being advantageous, requires caution.

Modern technology supports proliferation of digital signals and processing that provide the possibility of regenerating damaged information, high and independent of destabilizing factors accuracy of processing, superior versatility and flexibility, largest scale of integration, and reduced cost of equipment development and production. D&R are interfaces between the radios' digital portions and the analog world outside. Digitization comprises antialiasing filtering, generation of samples, quantization, and digital digitization-related operations. Reconstruction comprises digital reconstruction-related operations, analog decoding of digital signals, pulse shaping, and analog interpolating filtering. Although antialiasing filtering and generation of samples are currently performed separately, the sampling theorem's direct interpretation allows their combining. The situation with pulse shaping and interpolating filtering is identical.

Deterministic signals are used as test, pilot, reference, synchronization, and carrier signals, but only stochastic signals can be employed for information transmission. Interfering signals are also stochastic. The most exhaustive characteristics of random events are their probabilities, of random variables their probability distributions, and of random processes the joint probability distributions of their samples taken sufficiently often. Many probabilistic problems can be solved using only moments of random variables and moment functions of stochastic processes. The first two moments and moment functions are most important because they reflect the positions and spreads of random variables and stochastic processes, as well as linear statistical dependence (correlation) between different random variables, stochastic processes, and samples of the same stochastic processes. The absence of correlation means complete statistical independence only for Gaussian variables and processes.

In the general case, analysis of stochastic processes is very complex. However, many important classes of these processes satisfy certain constraints that simplify their analysis and practical use. Stationary and ergodic stochastic processes are among them. Strictly stationary stochastic processes have time-shift-invariant joint probability distributions, whereas wide-sense stationarity means time-shift invariance of just the first two moment functions. Strict stationarity implies wide-sense stationarity, but the inverse statement is correct only for Gaussian processes. A stochastic process is ergodic if any of its realizations reflects all its statistical properties. This makes the results of statistical and time averaging identical. Although ergodicity, in principle, does not presume stationarity, ergodic processes are stationary in most applications.

Expansion of signals over bases often simplifies their analysis and processing. The simplification is greatest when a basis is orthogonal or orthonormal. Such an expansion with Fourier coefficients minimizes the rms error of signal approximation. Trigonometric and complex exponential Fourier series, as well as classical sampling, are special cases of generalized Fourier series. In this book, the Fourier series and transform are widely used for D&R analysis because they can be applied to analog, discrete-time, and digital signals. Besides, their properties make operations on signals intuitively understandable and simplify the assessment of operations' results.

Input signals of Txs and output signals of Rxs are baseband. Due to the impossibility of their effective radiation by reasonable-size antennas, they are used for modulation of carriers. Bandpass real-valued signals obtained as a result of this modulation carry information over RF channels. In TDPs and RDPs of most digital radios, these signals are represented by digital samples of the I and Q components of their baseband complex-valued equivalents. The analog bandpass real-valued signals should be reconstructed from their equivalents at the TDPs' outputs and digitized at the RDPs' inputs. The D&R can be baseband or bandpass.

The most theoretically consistent and convenient for analysis definition of a signal's or circuit's bandwidth is its rms bandwidth. In particular, it allows deriving the duration-bandwidth uncertainty relation. However, this definition is not always applicable, and, even when applicable, it may not sufficiently characterize some practically important properties of signals and circuits. Therefore, many other definitions of bandwidth are used. For filters, the definitions of their passbands, stopbands, and transition bands are required. The major properties of signals, important for specific applications, are reflected by their essential bandwidths.

References

[1] Doob, J., *Stochastic Processes*, New York: John Wiley & Sons, 1953.
[2] Rosenblatt, M., *Random Processes*. Oxford, U.K.: Oxford University Press, 1962.
[3] Davenport Jr., W. B., and W. L. Root, *An Introduction to the Theory of Random Signals and Noise*, New York: IEEE Press, 1987.
[4] Gikhman, I. I., and A. V. Skorokhod, *Introduction to the Theory of Random Processes*, Mineola, NY: Dover Publications, 1996.
[5] Papoulis, A., and S. U. Pillai, *Probability, Random Variables, and Stochastic Processes*, 4th ed., New York: McGraw-Hill, 2002.
[6] Stark, H., and J. W. Woods, *Probability and Random Processes with Application to Signal Processing*, 3rd ed., Upper Saddle River, NJ: Prentice Hall, 2002.
[7] Karlin, S., and H. E. Taylor. *A First Course in Stochastic Processes*, 2nd ed., New York: Academic Press, 2012.
[8] Van Trees, H. L., K. Bell, and Z. Tian, *Detection, Estimation, and Modulation Theory*, Part I, New York, John Wiley & Sons, 2013.
[9] Lindgren, G., H. Rootzen, and M. Sandsten, *Stationary Stochastic Processes for Scientists and Engineers*, Boca Raton, FL: CRC Press, 2013.
[10] Florescu, I., *Handbook of Probability*, New York: John Wiley & Sons, 2013.
[11] Borovkov, A. A., *Probability Theory*, New York: Springer, 2013.

References

[12] Yates, R. D., and D. J. Goodman, *Probability and Stochastic Processes: A Friendly Introduction for Electrical and Computer Engineers*, 3rd ed., New York: John Wiley & Sons, 2014.

[13] Blitzstein, J., and J. Hwang, *Introduction to Probability*. Boca Raton, FL: CRC Press, 2015.

[14] Hancock, J., and W. Wade, "A Digital Wide-Band Nonlinear Receiver Capable of Near Optimum Reception in the Presence of Narrow-Band Interference," *IEEE Trans. Commun. Syst.*, Vol. 11, No. 3, 1963, pp. 272–279.

[15] Allen, W. B., and E. C. Westerfield, "Digital Compressed-Time Correlators and Matched Filters for Active Sonar," *J. Acoust. Soc. of America*, Vol. 36, No. 1, 1964, pp. 121–139.

[16] Poberezhskiy, Y. S., "Analysis of Time Compressors with Analog Representation of Samples" (in Russian), *Problems of Radioelectronics*, TRC, No. 3, 1968, pp. 18–27.

[17] Poberezhskiy, Y. S., "Spectral and Correlation Analysis of Time-Compressed Signals" (in Russian), *Problems of Radioelectronics*, TRC, No. 4, 1968, pp. 109–117.

[18] Poberezhskiy, Y. S., "Methods of Time-Compression of Digital Signals" (in Russian), *Problems of Radioelectronics*, TRC, No. 9, 1969, pp. 99–111.

[19] Goldman, S., *Frequency Analysis, Modulation, and Noise*, New York: McGraw-Hill, 1948.

[20] Fink, L. M., *Signals, Interference, Errors, …* (in Russian), 2nd ed., Moscow, Russia: Radio and Communications, 1984.

[21] De Coulon, F., *Signal Theory and Processing*, Norwood, MA: Artech House, 1986.

[22] Gonorovskiy, J. S., *Signals and Circuits in Radio Engineering* (in Russian), 4th ed., Moscow, Russia: Radio and Communications, 1986.

[23] Papoulis, A., *Circuits and Systems: A Modern Approach*, Oxford, U.K.: Oxford University Press, 1995.

[24] Siebert, W. M., *Circuits, Signals and Systems*, New York: McGraw-Hill, 1998.

[25] Vakman, D, *Signals, Oscillations, and Waves: A Modern Approach*, Norwood, MA: Artech House, 1998.

[26] Lathi, B. P., *Linear Systems and Signals*, 2nd ed., Oxford, U.K.: Oxford University Press, 2004.

[27] Sundararajan, D., *A Practical Approach to Signals and Systems*. New York: John Wiley & Sons, 2009.

[28] Hsu, H., *Schaum's Outline of Signals and Systems*, 3rd ed., New York: McGraw-Hill, 2014.

[29] Phillips, C. L., J. Parr, and E. Riskin, *Signals, Systems, and Transforms*, 5th ed., Harlow, U.K.: Pearson Education, 2014.

[30] Oppenheim, A. V., A. S. Willsky, and S. H. Nawad, *Signals and Systems*, 2nd ed., Harlow, U.K.: Pearson Education, 2015.

[31] Poberezhskiy, Y. S., *Digital Radio Receivers* (in Russian), Moscow, Russia: Radio & Communications, 1987.

CHAPTER 2
Radio Systems

2.1 Overview

Often, the technologies used in systems intended for different purposes have the same theoretical basis and/or technical solutions (e.g., synchronization in communication systems and target acquisition and tracking in radar). Yet the technologies used in systems intended for the same purpose may require different theoretical approaches and/or technical solutions. D&R provide examples of both possibilities because requirements for them are determined by their input signals regardless of the systems' purposes. Thus, systems with identical purposes require different D&R circuits if they use different signals and/or operate in different environments, whereas systems used for different purposes require the same D&R circuits if parameters of their input signals are similar. Therefore, theoretical results and technical solutions developed for D&R in radio systems are also useful for most other applications. Chapter 2 presents basic information on radio systems, shows the role of D&R, identifies factors that determine requirements for these procedures, and explains why the book is focused on communication radios.

Section 2.2 notes that although radio waves are utilized for transmitting both energy and information, D&R circuits are used only in the Txs and Rxs of the systems transmitting, receiving, and disseminating information. The general purpose of such systems is to reliably, accurately, and cost-effectively deliver information to end users, consuming minimum energy and bandwidth. D&R play a significant role in achieving this objective. An explanation of why communication radios are the best case study for the D&R analysis and further development in systems of all types is provided. The division of RF spectrum into frequency bands, radio wave propagation modes in these bands, and spectrum utilization by various systems are described.

Section 2.3 provides concise information on communication systems, identifies their development trends, and demonstrates their diversity. The signal processing operations performed in communication systems are analyzed from the perspective of their interdependence with D&R. Since channel encoding/decoding, modulation/demodulation, and spreading/despreading influence D&R, their basic principles are discussed, and more information on the subjects is provided in Chapters 3 and 4. Similarities between source coding and digitization are outlined.

Several types of radio systems are described in Section 2.4 with the purpose of comparing the requirements for their D&R circuits to those of communication systems. Broadcast systems are presented first, and the specifics of broadcast Txs and Rxs are emphasized. Since the principles of navigation, positioning, and geolocation significantly differ from those of communications, they are described in more detail. The attention to radar and electronic warfare (EW) systems is caused by the fact that some requirements for the digitization circuits of their Rxs can be higher than for these circuits in most communication Rxs.

2.2 Radio Systems and Radio Spectrum

2.2.1 Diversity of Radio Systems

Radio transmissions are based on the electromagnetic field theory, originated by J. C. Maxwell (United Kingdom) in the mid-1860s. This theory was rooted in the ideas of M. Faraday (United Kingdom) formulated during the 1830s and 1840s. After its experimental proof by H. R. Hertz (Germany) in the late 1880s, the efforts of many outstanding scientists and engineers resulted in the development of radio communications. Among those who contributed most to these efforts were É. Branly (France), O. Lodge (United Kingdom), A. S. Popov (Russia), G. Marconi (United Kingdom), N. Tesla (United States), and F. Braun (Germany). The subsequent progress of radio systems was unprecedented in human history.

Radio waves are electromagnetic waves with frequencies from 3 Hz to 3 THz. They are currently used for transmission of energy and information. Systems generating, transmitting, and delivering RF energy (microwave ovens, diathermy, directed energy weapons) should provide maximum efficiency and minimally affect unintended recipients. Although signals carrying RF energy usually have narrow bandwidths, their high power makes them dangerous to Rxs operating at neighboring frequencies. The major design objectives for the systems using radio waves for carrying information include minimum energy and/or bandwidth consumption. The minimum bandwidth consumption does not necessarily mean the minimum bandwidth of each transmitted signal because many orthogonal signals can be transmitted within the same frequency band without mutual interference. Still, several factors restrict absolute and relative bandwidths of radio signals: limited bandwidths of electronic components, complexity of simultaneously achieving wide relative bandwidths and high gains of antennas, and frequency-dependent radio wave propagation near Earth and other planets due to the influence of their surfaces and atmospheres. Therefore, most radio signals have relative bandwidths $B/f_0 \sim 0.01$ where B and f_0 are, respectively, the signal bandwidth and center frequency. Only in ultrawideband radios, used for see-through-the-wall radar, precision measurements, and some communications, $B/f_0 \geq 0.2$.

D&R are used exclusively in information-processing systems. The number and diversity of such radio systems are large and increasing. In all of them, information is carried by radio signals that are intercepted by antennas and sent to Rxs that extract this information. In all other respects, the systems may significantly differ. Indeed, although manmade Txs are used in most systems, they are not needed in passive systems (e.g., radio telescopes) investigating RF-emitting objects. In most radio systems with Txs, the information of interest is introduced there. In radar and sounding systems, however, such information is introduced during signal reflection by objects of interest. In communication, navigation, broadcasting, and many other systems, Rxs process signals of cooperative Txs, whereas Rxs of EW systems usually process signals of noncooperative Txs. Despite these dissimilarities, the requirements for D&R circuits can be similar even in very different systems.

All digital Rxs digitize their input analog bandpass signals and all digital Txs convert the digitally generated signals into bandpass analog ones prior to their transmission. In communication and a few other types of systems, Txs also digitize their analog baseband input signals, and Rxs also reconstruct them after reception. Thus, digital communication radios perform all types of D&R. Communication systems operate within the entire RF spectrum and in all types of RF environments. Their diversity is much larger than that of any other type of radio systems, and the same can be stated about their applications and source signals. Communication radios perform the biggest number of different signal processing operations that influence D&R. The requirements for the radios' bandwidths and dynamic ranges in communication systems are as high as or higher than in most other radio systems. Therefore, digital communication radios are the best case study for analysis and further development of D&R circuits and algorithms, and the results of these studies are applicable to virtually all other technical fields.

2.2.2 RF Spectrum and Its Utilization

The RF spectrum is divided into frequency bands. At first glance, the use of the highest frequencies is preferable since the higher f_0 is, the wider B can be achieved for a given B/f_0 ratio, and a wider B allows, for example, increasing the communication throughput and radar range resolution. According to the Shannon-Hartley theorem, enhancing the throughput of a communication channel with additive white Gaussian noise (AWGN) by increasing signal bandwidth is more efficient than by increasing its power. Indeed, a linear increase in signal bandwidth for a given signal power provides the same result as an exponential increase in signal power for a given signal bandwidth if SNR is sufficiently high. An increase in f_0 also improves the directivity and gain of antennas that are proportional to the ratio of antenna size to signal wavelength. Still, there are many factors that limit the use of high frequencies and make the utilization of low frequencies beneficial.

For instance, the frequencies from 3 Hz to 12 kHz (see Table 2.1) have a very high level of atmospheric noise, allow only very low-throughput communications, and require extremely large, yet still inefficient, transmit antennas. However, relatively large depth of these waves' penetration into seawater and ground make them effective for communications with submerged submarines and underground facilities. Another example is the short-wave or high-frequency (HF) band. These waves can propagate over great distances due to their reflection from and/or refraction in ionospheric layers. For this reason, HF systems were the major means of long-range communications from the mid-1930s to mid-1960s. Compared to the previously used long waves, these waves allowed communications with a higher throughput over longer distances using smaller antennas and transmitters. Yet multipath propagation of short waves, irregularity of ionospheric conditions, limited width of the band, and unavoidable mutual interference among HF radios stimulated the search for alternative solutions.

The development of radio relay, meteor burst, ionospheric scatter, troposcatter, fiber optic cable, and especially satellite communications diminished the HF band

Table 2.1 ITU Classification of Radio Bands

Band	Frequency and Wavelength	Propagation Modes	Examples of Applications
Extremely low frequency (ELF)	3–30 Hz, 100,000–10,000 km	Ground wave caused by the diffraction around the Earth curvature and supported by different refractive indexes of the Earth and its atmosphere; penetrate seawater and ground; high atmospheric noise	Communications with submerged submarines
Super low frequency (SLF)	30–300 Hz, 10,000–1,000 km		
Ultralow frequency (ULF)	300 Hz–3 kHz, 1,000 km–100 km		Communications with submerged submarines and underground facilities, communications within mines
Very low frequency (VLF)	3 kHz–30 kHz, 100 km–10 km	The same mode plus effect of waveguide between the Earth and ionosphere; high atmospheric noise	Communications with submerged submarines, navigation and timing signals, geophysics, wireless heartbeat monitors
Low frequency (LF)	30 kHz–300 kHz, 10 km–1 km	The same mode plus strong effect of waveguide between the Earth and ionosphere; high atmospheric noise	Navigation, time signals, AM broadcasting, RFID, amateur radio
Medium frequency (MF)	300 kHz–3 MHz, 1 km–100m	Ground wave, skywave at night; still high atmospheric noise	AM broadcasting, amateur radio, communications in forests, avalanche beacons
High frequency (HF) or short wave	3 MHz–30 MHz, 100m–10m	Ground wave (minimum range), skywaves due to refraction and reflections from ionosphere; strong unintentional interference	Long-range communications, AM broadcasting, over-the-horizon radar, amateur radio, RFID, citizens band radio
Very high frequency (VHF)	30 MHz–300 MHz, 10m–1m	Generally direct wave, meteor scatter, ionospheric scatter	FM and television broadcasting, line-of-sight communications, meteor burst communications, amateur radio, weather radio
Ultrahigh frequency (UHF)	300 MHz–3 GHz, 1m–10 cm	Direct wave, tropospheric scatter, irregular tropospheric ducting	Television broadcasting, mobile phones, wireless LAN, GNSS, RFID, satellite radio, amateur radio, general mobile radio service, family radio service, troposcatter communications, microwave oven
Super high frequency (SHF)	3 GHz–30 GHz, 10 cm–1 cm	Direct wave, tropospheric scatter, irregular rain scatter; absorption by water vapor	Satellite communications, RFID, wireless LAN, cable and satellite radio and television broadcasting, radar, amateur radio, radio astronomy
Extremely high frequency (EHF)	30 GHz–300 GHz, 1 cm–1 mm	Direct wave; absorption by water vapor and oxygen; rain scatter	Space and satellite communications, microwave radio relay, remote sensing, radio astronomy, radar, amateur radio, directed-energy weapon, millimeter-wave scanner
Tremendously high frequency (THF)	300 GHz–3 THz, 1 mm–100 μm	Direct wave in free space; virtually nontransparent atmosphere	Space communications, remote sensing, amateur radio, medical imaging, condensed matter physics, terahertz time-domain spectrography

role. Satellite communications, capable of connecting any geographically remote areas with very high throughput and independently of ionospheric conditions, gradually became dominant. Later, however, it was found that these communications are not reliable enough for military and some other users due to the vulnerability to jamming and physical destruction. This revitalized the interest in the HF band, which is currently widely used for international broadcasting, various communications, over-the-horizon radar, and scientific equipment. In the future, it can provide low-cost communications among unmanned and manned stations on other planets with ionospheres (e.g., Mars and Venus).

While the radio waves of lower-frequency bands have advantages despite the bandwidth limitations and high noise levels, the radio waves of higher-frequency bands, allowing wider bandwidths, have drawbacks related to their absorption (mainly by oxygen and water vapor) and scattering (by raindrops and hail). The absorption becomes significant at the frequencies of several gigahertz. The largest absorption corresponds to the resonant frequencies of oxygen (60 GHz and 119 GHz) and water vapor (22 GHz and 183 GHz) molecules. Although the absorption peaks are widened by molecular collisions, the frequency intervals between neighboring peaks are used for signal transmission. In the tremendously high frequency (THF) band, the absorption is so significant that the atmosphere becomes virtually nontransparent. Still, this band can be utilized for space communications and radar. It is also used for other space- and ground-based applications. Since the radio spectrum is a valuable but limited resource, the International Telecommunication Union (ITU) allocates its parts to different applications and services to prevent or minimize the interference among them. This allocation and radio wave propagation modes are reflected in the classification shown in Table 2.1. Here, each band's highest frequency is 10 times larger than the lowest one. Table 2.2 shows an alternative IEEE classification of microwave frequency bands (commonly used in radar, navigation, and satellite communications). More information on radio wave propagation can be found in [1–7] and in publications on communications, navigation, and radar. Some of them also characterize the RF interference.

Natural phenomena cannot be unambiguously negative or positive because, being undesirable for some applications, they are beneficial to others. For instance, the atmospheric phenomena, inhibiting communications and radar operations in UHF and SHF bands, enabled weather radar development. Similarly, the ionospheric layers, allowing long-distance HF communications among the radios on the Earth surface and close to it, prevent communications between terrestrial and space radios in this band. The utilization of frequency bands depends on radio wave propagation, RF interference, and contemporary technological level. The first two factors form the RF environment. The RF interference sources can be natural (lightning, Sun radiation) and man-made (undesired Txs, automobile ignition systems, electric

Table 2.2 IEEE Classification of Microwave Bands

Band	L	S	C	X	Ku	K	Ka	V	W	Millimeter (mm) band
Frequency, GHz	1–2	2–4	4–8	8–12	12–18	18–27	27–40	40–75	75–110	110–300

welding). The interference can be intentional (jamming) or unintentional, wideband or narrowband. The wideband interference can be pulsed or continuous.

The impact of RF interference on signal reception ranges from the reception degradation to its termination and even to the Rx destruction. The approaches to dealing with RF interference combine regulatory and technical measures. The regulatory measures include the allocation of separate RF spectrum parts to different users as well as restrictions on generation and emission of radio waves by industrial, medical, scientific, and other devices and installations. These measures, being important, are often insufficient and may be difficult to enforce. For example, poor predictability of ionospheric conditions causes mutual interference among HF radios despite the regulations. Besides, many regulations are ignored during international conflicts. Therefore, technical protection is always implemented. It heightens the requirements for D&R circuits. The intensity of the RF spectrum utilization constantly increases, necessitating the improvement of the dynamic range, adaptivity, and reconfigurability of digital radios. These characteristics are decisively influenced by D&R techniques.

2.3 Radio Communication Systems

2.3.1 General

The diversity of radio systems is outlined in Section 2.2. Radio communication systems form their most diverse and largest type. They operate in all frequency bands and in all kinds of RF environments. The variety of their applications, source signals, capabilities, structures, modes of operation, methods of signal processing, and ways of implementation is much larger than that of any other type of radio systems. These systems are widely used for civilian and military purposes. Depending on specific applications, frequency bands, and required capabilities, the length of the radio links varies from several centimeters to billions of kilometers, the power of transmitted signals from picowatts to hundreds of kilowatts, the signals' bandwidth from fractions of hertz to several gigahertz, the size of antennas from millimeters to kilometers. Some communication systems operate independently, whereas others are parts of more complex systems, such as radar or higher-level communication systems. Communication systems can operate in simplex, half-duplex, and duplex modes (according to the ITU definition, half-duplex mode is a version of simplex one). Half-duplex and duplex modes are used most widely. Simplex mode dominates in broadcast systems, communications with submerged submarines, that carry only receive VLF-ELF antennas (due to extremely large sizes of transmit VLF-ELF antennas), and a few other systems. Figure 2.1 presents a simplex communication system. It includes a Tx, a Rx, and an RF communication channel that provides a medium for signals transmission but also weakens and distorts signals and introduces interference.

The first radio telegraph messages were transmitted in the mid-1890s. Voice was transmitted over radio channels in the early 1900s, and experimental television (TV) transmissions started in the mid-1920s. The invention and implementation of frequency-selective circuits, vacuum tubes, effective antennas, new modulation

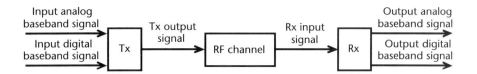

Figure 2.1 Simplex radio communication system.

techniques, and superheterodyne receivers played a great role in the evolution of communication and broadcast systems during the pre-World War II period. Initially, the LF band was used for long-range radio communications. However, after the experimental discovery of ionospheric layers in the mid-1920s and successful establishment of long-range HF links by ham radio enthusiasts, this band was utilized for commercial and military communications and broadcasting. The development of communication and broadcast systems stimulated the progress of electronic industry that became the foundation for advances in radar, navigation, and EW, as well as for the further development of communications on the eve of and throughout World War II. By the war's end, every military aircraft, tank, and even smallest navy vessel had its own radio, and many of them could transmit both voice and data in the HF and VHF bands. VLF-band radio stations could transmit information to submerged submarines. In addition, the efforts, related mainly to the military communication and radar systems advancement, produced revolutionary concepts such as information and detection theories.

The industrial base, technology, and theoretical results developed throughout World War II, as well as a large number of specialists trained at that period, contributed to the subsequent fast progress of radio systems and expansion of their civilian and military applications. This progress, accelerated by the Cold War military R&D and space exploration, was characterized by expeditious implementation of innovations and intensive exchange of ideas among the researchers and engineers working in different areas. Two factors simplified this exchange: close relations between theorists and practical engineers, established during World War II, and common (related to communication and/or broadcast systems) background of most leading specialists. Many currently used energy-efficient modulation techniques, spread spectrum (SS) signals, first effective channel codes, optimal demodulation and diversity combining techniques, transmission of analog signals using pulse-code modulation (PCM), as well as effective methods of sampling, quantization, source coding, and encryption were implemented during the first two post-war decades. The invention of transistors accelerated this implementation.

New types of communication systems, such as meteor burst, ionospheric scatter, troposcatter, and, most importantly, first space communication systems, were introduced, and numerous microwave relay point-to-point communication systems were built. Directional antennas were widely employed in these systems. Some innovations (e.g., orthogonal frequency-division multiplexing (OFDM), adaptive antenna arrays, packet-switched networks, and many DSP-based techniques) proposed and prototyped at that period were implemented later using newer technologies. The use of communication and sampling theories, probabilistic approach, and formal optimization methods became a common practice. Methods of channel coding and modulation developed at that period became more oriented towards specific communication channels. Computers began to be used in communication systems design. Some major trends in the development of communication systems became apparent: increase in the systems diversity and complexity; improvements in their throughput, reliability, and quality of service; and more intensive use of already adopted frequency bands combined with utilization of new ones.

These trends still remain, but new ones emerged as a result of more recent scientific and technological achievements such as integrated circuits (ICs); DSP chips and field programmable gate arrays (FPGAs); advanced aerospace technology; the internet and its applications; digital, software-defined, and cognitive radios; adaptive antenna arrays; new devices and systems based on photonics, ultrasound technology, and microelectromechanical systems (MEMS); novel modulation and coding techniques; multiple-input multiple-output (MIMO) systems; and computer-aided design (CAD) and related software. These achievements enabled a radical increase in communication systems throughput, diversity, flexibility, and self-organization. The last decades have also changed the mindset and skillset of researchers and engineers, increasing their tolerance to the design complexity; the ability to use computer simulations in R&D and CAD tools in design; proficiency in realizing technical solutions in application-specific ICs (ASICs), DSP chips, and/or FPGAs; and capability of finding intelligent solutions to the problems previously solved by increasing power consumption. Some of these changes have negative implications along with positive ones. For instance, the use of computer simulations sometimes produces neglect of fundamental theories. Specialization caused by diversity of radio systems created information gaps among researches and engineers.

Some of the later trends of the communication systems development are: rising significance of software; diminishing role of analog signal processing and growing role of DSP and mixed-signal processing; ongoing enhancement of adaptive, cognitive, and self-organizing capabilities of the systems and radios; widening employment of adaptive coding and modulation, dynamic bandwidth and resource allocation, and intelligent management of radios' dynamic ranges and signal powers; growing influence of D&R on the performance of radios; increasing functionality and capabilities of TDPs, RDPs, and fully integrated digital radios; emergence of mixed-signal field programmable ICs (FPICs); fast expansion of energy-efficient, ultralow-power, and energy-harvesting radios; implementation of radio platforms with intelligent hardware; systematic use of digital correction of distortions introduced in the analog and mixed-signal domains; combined rejection of strong interference in analog and digital domains; joint employment of frequency, code, and spatial selectivities stimulated by broadening use of adaptive antenna arrays; incorporation of sensing and navigation functions in communication radios; and self-organizing and self-configuring ad hoc networks, including self-organizing sensor networks formed by a large number of miniature radios and capable of operating as a single spatially distributed radio when communicating with external radios or networks. This diversity of trends shows, in particular, that the variety of requirements for D&R circuits and algorithms in digital communication radios is larger than that in any other technical area. More information on communication systems can be found in [8–36].

Mobile phone technology illustrates the continuing improvement in communication systems capabilities. The deployment of the first-generation (1G) commercial mobile phone systems started in Japan in 1979, in the Nordic countries in 1981, and in North America in 1983. They transmitted voice using analog technology. First-generation systems are no longer in use. The second-generation (2G) mobile communications, brought into service in the early 1990s, were digital. Initially, 2G was oriented toward the transmission of voice and low-bit-rate data (tens of kilobits per second). Evolutionary upgrades increased the maximum data rates to hundreds of kilobits per second per user. The most widespread 2G standard was GSM together with its upgrades GPRS and EDGE. In 2017, it was the only surviving 2G standard. GSM uses Gaussian minimum shift keying (GMSK) modulation. It combines frequency-division multiple access (FDMA) for channel separation with time-division multiple access (TDMA) for sharing each frequency channel among several users. At the beginning of the twenty-first century, the third generation (3G) mobile networks were deployed. Both major 3G standards (WCDMA and CDMA2000) use code-division multiple access (CDMA). The initial maximum data rates of a few megabits per second per user were increased to tens of megabits per second by the subsequent upgrades.

Still, as became clear by 2009, 3G could no longer satisfy high-data-rate applications. To radically increase data rates and provide data services more efficiently, the industry developed the fourth-generation (4G) technologies. These technologies were included in the LTE (Long-Term Evolution) standard and its upgrades LTE Advanced (LTE-A) and LTE-A Pro. These standards support data rates up to hundreds and even thousands of megabits per second per user. While the LTE standard falls short of some 4G benchmarks, LTE-A and LTE-A Pro meet all 4G requirements. The data rate increase is achieved partly by widening the signal spectrum, but mostly by using more bandwidth-efficient signals. The standards use orthogonal FDMA (OFDMA) with quadrature PSK (QPSK) and QAM subcarrier modulation. Similar multiple access and modulation techniques are used in the 4G wide area network standard WiMAX Advanced. Fifth-generation (5G) standards are aiming to provide even higher data rates, lower latency, and capability of accommodating higher density of mobile broadband users. The astonishing improvement in the throughput and universality demonstrated by mobile radios is typical for most communication systems.

2.3.2 Communication Txs and Rxs

Since the requirements for D&R circuits are influenced by the Txs' and Rxs' structures and signal processing operations, these structures and operations are concisely discussed in this and the next sections. The block diagrams in Figure 1.19 and most other block diagrams in the book correspond to multipurpose and/or multistandard radios where requirements for D&R circuits are most challenging. Since modern technology has made DSP advantageous for specialized radios as well, their specifics are also addressed.

In the Tx shown in Figure 2.2(a), its input baseband signals first undergo source encoding that reduces their redundancy. The redundancy could be necessary or useful

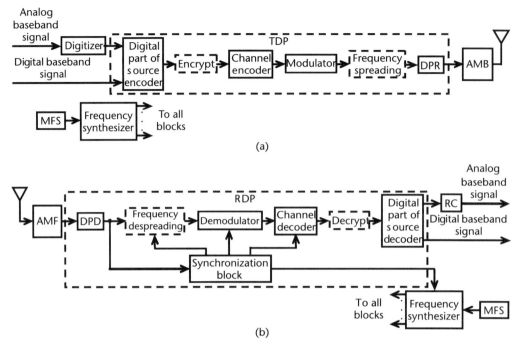

Figure 2.2 Digital communication: (a) Tx and (b) Rx.

for perception of the information carried by signals or for its more reliable transmission to the Tx input. Despite its prior usefulness, the redundancy, which cannot be effectively utilized for improving the transmission quality, should be removed to increase the system throughput. While source encoding of digital input signals is performed digitally, a part (sometimes significant) of analog signals' source encoding can be performed during their digitization (i.e., in the mixed-signal domain). For instance, analog input signals can be compressed using nonuniform quantization if they have a high dynamic range and their undesired components, such as noise, are negligible. The obtained compressed signals are represented by a smaller number of bits throughout their transmission and reception. After the reception, they are expanded in the Rx RC (see Figure 2.2(b)) for their correct perception by end users. Such processing is called companding. When samples of input signals are correlated, predictive quantization reduces the number of bits needed for their representation. There are other methods of the mixed-signal domain source encoding.

In some radios, digital bit streams from the source encoders are encrypted. Since this operation is optional, the block executing it is shown with a dashed line. Mutual influence of encryption/decryption and D&R is negligible. In contrast, channel encoding, modulation, optional spreading, and optional multiplexing (not shown in Figure 2.2(a)) affect signal reconstruction in Txs. Actually, channel encoding, modulation, and spreading are different stages of the same procedure that can be called generalized modulation. Its purpose is varying the carrier signal parameters in a way that allows optimal transmission of information over communication channels.

The earliest communication systems typically transmitted sensible texts with separately encoded letters using on-off keying. Most errors could be detected and corrected by recipients due to the redundancy of sensible texts. Two techniques were used to increase the reliability of transmissions: the most important messages were transmitted several times, and numbers, because of their insufficient redundancy, were replaced with the corresponding words, for instance, "26" was transmitted as "twenty-six." Since then, the situation has changed dramatically. Now channel codes with lower redundancy than a sensible text allow better, faster, and less expensive error correction than that provided by humans. Therefore, channel encoding and decoding are currently used in virtually all radios. Channel encoding, modulation, optional frequency spreading, and the reasons for separating the generalized modulation stages are discussed in the next section. Here, it is worthwhile to mention that frequency spreading in communication systems improves resistance to jamming, narrowband interference, and multipath propagation; increases security; reduces signal detection and interception probabilities; and/or allows CDMA implementation.

As noted in Section 1.4, it is preferable to represent analog bandpass real-valued signals, used for transmissions over radio channels, by the *I* and *Q* components of their digital baseband complex-valued equivalents (complex envelopes) in the TDPs and RDPs of most radios. In this case, the analog bandpass real-valued signals should be generated from their equivalents at the TDP output and prepared for transmission in the AMB. This generation is a two-stage procedure comprising the conversion of digital signals into analog ones and forming bandpass real-valued signals from their baseband complex-valued equivalents. If the conversion is executed first, the reconstruction is baseband. Otherwise, it is bandpass. When the reconstruction is baseband, most operations related to it are performed in the Tx AMB. Still, some of them are carried out in the digital part of reconstruction (DPR) circuit. When the reconstruction is bandpass, digital bandpass real-valued signals are formed in the DPR, which carries out most operations related to the reconstruction. Yet the conversion of digital bandpass real-valued signals into the analog domain is performed in the AMB. The digital forming of bandpass real-valued signals is outlined in Section 1.4.2 and illustrated by Figure 1.22. The reconstruction quality decisively influences the accuracy of the signal generation in Txs. Independently of the reconstruction type, the analog bandpass RF signals of sufficient power should be produced at the AMB output and sent to the Tx antenna.

Digital radios require multiple internally generated frequencies for tuning and signal transformations in Txs and Rxs. The frequencies are produced by frequency synthesizers, and their accuracy and stability are ensured by deriving them from a master frequency standard (MFS) (see Figure 2.2). Most MFSs are high-precision crystal oscillators that currently encounter strong competition from MEMS oscillators. Atomic standards, used as MFSs, provide exceptional accuracy of synthesized frequencies.

Signals from the Rx antenna (see Figure 2.2(b)), even after their filtering in the AMF, contain, besides desired signals, noise and interference that can be much stronger than the desired signals. Therefore, the requirements for the Rxs' digitization circuits of different radio systems can be identical when the Rxs have similar bandwidths and operate in similar RF environments. The received analog bandpass real-valued signals, independently of the AMF processing specifics, are converted into their digital baseband complex-valued equivalents at the RDP input, as a result of a two-stage procedure reciprocal to that in Txs. The digitization is bandpass if the conversion to the digital domain is executed first and then the digital baseband complex-valued equivalents are formed in the digital part of digitization circuit (DPD), as outlined in Section 1.4.2 and illustrated by Figure 1.23. If the received signals are first converted into their analog baseband complex-valued equivalents and then the equivalents are digitized, the digitization is baseband.

From a general standpoint, digitization is a special case of source encoding that reduces the redundancy of analog signals prior to their digital processing. Indeed, as shown in Figure 1.3(a) and described in Section 1.2.1, antialiasing filtering rejects most of the analog signal's redundant frequency components, sampling excludes most of its instantaneous values, preserving only the samples necessary for the analog signal reconstruction, and quantization reduces the number of levels required for the signal representation. Finally, subsequent digital filtering and decimation remove the remaining unneeded redundancy. Thus, digitization preserves the essential information contained in an analog signal while compressing it. This standpoint allows sharing technical ideas and solutions between designers of digitization and source encoding circuits and algorithms.

In the RDP, multiplexed signals are demultiplexed, spread ones are despread, and all signals undergo demodulation and channel decoding. Despreading, demodulation, and channel decoding are the stages of generalized demodulation, performed on the signals' baseband complex-valued equivalents. Optimal demodulation of digital signals is executed by correlators or matched filters in AWGN channels. When the noise is Gaussian but nonwhite with the PSD $N(f)$, optimal demodulation requires a linear filter with the transfer function

$$H(f) = \frac{1}{N(f)} \exp(-j2\pi f t_0) \tag{2.1}$$

prior to the correlators or matched filters (here, t_0 is the filter delay). When the additive noise is non-Gaussian, optimal demodulators become more complex and their structures depend on the noise distribution, but they still include correlators or matched filters. The demodulated symbols are sent to the channel decoder for error detection or correction. Although conceptually different operations of generalized modulation and demodulation are shown as separate blocks in Figure 2.2, some of them can be performed jointly. If the received information is encrypted, it is decrypted after the channel decoding. Finally, the retrieved information is converted into a form convenient for recipients by source decoding. The signals expected to be received in analog form are reconstructed in the RC.

Most signal processing operations in RDPs require synchronization [17, 21–23, 31–33]. In direct-sequence (DS) SS systems, the received and reference pseudonoise (PN) sequences should be synchronized. Frequency-hopping (FH) SS systems also require synchronization. To fully accumulate symbols' energies in correlators or matched filters, symbol synchronization is needed. Word synchronization allows correct decoding of block codes. When transmitted information is organized in frames, frame synchronization is also necessary. The PN, symbol, word, and frame synchronizations can be combined in some cases [32]. Most noncoherent demodulators need frequency synchronization, and coherent demodulators additionally require phase synchronization. Although all types of synchronization are symbolically represented by a specialized block in Figure 2.2(b), actual synchronization systems are usually distributed among various Rx blocks and their input signals are generated in these blocks. In some communication systems, mostly in systems with adaptive directional antennas and/or TDMA [23, 32], Txs are also involved in synchronization. Note that no synchronization is needed between the D&R procedures in Txs and Rxs.

Complexity of D&R is determined by the statistical properties of their input signals (mainly by the signals' crest factors, bandwidths, and spectral positions) and by the needed accuracy of signal generation in Txs and processing in Rxs. The crest factor, defined as the ratio of the signal's maximum and rms values, affects the required dynamic range. Larger crest factors, wider bandwidths, and higher center frequencies of signals usually necessitate more complex D&R. Knowledge of the signals' statistical properties can be utilized to improve or simplify the D&R. The signal digitization complexity at the Rxs' inputs is much higher than that at the Txs' inputs because the Rxs' input signals are bandpass, whereas the Txs' input signals are baseband (digitization of bandpass signals is, in general, more complex than that of baseband ones), crest factors of the Rxs' input signals are usually much larger than those of the Txs' input signals due to multipath propagation and interference, and statistical properties of the Txs' input signals are known better than those of the Rxs' input signals. For similar reasons, digitization of the Rxs' input signals is also more complex than reconstruction of the Txs' output signals.

Thus, digital communication radios perform D&R of both baseband and bandpass signals, and many signal processing operations in the radios are interdependent with D&R. These facts confirm the validity of using communication radios as a case study for analysis and further development of D&R circuits and algorithms.

2.3.3 Channel Coding, Modulation, and Spreading

As stated above, channel encoding, modulation, and frequency spreading are different stages of generalized modulation. Since frequency spreading is optional, let us first clarify why channel encoding and modulation are separated, despite their common purpose. To this end, recall that there are two types of modulations: energy-efficient and bandwidth-efficient. Energy-efficient modulations minimize energy consumption per bit of transmitted information at the expense of wider signal bandwidth. They are used when energy and/or power are more limited than bandwidth. Battery capacity limits both energy and power. Power can also be limited by the constraints on the size or power dissipation of radios. Bandwidth-efficient modulations minimize signal bandwidth at the expense of higher energy consumption per bit. They are used when bandwidth is more limited than energy and power.

Most energy-efficient modulations use orthogonal, biorthogonal, and simplex signals [21–23]. Orthogonal signals are produced, for example, by FSK, pulse position modulation (PPM), or binary PSK (BPSK) sequences modulated by Walsh-Hadamard functions. Biorthogonal signals, formed by augmenting orthogonal signals with the negative of each signal, allow doubling the alphabet size M without increasing bandwidth. Simplex (transorthogonal) signals are generated from orthogonal ones by subtracting the alphabet mean value (centroid). Simplex signals have uniform negative cross-correlation. BPSK forms the smallest set of simplex signals, which are antipodal in this case (see Figure 2.3(a)). The smallest set of biorthogonal signals

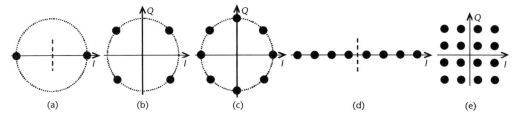

Figure 2.3 Signal constellations of different digital modulations: (a) BPSK, (b) QPSK, (c) 8PSK, (d) 8PAM, and (e) 16QAM.

corresponds to QPSK (see Figure 2.3(b)). Recall that the I component of a signal corresponds to its real part, whereas the Q component corresponds to its imaginary part. Coherently demodulated BPSK and QPSK have the same energy efficiency, which is two times higher than that of coherently demodulated binary FSK (BFSK). However, the energy efficiency of simplex and biorthogonal signals insignificantly exceeds that of orthogonal signals when M is large. Still, biorthogonal signals are the best choice for large M because, having only slightly better energy efficiency, they provide two times better bandwidth efficiency.

Signal constellations of most widely used bandwidth-efficient modulations: M-ary PSK, pulse-amplitude modulation (PAM), and QAM are shown in Figure 2.3(c, d, e), respectively [21–23]. These modulations become bandwidth-efficient only when M is sufficiently large, and their bandwidth efficiency improves as M grows. When $M = 2$, they turn into BPSK, and when $M = 4$, M-ary PSK and QAM turn into QPSK. Among bandwidth-efficient modulations, QAM provides the highest bit rate for given bandwidth and average energy per symbol because it, in contrast with M-ary PSK and PAM, utilizes both amplitude and phase for increasing M. When a priori probabilities of all symbols are the same, the number of bits per symbol is

$$k = \log_2 M \qquad (2.2)$$

Increasing k for a given symbol length τ_{sym} proportionally increases the bit rate for all types of signals. However, this increase has opposite effects on the bit error probabilities P_b of energy-efficient and bandwidth-efficient signals if the average symbol energy E_s remains the same. It raises P_b for bandwidth-efficient signals and reduces P_b for energy-efficient ones if $P_b \leq 10^{-1}$ for $k = 2$ [23]. The energy gain, provided by increasing k without changing E_s in energy-efficient signals is obtained at the expense of widened signal bandwidth. This gain quickly diminishes as k becomes larger. For biorthogonal signals, for instance, the transition from $k = 2$ to $k = 4$ provides higher gain than that from $k = 10$ to $k = 20$. If k of energy-efficient signals is increased to minimize P_b for given bit rate and energy per bit E_b, τ_{sym} should be extended proportionally to k. This also increases E_s proportionally to k for the same Tx power. As follows from (2.2), an n-fold increase in k (and, consequently, in E_s) requires N-fold increase in M where

$$N = 2^{(n-1)k} \qquad (2.3)$$

This increase in M proportionally increases the required bandwidth and the number of correlators or matched filters. Thus, a linear increase in k and E_s corresponds to an exponential increase in the bandwidth and system complexity. Therefore, improving the reliability of communications by enlarging M is reasonable only when k is relatively small ($k \leq 5$) because trading the signal bandwidth and system complexity for the symbol energy becomes wasteful for $k > 5$. Somewhat larger k can be justified if, for example, the processing gain of DS frequency spreading, performed after the modulator, permits it. Indeed, since the bandwidth is already allocated for the signal spreading, its utilization (even less efficient) for raising E_s can be practical. Besides the limitation caused by wasteful trading for large k, increasing E_s is often ineffective, for instance, in multipath channels.

These weaknesses of energy-efficient modulations are compensated by augmenting them with channel encoding that employs different and more flexible approaches to the communications reliability improvement. In the systems with limited energy or power, channel encoding continues trading bandwidth for energy, but does it more efficiently than modulation for large k and allows adjusting this trade to a specific communication channel. Soft-decision demodulation and decoding prevent information loss and noise immunity reduction during transition from demodulation to channel decoding. Thus, the conceptual separation of channel encoding and modulation in generalized modulation allows bringing together different approaches and techniques for improving reliability of communications. Spreading is included in generalized modulation as its last stage for the same reasons. Division of generalized modulation into stages does not preclude combining the stages when channel encoding and modulation or modulation and spreading can be performed jointly.

The situation with bandwidth-efficient signals is similar. In systems with QAM, for instance, the modulation improves communication reliability for a given average energy per symbol by optimizing the constellation boundaries and densest packing of signal points within these boundaries. A channel encoding technique called trellis coding [23] achieves this goal by increasing the minimum distance between the allowable code sequences without changing the symbol rate or average energy per symbol. In contrast with encoding of energy-efficient signals where coding gain is attained at the cost of increased bandwidth, trellis coding provides it at the cost of increased complexity of encoding and decoding. Trellis encoding is performed jointly with modulation.

High energy efficiency of signals, besides minimum E_b for a given P_b, requires a maximum utilization of Tx power. The simplest way to meet this requirement is allowing a Tx to operate in saturation mode. At first glance, the energy-efficient modulation techniques with constant envelopes meet it. However, their high spectral sidelobes interfere with neighboring channels. Filtering, which suppresses the sidelobes, increases the signals' crest factors if 180° phase transitions between adjacent symbols are possible. Hard limiting of the filtered signals restores their near-constant envelopes at the cost of sidelobe regeneration. Chapter 3 describes several crest-factor-minimizing techniques that improve Tx power utilization and simplify signal reconstruction. Among bandwidth-efficient signals, only M-ary PSK provides constant envelope, but its efficiency is lower than that of QAM when $M > 8$ because PSK does not utilize amplitude. However, amplitude utilization leads to high crest factors of QAM signals that complicate reconstruction and power utilization in Txs. PAM, CDM, CDMA, FDM, FDMA, and OFDMA as well produce signals with high crest factors. Many of these signals are also intolerant to distortion of their constellations. The ways to solve these problems are discussed in Section 3.4.3.

When the first DS SS communication systems were realized in the 1950s, a parameter called processing gain G_{ps} was introduced as a measure of their antijam (AJ) capability [17, 23]. Since spreading distributes a low-dimensional modulated signal $u_m(t)$ in a high-dimensional space of the resulting SS signal $u_{ss}(t)$, G_{ps} was defined as:

$$G_{ps} = \frac{D_{ss}}{D_m} \tag{2.4}$$

where D_{ss} and D_m are the dimensionalities of u_{ss} and u_m, respectively. At that time, (2.4) adequately characterized the AJ capability of SS communication systems because the influence of modulation and channel coding was negligible. Since then, many effective modulation/demodulation and encoding/decoding techniques, contributing to the systems' AJ capabilities, were implemented. Therefore, the total processing gain of a DS SS system is

$$G_p = G_{pc} \cdot G_{pm} \cdot G_{ps} \tag{2.5}$$

where G_{pc} and G_{pm} are, respectively, coding and modulation gains of the system. Thus, G_{ps}, determined by (2.4), is only one of the factors characterizing AJ capability of SS systems. It is also one of the factors characterizing the probability of detection and interception, capability to cope with multipath propagation, and security of transmitted information. Since channel coding, modulation, and spreading differently tackle various adverse phenomena in communication channels, adaptive allocation of available system bandwidth to optimally adjust G_{pc}, G_{pm}, and G_{ps} is advantageous. When thermal noise is the only interference, it is better to allocate most of the available bandwidth to maximize G_{pm} and G_{pc}, minimizing G_{ps} that does not improve reception reliability in this case. In the presence of jamming, G_{ps} should be significantly increased, although, as mentioned above, G_{pc} and G_{pm} can also contribute to the system AJ capability. In channels with multipath propagation, G_{ps} and G_{pc} play larger roles than G_{pm}.

The dimensionalities D_{ss} and D_m of u_{ss} and u_m used in (2.4) require clarification. In principle, the dimensionality D of a signal $u(t)$ is determined by the number of samples needed for its discrete-time representation. Bandpass signals, for example, are most often represented by the samples of their instantaneous values, I and Q components, or envelopes and phases. According to the sampling theorem, $u(t)$ with length T and bandwidth B requires $2BT$ samples for any of these representations in the general case. Consequently, $D = 2BT$. However, some modulation and spreading techniques restrict $u(t)$, reducing the required number of samples. For instance, a constellation of a BPSK signal can be oriented: (1) along the I-axis (see Figure 2.3(a)), (2) along the Q-axis, or (3) between these axes. The representation of $u(t)$ by the samples of I and Q components requires only BT samples in the first two cases, and $2BT$ samples in the third case. Since the amplitudes of all PSK signals (see Figure 2.3(b, c)) are constant, they can be represented by only $BT + 1$ samples of the amplitude and phase, but their representation by I and Q components still requires $2BT$ samples.

The dependence of signal dimensionality on the modulation and/or spreading type, initial phase, and coordinate system creates ambiguities. The fact that phase shifts in communication channels make it impossible to align the phases of jamming and desired signals resolves the ambiguities. Indeed, I and Q components of the desired signal should be jammed with the same intensity. Similarly, distortions, noise, and interference in communication channels make constant amplitudes of Rx input signals impossible. Therefore, the dimensionalities in (2.4) should be calculated as

$$D_{ss} = 2B_{ss}\tau_{sym} \quad \text{and} \quad D_m = 2B_m\tau_{sym} \qquad (2.6)$$

where B_{ss} and B_m are the bandwidths of u_{ss} and u_m, respectively, and conventional formula

$$G_{ps} = \frac{B_{ss}}{B_m} \qquad (2.7)$$

is true, despite the ambiguities mentioned above. Although the signal dimensionality reduction does not influence G_{ps}, it can be utilized during DSP in TDPs and RDPs.

2.4 Other Radio Systems

2.4.1 Broadcasting Systems

Broadcasting is unidirectional transmission of radio signals carrying video, audio, and/or data intended for wide and dispersed audiences. Typically, broadcast stations use high-power Txs with effective antennas to provide sufficient reception quality and coverage area even for inexpensive Rxs. Broadcast transmission of speech and music, which gained popularity in the 1920s, was originally performed with AM signals in the LF and MF bands. Subsequent utilization of the HF band significantly increased the number of broadcast stations and widened the audience. Although experimental TV broadcasting started in 1925 and commercial one in the 1930s, the TV and FM audio broadcasts became common after World War II when the VHF and UHF bands were allocated for them. Later, cable TV became a competitor to over-the-air broadcasting. In radio and TV studios, as well as several European urban sites, cables were used for transmitting radio since the 1920s and TV since the early 1930s. From the 1950s, cables connected individual houses and apartments to effective community antennas, extending over-the-air TV reception to the areas where it was limited by terrain or distance from TV stations. For decades, the number of cable TV users was increasing slowly. However, the implementation of optical fiber trunk lines between cable distribution hubs, emergence of the internet, and multiple computer applications significantly accelerated this process. Cable allows transmitting hundreds of TV channels and high-speed data streams, provides uplinks and downlinks, and is less prone to interference than over-the-air transmissions. Its major limitation is the uselessness for mobile platforms.

Satellite radio and TV broadcasting is free of this limitation and covers larger area than terrestrial over-the-air broadcasting. The possibility of worldwide communications using three satellites equidistantly positioned on the geostationary orbit was shown by A. C. Clarke (United Kingdom) in 1945. The first commercial TV programs from Europe to the United States and vice versa as well as from the United States to Japan were transmitted in 1962 to 1963. Later in the 1960s and 1970s, several satellite systems with TV and FM audio programs were launched. By the 1980s, satellite TV became common, but the high cost of satellite TV sets with large antennas was still an obstacle. In the 1980s, the implementation of direct-broadcast satellite TV in the Ku-band reduced the cost of TV sets with smaller antennas and substantially increased the number of satellite TV viewers. Since the 1990s, the development of direct-broadcast satellite TV and audio was connected to the digital technology. Several digital satellite broadcasting standards with increasingly high quality have been implemented. For example, satellite TV standard DVB-S2 specifies 7 grades of video signal quality. The lowest grade with 6-MHz bandwidth provides 10.8-Mbps data rate, whereas the highest grade with 36-MHz bandwidth provides 64.5-Mbps data rate. This standard uses 8PSK modulation.

The cable and terrestrial over-the-air broadcasting have also moved to digital transmission technology, allowing streaming video and audio programs over the internet. Three major digital TV standards for terrestrial broadcasting in the VHF and UHF bands currently exist: ATSC (Advanced Television Systems Committee) implemented in North America, South Korea, most of the Caribbean, and some other countries; DVB-T (Digital Video Broadcasting-Terrestrial) implemented in Europe and most of Asia, Africa, and Oceania; and ISDB-T (Integrated Services Digital Broadcasting-Terrestrial) used in Japan, Philippines, most of South America, and a few other countries. ATSC utilizes 8-level vestigial sideband modulation (8VSB). Its 3-bit symbol is obtained from two data bits by trellis encoding. Thus, the 10.76-Mbaud channel symbol rate corresponds to the gross and net bit rates of 32 Mbps and 19.39 Mbps, respectively. The resulting signal is filtered with a Nyquist filter to obtain 6-MHz channel bandwidth. DVB-T utilizes coded OFDM (COFDM), which separates a digital data stream into a large number of slower streams modulating closely spaced subcarrier frequencies. DVB-T allows a choice of 1,705 or 6,817 subcarriers that are approximately 4 kHz or 1 kHz apart. The subcarrier modulations are QPSK, 16QAM, or 64QAM. The channel bandwidth (6 to 8 MHz) depends on a DVB-T version. The inner code is a punctured convolutional code with one of five coding rates: 1/2, 2/3, 3/4, 5/6, or 7/8. The optional outer code is Reed-Solomon (204, 188). ISDB-T modulation, coding, and bandwidths are very similar to those of DVB-T. ISDB-T also has the time-interleave capability. ATSC performs slightly better than other standards in rural areas with insignificant multipath. To achieve the same performance as DVB-T and ISDB-T in areas with severe multipath, ATSC Rxs need advanced equalizers. ISDB-T is least susceptible to impulse interference and provides the highest mobility.

Digital Rxs can, in principle, process both digital and analog signals, carrying any kind of information and modulated according to current or legacy standards. Cost constraint is a major obstacle to their universality, and this constraint is particularly important for mass-produced broadcast Rxs. Therefore, digital broadcast Rxs are somewhat specialized. The desire to make Rxs inexpensive forces the designers to minimize the cost, size, and power consumption of their D&R circuits. In most digital Rxs intended for the reception of video, audio, and data signals within the bandwidth of 6–8 MHz, baseband digitization with the sampling rate of 10 Msps and 12-bit A/D resolution in I and Q channels is currently used. High-end TV Rxs with wider bandwidths require a proportionally higher sampling rate, and, as shown in the subsequent chapters, their reception quality can be improved by using bandpass digitization with novel sampling. Still, the requirements for their D&R circuits are well within the limits of what is required for them in advanced communication systems. Txs and Rxs for transmission and reception of full-band cable or satellite signals impose the highest requirements for their D&R circuits. However, these Txs and Rxs should be considered communication rather than broadcast ones. Besides that, most broadcast systems operate in well-regulated and friendly RF environments. For these reasons, the progress of D&R circuits and algorithms in digital communication radios also solves problems of D&R in broadcast radios.

2.4.2 Radio Navigation and Positioning Systems

Radio navigation systems determine the locations and velocities of navigated objects using radio beacons. Originated to navigate ships and aircraft, they are now used in virtually every platform from cell phones and cars to spacecraft. Their principles and purposes are somewhat similar to those of geolocation, radar (especially passive), and EW. A variety of radio navigation systems have been developed since their origination prior to World War I. The most important systems are concisely described below, while their methods are discussed in Section 2.4.3.

The earliest systems determined the directions to radio beacons with known locations and then performed triangulation. Initially, nondirectional beacons were employed, and the directions towards them were found using directional antennas (rotating small vertical loops) mounted on the navigated vehicle. The directions were indicated by sharp drops in reception caused by the antenna nulls. The intersection of bearing lines on a map revealed the vehicle location. These systems could use high-power commercial broadcast stations as beacons along with lower-power nondirectional beacons specially built to mark airways and approaches to harbors. The necessity to install rotating antennas on aircraft and complexity of onboard electronic equipment were the method's major drawbacks. To avoid them, several navigation systems, whose beacons had rotating directional antennas, were developed. When the antenna beam pointed in a certain direction (e.g., north), each beacon transmitted its identifier. The time between the identifier reception and the drop in the signal level allowed calculating the beacon's bearing. The first such systems were introduced prior to World War I. This approach, being significantly improved, is employed in the aircraft radio navigation system called VHF omnidirectional radio range (VOR).

Despite the continuing development of the systems with rotating ground antennas, the use of nondirectional beacons was revitalized in the second half of the twentieth century when technological progress permitted determining the wavefront arrival angle by a single directional solenoid or by comparing the signal phases on two or more small antennas using portable equipment. Since nondirectional beacons operated at 190–1,750 kHz (LF and MF bands), their signals followed the Earth curvature and, therefore, could be received at greater distances and lower altitudes than VOR signals. Signals at these frequencies are affected by ionospheric conditions, terrain, and interference from stations operating at the same or close frequencies. Navigation Rxs mitigate some of these adverse effects.

As to VOR, it uses the VHF band (108–117.95 MHz) and has a relatively short range (practically up to 200 km). While the earliest systems with rotating ground antennas transmitted a single signal, VOR stations transmit three different signals over three channels. The voice channel transmits the station identifier and voice signal that may contain the station name, recorded in-flight advisories, or live flight service broadcasts. The second channel omnidirectionally broadcasts a continuous signal, whereas the third one uses an antenna array to transmit a highly directional signal that rotates clockwise at 30 rpm. The phase of the rotating signal changes synchronously with its rotation so that it is equal to 0° when pointed north and 90° when pointed east. The direction to the station can be determined by comparing the phase of the received directional signal with that of the continuous omnidirectional one. Military UHF tactical air navigation system (TACAN) uses similar methods. VOR system was developed in the United States in 1937 and deployed by 1946. In a few decades, it became the world's most widely used aircraft radio navigation system. In the twenty-first century, however, the number of VOR stations has been decreasing due to the growing use of global navigation satellite systems (GNSSs). Yet this decrease has a limit because the concerns about GNSS vulnerability to jamming and physical destruction require functioning VOR stations as a backup. The digitization circuits of onboard VOR Rxs should withstand strong interference from the FM broadcast stations located near airports because the VOR subband is adjacent to their subband (87.5–108.0 MHz).

Between two world wars, several navigation systems that kept aircraft centered within the radio stations' beams were built. This approach required very simple onboard Rxs. The progress of electronics, however, made this advantage unimportant, whereas the impossibility of supporting navigation outside the beams caused gradual replacement of such systems after World War II. The only remaining beam systems are the ones controlling aircraft landing.

Among other implemented land-based radio navigation systems, LF maritime systems comprising multiple beacons are most important. The first such system, Decca (United Kingdom), developed and deployed during World War II, was shut down by 2001. Similar systems still in use are Loran-C (United States) and Chayka (Russia). Since the 1990s, these systems were in decline due to the growing use of GNSSs. However, a radical Loran-C upgrade, eLoran (Enhanced Loran) is currently planned due to the concerns about GNSS vulnerabilities.

Satellite-based navigation systems are the most successful and widely used ones. Initially intended for military, aviation, and maritime applications, they have become ubiquitous. The development of the first GNSS called Global Positioning System (GPS) NAVSTAR started in 1973 in the United States. Initially, NAVSTAR was a name of the system, whereas the acronym GPS indicated the system's purpose. Now, the name NAVSTAR is used rarely, while GPS has become the system's name. GPS achieved its initial and full operational capabilities in 1993 and 1995, respectively. Its satellites have circular medium Earth orbits. Each satellite has a high-precision atomic clock and broadcasts DS SS signals carrying its navigation message containing the satellite's time and orbital parameters, as well as other data for Rx positioning. Being a dual-use system, GPS provides two services: unrestricted Standard Positioning Service (SPS) open to all users and restricted Precise Positioning Service (PPS) with encrypted signals available to the U.S. government-authorized users. Originally, GPS provided one SPS signal at 1,575.42 MHz and two identical PPS signals at 1,575.42 MHz and 1,227.60 MHz. Since the early 2000s, GPS has been undergoing modernization that includes adding three different SPS signals at 1,575.42 MHz, 1,227.60 MHz, and 1,176.45 MHz, as well as two new PPS signals at 1,575.42 MHz and 1,227.60 MHz.

GPS has become a de facto standard-setter for the GNSSs developed by other countries (see Table 2.3), which are based on the same principles, transmit navigation messages using SS signals at several L-band frequencies, and provide open and restricted positioning services. All GNSSs evolve towards better interoperability by coordinating their frequencies, waveforms, and navigation messages. Besides the GNSSs, India and Japan are developing their regional navigation satellite systems using geostationary and geosynchronous satellites.

A GNSS Rx with an ideal clock can calculate its three-dimensional position (x, y, and z) by trilateration using the signals of at least three satellites. Since Rxs' clocks of most users are imprecise, their positioning requires signals from at least four satellites (multilateration) to solve equations for four unknowns: coordinates x, y, z, and Rx clock bias τ. Typically, Rxs also calculate velocity components and clock drift $v_x = x'$, $v_y = y'$, $v_z = z'$, and τ'. Increasing the number of satellites improves positioning and timing accuracy. Modern GNSS Rxs have tens or even hundreds of satellite tracking channels and often utilize signals of several GNSSs. A typical unaided GNSS Rx has positioning error of a few meters.

Kalman filters are typically used for calculating the Rx position because they can take into account prior measurements and combine GNSS measurements with those from other sensors, thus increasing the accuracy and reliability. Integrating GNSS Rxs with inertial measurement units (IMUs) is especially beneficial because they are complementary: GNSSs have noticeable random errors almost without biases, whereas IMUs have inherent biases but low random errors. Their combining compensates these drawbacks and sustains navigation during signal reception disruptions, albeit with gradual degradation. Other sensors usually integrated with GNSS Rxs are altimeters, electro-optical sensors, and Wi-Fi positioning. Even small consumer devices like tablets and cell phones currently contain multiconstellation GNSS Rxs combined with miniature low-cost IMUs and other sensors.

Applying differential corrections from local, regional, and/or global augmentation systems improves the GNSS Rx accuracy. Local systems are usually terrestrial and may support approximately 1-cm accuracy. Regional and global augmentation systems (supporting submeter accuracies) are mostly satellite-based, commonly called SBAS (Satellite-Based Augmentation System). Major examples are the U.S.-operated Wide Area Augmentation System (WAAS) covering North America and Hawaii, the EU-operated European Geostationary Navigation Overlay Service (EGNOS), the

Table 2.3 Selected Parameters of GNSSs

GNSS	Country/Area	Multiple Access Method	Center Frequencies (MHz)	Constellation, Satellites	Orbital Radius (km)	Orbital Inclination	Orbital Period	Ground Track Repetition Period
GPS	United States	CDMA	1,176.45, 1,227.60, 1,575.42	24–32 (31 as of October 2018)	26,600	55°	11 hr, 58 min	2 orbits (1 sidereal day)
GLONASS	Russia	FDMA, CDMA	1,246.0 and 1,602.0 (FDMA), 1,202.025 (CDMA)	24 or more (24 as of October 2018)	25,500	64.8°	11 hr, 16 min	17 orbits (8 sidereal days)
Galileo	European Union	CDMA	1,191.795, 1,278.75, 1,575.42	30 (planned in 2020)	29,600	56°	14 hr, 5 min	17 orbits (10 sidereal days)
BeiDou-2 (Compass)	China	CDMA	1,207.14, 1,268.52, 1,561.098, 1,589.742	27 and 3 geosync., 5 geostat. (planned in 2020)	27,800	55°	12 hr, 52 min	13 orbits (7 sidereal days)

Japanese Multi-Functional Satellite Augmentation System (MSAS), and the Indian GPS-Aided Geo Augmented Navigation (GAGAN). Two commercial global SBASs are StarFire and OmniSTAR.

The front-end bandwidths of GNSS Rxs vary from a few megahertz (sufficient for most unrestricted signals) to 20–40 MHz to receive higher-precision signals. Some Rxs have even wider front-end passbands to receive signals from more than one GNSS and/or SBAS constellations. Since the power of GNSS signals at the Rx input is well below thermal noise, the dynamic range of a GNSS Rx in interference-free conditions can be low. However, unintentional and intentional interference are quite common. Unintentional interference can be caused by any strong signal within the GNSSs' or adjacent bands. While the 1,559–1,610-MHz band is reserved exclusively for satellite navigation, this is not true for other GNSS frequency bands. For example, the 1,215–1,240-MHz band is also reserved for radar and the 1,164–1,215-MHz band for other aeronautical navigation aids. In addition, power leakage from adjacent bands and nonlinear products from strong signals in other bands can interfere with GNSS signals. GNSS signals are also subject to jamming. For example, drivers of vehicles equipped with GNSS trackers sometimes use low-power jammers to disable the trackers for privacy reasons. Such jammers can usually affect GNSS Rxs within a few meters, but some of them are powerful enough to interrupt GNSS signal reception within tens or hundreds of meters. While interference is a problem for civilian users, it is a much more acute problem for military users whose GNSS Rxs must be resistant to very strong and sophisticated jamming. This requires a high dynamic range of the Rxs and their digitization circuits. Additional information on navigation systems can be found, for instance, in [37–41].

2.4.3 Radio Methods in Positioning and Geolocation

Since radio methods of navigation are also used in geolocations that determine positions and velocities of radio-wave-emitting objects (e.g., Txs, mobile phones, internet-connected computers), direction finding (DF), radar, and EW systems, additional information on them is provided below. Triangulation, trilateration, and multilateration, mentioned in Section 3.4.2, as well as related technical procedures, are discussed here. Triangulation determines an object location by forming triangles to it from reference objects with known locations. In navigation, triangulation is performed after finding the directions to several radio beacons with known locations by the navigated vehicle's Rx. In geolocation, it is performed after finding the directions to a radio wave emitter from several known geographic locations. While triangulation itself is a geometric operation, DF is a radio procedure. It can be accomplished sequentially using a single moving platform (single-platform geo-location) or simultaneously using several platforms (multiplatform geolocation). The second technique, being more complex and expensive, can locate moving and burst emitters.

Rx antenna directivity is important for DF. There are two types of directional Rx antennas: those that provide maximum gain in the direction of a Tx of interest and those that provide a sharp null in that direction. The second type is more suitable for DF because changes in the Tx direction cause larger variations of the antenna output signals near the antenna pattern nulls than near the main lobe maximum. Vertical small loops (i.e., loops with the diameters of a tenth of a signal wavelength or smaller) were the earliest second-type antennas developed for DF. They have a figure-eight antenna pattern with sharp nulls in the directions normal to the loop plane, caused by the subtraction of the voltages induced on the opposite sides of the loop. The direction ambiguity can be eliminated by finding the direction to the Tx of interest from the second point or by combining a small vertical loop with a whip antenna. The combined antenna has a cardioid pattern in the horizontal plane with just one null, albeit less sharp than the nulls of a figure-eight pattern. While the cardioid pattern gives the initial direction estimate, switching off the whip antenna restores the figure-eight pattern that refines this estimate. The first DF loop antennas required mechanical rotation. Later, antenna systems with electronic steering were developed.

In small loops, the sharp antenna pattern nulls are created by the difference of voltages induced on the opposite sides of the loop. This idea was later transformed into a more general approach based on the subtraction or comparison of the output signals of two separate antennas or oppositely deflected antenna beams. This approach provides better DF accuracy than absolute measurements because the difference obtained as a result of the subtraction reflects a sum of changes in both antennas and beams produced by the Tx movement. In radar, this approach is applied to signals reflected by targets. For instance, phase-comparison monopulse and amplitude-comparison monopulse techniques are based on it. Due to the use of a single pulse and simultaneous comparison of received signals, these techniques allow avoiding problems caused by rapid changes in signal strength. Similar techniques are effective in interception of burst transmissions. Interestingly, the differential approach is also widely used in communications and signal processing for entirely different purposes [18, 21–23, 36].

Many types of DF antennas for different frequency ranges have been developed. Antennas based on the intentionally created Doppler effect are among them. The first versions of such antennas were used for navigation and signal intelligence during and after World War II (see, for example, [42, 43]). Initially, this effect was created by circular mechanical motion of a single antenna in the horizontal plane, as illustrated in Figure 2.4(a). The Doppler shift f_d reaches its maximum absolute value $|f_{d\max}|$ at points B and D where the antenna velocity vector $\mathbf{v}(t)$ is collinear with the direction to an emitter. The Doppler shift becomes equal to zero at points A and C where $\mathbf{v}(t)$ is perpendicular to this direction. This phenomenon allows determining the direction to a Tx of interest. The attainable Doppler shift and, consequently, the DF accuracy was radically increased when mechanical rotation was replaced with virtual antenna rotation created by sequential switching of the antenna elements (AEs) of a stationary uniform circular array (UCA) shown in Figure 2.4(b). Here, the AEs of the UCA are connected to the Rx by an electronic cyclic switch (ECS) that cycles through the AEs, creating the rotation effect.

Modern developments in DF technology are characterized by utilizing antenna arrays and advanced DSP algorithms. The major objectives are increasing the angular resolution, speed, and interference immunity of DF and improving its capabilities of resolving and finding the directions to several emitters simultaneously.

2.4   Other Radio Systems   67

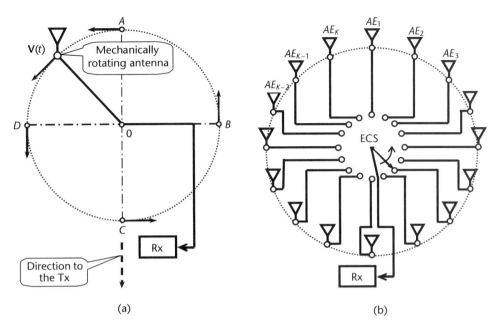

Figure 2.4 Antennas creating Doppler effect: (a) with mechanical rotation and (b) with virtual rotation.

While triangulation requires determining the angles of arrival, trilateration is based on determining the times of arrival. Multiplying this time by the speed of radio wave propagation gives the distance between the Tx and Rx. In the case of two-dimensional positioning and navigation, trilateration allows determining the location of a Rx using the signals of two Txs with known locations. In geolocation, it allows determining the location of a Tx of interest using two Rxs with known locations, capable of estimating the Tx signals' times of arrival. The major problem of the trilateration is the necessity to have precise synchronized clocks in all Txs and Rxs. Multilateration avoids this problem by using the time-of-arrival differences. It requires at least three stations for two-dimensional positioning, and at least four stations for three-dimensional positioning. Despite the increased number of the reference stations, multilateration is attractive for navigation due to moderate requirements for the accuracy and stability of the users' clocks. Therefore, it is employed in most popular navigation systems (e.g., GNSSs, Loran-C, and Chayka). Its major advantage for geolocation is that multilateration does not require any cooperation from a radio wave emitter.

Instead of measuring time of arrival or time difference of arrival, phase differences of arrived signals can be measured. Although this method requires accurate knowledge of the signal carrier frequency, it allows placing the antennas and Rxs of an interferometer needed for this operation on the same relatively small platform.

The accuracy of geolocation with airborne or space-based platforms can be improved by combining the time and frequency differences of arrival estimations. The latter technique estimates the differences of Doppler shifts of the signal copies received by platforms approaching a Tx of interest with different radial velocities.

The material of this and previous sections shows that, in principle, radio navigation, positioning, and geolocation do not require wider Rxs' passbands than those of wideband communication Rxs. An exception is EW geolocation scenarios with unknown spectral positions of signals of interest that should be quickly determined. Since the requirements for the Rxs' dynamic ranges mostly depend on the RF environment and the Rxs' bandwidths, these requirements in navigation, positioning, and geolocation Rxs are close to those for other types of Rxs with comparable bandwidths and operating in similar RF environments.

2.4.4 Radar and EW Systems

Both radar and EW systems are vast technical areas that cannot be even concisely described in this section. Fortunately, many excellent books have been published on these subjects (e.g., [44–66]). The material below provides examples of the most widely used radar systems and their frequency bands, mentions several facts rarely reflected in the technical literature, and outlines the highest requirements for the D&R circuits in these systems.

Radar uses the reflection or scattering of radio waves by conductive objects to determine their presence, locations, and velocities. The term "radar" was coined as an acronym for "radio detection and ranging" or "radio direction and ranging" in the United States in 1940. C. Hülsmeyer (Germany) contributed most to the discovery of the pulsed radar principles in 1904. The contemporary technology did not allow effective practical realization of these principles then, and they were forgotten for some time. In the 1920s and later, the idea of radio wave reflection from conductive media was used by E. V. Appleton (United Kingdom) for the discovery of ionospheric layers. In the 1930s, the intensive development of radar technology was initiated in the United Kingdom, United States, Germany, Russia, France, Japan, and some other countries. During World War II, radar was a decisive weapon in several military campaigns including the Battle of Britain. After that, fast expansion of radar systems was stimulated by their numerous military and civilian applications (see Table 2.4). Radar is currently used for safe landing of aircraft and passage of ships in poor visibility; preventing collisions on the roads; weather forecasting; exploring Earth and other planets; and detecting, locating, and characterizing various objects from animals and birds to satellites and asteroids.

The first radar systems, implemented in the 1930s, took advantage of the theoretical results, components, and industrial base created earlier for communications, broadcasting, and DF. However, military necessity forced the advanced development of radar-related theory and technology, which stimulated many discoveries and innovations. Statistical methods of signal processing, new types of directional antennas and antenna arrays, novel electronic devices, SS signals, MIMO techniques, and many other innovations were first implemented in radar and later found their ways to other radio systems as well as to sonar, lidar, seismological, medical, measurement, and other equipment. The emergence of IC and DSP technologies significantly accelerated the radar progress. The following example illustrates the mutual influence of different fields in electrical engineering. The adverse effect of linear distortions in Txs and Rxs first became apparent in TV, and a paired echoes approach [67] was developed to study this effect. When it was noticed that the same distortions were

Table 2.4 Types and Frequency Bands of Radar

Radar Type	Frequency Band
Over-the-horizon radar	VLF, LF, MF, HF
Very long range radar	VHF, UHF (< 1 GHz)
Ground-penetrating radar (including radar astronomy)	VHF, UHF (< 1 GHz)
Foliage-penetrating radar	UHF (< 1 GHz)
Airport radar (long-range air traffic control and surveillance)	UHF (L-band)
Airport radar (terminal traffic control and moderate range surveillance)	UHF, SHF (S-, X-bands)
Airport radar (short-range surveillance)	SHF, EHF (Ka-band)
Mapping radar (including radar astronomy)	UHF, SHF, EHF (S-, C-, X-bands)
Marine radar	UHF, SHF (S-, X-bands)
Weather radar (long and medium range)	UHF, SHF (S-, C-bands)
Weather radar (short range)	SHF (X-band)
Weather radar (cloud and fog detection)	SHF (K-band)
See-through-the-wall radar	UHF, SHF (S-, C-bands)
Tracking radar (long range)	SHF (C-band)
Tracking radar (short range)	SHF (X-band)
Tracking radar (high resolution)	SHF (Ku-band)
Police radar guns	SHF (K-band)
Police photo enforcement radar	EHF (Ka-band)
Missile guidance radar and fire control radar	SHF (X-band)
Missile guidance radar (active homing) and fire control radar (short range)	EHF (W-band, millimeter-band)

reducing the range resolution and causing false targets in radar, this approach was modified and refined in [68, 69] to determine the distortions level acceptable for radar. The implementation of SS signals in communication and sounding systems coupled with the necessity of suppressing the parts of signal spectrum affected by narrowband interference produced similar problems. To solve them, the additionally enhanced paired echoes approach was combined with the optimal filtering theory in [70–74]. Thus, ideas originated to solve specific problems in particular fields become more general and productive after applying them to other fields.

Radar Txs and Rxs are currently digital. As discussed in Section 2.2, the requirements for the dynamic range and bandwidth of D&R circuits determine their complexity. Since the range resolution of radar depends on the signal bandwidth, the latter is usually quite wide. Bandwidths of 50–200 MHz are common, and they can reach several gigahertz in see-through-the-wall radar. Still, the complexity of reconstruction circuits in radar Txs is moderate because the required dynamic range is not high. The complexity of digitization circuits in radar Rxs is usually much higher since they must support high dynamic range and sensitivity in addition to wide bandwidth. The required dynamic range is particularly large in military systems because, besides the necessity to detect targets with small radar cross-sections in the presence of strong clutter (that must be rejected in the RDP), their Rxs can become victims of brute-force jamming. In such scenarios, even 80-dB dynamic range can be insufficient, and there is no upper limit for the required dynamic range. In practice, the Rx dynamic range is restricted by technological limitations as well as constraints on power consumption, size, weight, and/or cost of equipment. Thus, the requirements for digitization circuits in military radar can be the same or even higher than in the most advanced communication systems operating in hostile RF environments.

EW Rxs may operate in various conditions and have different tasks. Some of them are compact, fairly simple, and specialized in fast recognition of expected immediate threats in order to avoid or eliminate them. More complex multipurpose Rxs are capable of performing not only immediate threat recognition but also many other functions including intelligence collection. Their superior capabilities are usually achieved at the expense of larger size, weight, power consumption, and cost. Being multipurpose, such Rxs are still rather specialized on communication or radar signals because different carrier frequencies, bandwidths, purposes, and structures of these signals require different processing. The instantaneous bandwidths of intercept and intelligence-gathering Rxs should be wider than the spectra of their signals of interest because the spectral positions and widths of these signals may be unknown. The wide required instantaneous bandwidths and uncertainty about the input signal levels necessitate high dynamic ranges of these Rxs. Modern technology allows the development of single-channel high-dynamic-range intercept Rxs with the instantaneous bandwidth of about 4 GHz [66, 75]. Both AMFs and RDPs of such Rxs are expensive.

The implementation of novel digitization circuits described in Chapter 6 will reduce the cost and improve the parameters of the AMFs. Simultaneously, the progress of DSP and IC technologies will reduce the cost of the RDPs. Currently, a cost-benefit analysis may show that, is some cases, it is more beneficial to employ several parallel lower-cost Rxs to achieve the combined instantaneous bandwidth of about 4 GHz. When the required instantaneous bandwidth of an intercept Rx is much wider than 4 GHz, parallel structures are unavoidable. Note that intercept Rxs have some advantages over the own Rxs of radar and communication systems of interest. For instance, the signal power at the input of a Rx intercepting radar signals is typically much higher than that of reflected signals at the inputs of the radar's own Rx. This is also true for a Rx intercepting communication signals if it is located much closer to a Tx of interest than the communication system's own Rx.

2.5 Summary

Radio waves are electromagnetic waves with frequencies from 3 Hz to 3 THz, currently utilized for transmitting energy and information. D&R circuits are used in the systems transmitting information. The variety of such systems is large, and communication systems form their most diverse class.

Digital communication radios perform all types of D&R and the largest number of signal processing operations interdependent with D&R. The requirements

2.5 Summary

for D&R in communication radios are as high as or higher than in radios of most other systems. Thus, communication radios are the best case study for analysis and development of D&R circuits and algorithms.

RF spectrum is divided into frequency bands (see Tables 2.1 and 2.2). Each band has unique advantages that are more or less important for various applications. Since the spectrum is a valuable but limited resource, the ITU allocates its nonoverlapping parts to different applications and services.

Section 2.3 identifies the major trends in communication systems development. Many of the trends influence D&R circuits. Virtually all communication radios are currently digital, and representation of radio signals by the I and Q components of their digital baseband complex-valued equivalents is advantageous in most radios.

Source encoding in Txs removes the redundancy that cannot be effectively utilized for improving the radio communications reliability. Performing a part of analog signals' source encoding during their digitization is often efficient. Considering digitization a special case of source encoding allows sharing technical solutions between the designers of digitization and source encoding circuits and algorithms.

The complexity of D&R is determined by the properties of their input signals (mainly by the signals' crest factors, bandwidths, and spectral positions) as well as by the required accuracy of signal generation in Txs and processing in Rxs. The complexity of digitization at Rxs' inputs is higher than that of digitization at Txs' inputs and reconstruction at Txs' outputs.

Channel encoding, modulation, and spreading actually are the stages of generalized modulation whose purpose is varying the carrier signal parameters in a way allowing optimal transmission of information over communication channels. Despreading, demodulation, and channel decoding are the stages of generalized demodulation. The conceptual separation of the generalized modulation and demodulation operations allows combining different approaches and techniques for achieving the common goal.

This separation still allows joint realization of channel encoding and modulation or modulation and spreading as well as channel decoding and demodulation or demodulation and despreading. When demodulation and channel decoding are performed separately, soft-decision demodulation and decoding are preferable.

Processing gain, introduced in the 1950s, characterizes only the impact of spreading on the system's AJ capability. Currently, this and some other capabilities are characterized by the product of gains provided by channel coding, modulation, and spreading. Since these operations differently affect various adverse phenomena in communication channels, adaptive distribution of available system bandwidth among them is advantageous.

From a technical standpoint, broadcasting systems can be considered a special case of communication systems. The specifics of broadcast radios in terms of the requirements for their D&R circuits and algorithms are well within the limits of what is required for other communication systems.

A variety of radio navigation systems have been developed since their origination. GNSSs, which are the most successful and widely used among them, have fully or partially replaced many earlier deployed systems. However, concerns about vulnerability of GNSSs limit this replacement.

The principles of navigation, positioning, geolocation, and DF have been outlined because they significantly differ from those of communications. The requirements for D&R circuits and algorithms in all these systems are not higher than those in communication systems, except for some EW scenarios. In radar, the highest requirements for D&R circuits and algorithms are also related to EW.

References

[1]   Boithais, L., *RadioWave Propagation*, New York: McGraw-Hill, 1987.

[2]   Jacobs, G., T. J. Cohen, and R. B. Rose, *The New Shortwave Propagation Handbook*, Hicksville, NY: CQ Communications, 1997.

[3]   Bertoni, H. L., *Radio Propagation for Modern Wireless Systems*, Upper Saddle River, NJ: Prentice Hall, 2000.

[4]   Saakian, A., *Radio Wave Propagation Fundamentals*, Norwood, MA: Artech House, 2011.

[5]   Picquenard, A., *Radio Wave Propagation*, New York: Palgrav, 2013.

[6]   Poberezhskiy, Y. S., "On Conditions of Signal Reception in Short Wave Channels," *Proc. IEEE Aerosp. Conf.*, Big Sky, MT, March 1–8, 2014, pp. 1–20.

[7]   Ghasemi, A., A. Abedi, and F. Ghasemi, *Propagation Engineering in Wireless Communications*, 2nd ed., New York: Springer, 2016.

[8]   Lindsey, W. C., and M. K. Simon, *Telecommunication Systems Engineering*, Englewood Cliffs, NJ: Prentice Hall, 1973.

[9]   Jakes, W. C. (ed.), *Microwave Mobile Communications*, New York: John Wiley & Sons, 1974.

[10] Spilker Jr., J. J., *Digital Communications by Satellite*, Englewood Cliffs, NJ: Prentice-Hall, 1977.

[11] Holmes, J. K., *Coherent Spread Spectrum Systems*, New York: John Wiley & Sons, 1982.

[12] Smith, D. R., *Digital Transmission Systems*, New York: Van Nostrand Reinhold, 1985.

[13] Korn, I., *Digital Communications*, New York: Van Nostrand Reinhold, 1985.

[14] Benedetto, S., E. Bigiliery, and V. Castellani, *Digital Transmission Theory*, Englewood Cliffs, NJ: Prentice Hall, 1987.

[15] Ivanek, F., *Terrestrial Digital Microwave Communications*, Norwood, MA: Artech House, 1989.

[16] Schwartz, M., *Information, Transmission, Modulation, and Noise*, 4th ed., New York: McGraw-Hill, 1990.

[17] Simon, M. K., et al., *Spread Spectrum Communications Handbook*, New York: McGraw-Hill, 1994.

[18] Okunev, Y., *Phase and Phase-Difference Modulation in Digital Communications*, Norwood, MA: Artech House, 1997.

[19] Garg, V. K., K. Smolik, and J. E. Wilkes, *Applications of CDMA in Wireless/Personal Communications*, Upper Saddle River, NJ: Prentice-Hall, 1997.

[20] Van Nee, R., and R. Prasad, *OFDM for Wireless Multimedia Communications*, Norwood, MA: Artech House, 2000.

[21] Xiong, F., *Digital Modulation Techniques*, Norwood, MA: Artech House, 2000.

[22] Proakis, J. G., *Digital Communications*, 4th ed., New York: McGraw-Hill, 2001.

[23] Sklar, B., *Digital Communications, Fundamentals and Applications*, 2nd ed., Upper Saddle River, NJ: Prentice Hall, 2001.

[24] Rappaport, T. S., *Wireless Communications*, 2nd ed., Upper Saddle River, NJ: Prentice Hall, 2002.

References

[25] Calhoun, G., *Third Generation Wireless Systems, Post-Shannon Signal Architectures*, Vol. 1, Norwood, MA: Artech House, 2003.

[26] Haykin, S., *Communication Systems*, 5th ed., New York: John Wiley & Sons, 2009.

[27] Kalivas, G., *Digital Radio System Design*, New York: John Wiley & Sons, 2009.

[28] Wyglinski, A. M., M. Nekovee, and Y. T. Hou (eds.), *Cognitive Radio Communications and Networks: Principles and Practice*, New York: Elsevier, 2010.

[29] Lathi, B. P., and Z. Ding, *Modern Digital and Analog Communication Systems*, 4th ed., Oxford, U.K.: Oxford University Press, 2012.

[30] Furman, W. N., et al., *Third-Generation and Wideband HF Radio Communications*, Norwood, MA: Artech House, 2013.

[31] Torrieri, D., *Principles of Spread-Spectrum Communication Systems*, 3rd ed., New York: Springer, 2015.

[32] Poberezhskiy, Y. S., I. Elgorriaga, and X. Wang, "System, Apparatus, and Method for Synchronizing a Spreading Sequence Transmitted During Plurality of Time Slots," U.S. Patent 7,831,002 B2, filed October 11, 2006.

[33] Poberezhskiy, Y. S, "Method and Apparatus for Synchronizing Alternating Quadratures Differential Binary Phase Shift Keying Modulation and Demodulation Arrangements," U.S. Patent 7,688,911 B2, filed March 28, 2006.

[34] Poberezhskiy, Y. S., "Alternating Quadratures Differential Binary Phase Shift Keying Modulation and Demodulation Method," U.S. Patent 7,627,058 B2, filed March 28, 2006.

[35] Poberezhskiy, Y. S., "Apparatus for Performing Alternating Quadratures Differential Binary Phase Shift Keying Modulation and Demodulation," U.S. Patent 8,014,462 B2, filed March 28, 2006.

[36] Poberezhskiy, Y. S., "Novel Modulation Techniques and Circuits for Transceivers in Body Sensor Networks," *IEEE J. Emerg. Sel. Topics Circuits Syst.*, Vol. 2, No. 1, 2012, pp. 96–108.

[37] Hofmann-Wellenhof, B., K. Legat, and M. Wieser, *Navigation: Principles of Positioning and Guidance*, New York: Springer, 2003.

[38] Dardari, D., M. Luise, and E. Falletti (eds.), *Satellite and Terrestrial Radio Positioning Techniques: A Signal Processing Perspective*, Waltham, MA: Academic Press Elsevier, 2012.

[39] Nebylov, A. V., and J. Watson (eds.), *Aerospace Navigation Systems*, New York: John Wiley & Sons, 2016.

[40] Betz, J. W., *Engineering Satellite-Based Navigation and Timing: Global Navigation Satellite Systems, Signals, and Receivers*, New York: John Wiley & Sons, 2016.

[41] Kaplan, E. D., and C. J. Hegarty (eds.), *Understanding GPS/GNSS: Principles and Applications*, 3rd ed., Norwood, MA: Artech House, 2017.

[42] Hansel, P. G., "Navigation System," U.S. Patent No. 2,490,050; filed November 7, 1945.

[43] Hansel, P. G., "Doppler-Effect Omnirange," *Proc. IRE*, Vol. 41, No. 12, 1953, pp. 1750–1755.

[44] Sherman, S., *Monopulse Principles and Techniques*, Norwood, MA: Artech House, 1984.

[45] Blake, L., *Radar Range Performance Analysis*, Norwood, MA: Artech House, 1986.

[46] Wehner, D., *High-Resolution Radar*, Norwood, MA: Artech House, 1987.

[47] Levanon, N., *Radar Design Principles*, New York: John Wiley & Sons, 1988.

[48] Brookner, E. (ed.), *Aspects of Modern Radar*, Norwood, MA: Artech House, 1988.

[49] Nathanson, F., *Radar Design Principles*, 2nd ed., New York: McGraw-Hill, 1991.

[50] Stimson, G. W., *Introduction to Airborne Radar*, Raleigh, NC: SciTech Publishing, 1998.

[51] Skolnik, M. I., *Introduction to Radar Systems*, 3rd ed., New York: McGraw Hill, 2001.

[52] Shirman, Y. D. (ed.), *Computer Simulation of Aerial Target Radar Scattering, Recognition, Detection, and Tracking*, Norwood, MA: Artech House, 2002.

[53] Sullivan, R. J., *Radar Foundations for Imaging and Advanced Concepts*, Edison, NJ: SciTech Publishing, 2004.

[54] Barton, D., *Radar System Analysis and Modeling*, Norwood, MA: Artech House, 2005.

[55] Willis, N. J., and H. D. Griffiths (eds.), *Advances in Bistatic Radar*, Raleigh, NC: SciTech Publishing, 2007.

[56] Skolnik, M. I. (ed.), *Radar Handbook*, 3rd ed., New York: McGraw Hill, 2008.

[57] Meikle, H., *Modern Radar Systems*, 2nd ed., Norwood, MA: Artech House, 2008.

[58] Richards, M. A., *Fundamentals of Radar Signal Processing*, 2nd ed., New York: McGraw-Hill, 2014.

[59] Budge Jr., M. C., and S. R. German, *Basic Radar Analysis*, Norwood, MA: Artech House, 2015.

[60] Schleher, D. C., *Electronic Warfare in the Information Age*, Norwood, MA: Artech House, 1999.

[61] Adamy, D. L., *EW 101: A First Course in Electronic Warfare*, Norwood, MA: Artech House, 2001.

[62] Adamy, D. L., *EW 102: A Second Course in Electronic Warfare*, Norwood, MA: Artech House, 2004.

[63] Adamy, D. L., *EW 103: Tactical Battlefield Communications Electronic Warfare*, Norwood, MA: Artech House, 2009.

[64] Poisel, R. A., *Modern Communications Jamming: Principles and Techniques*, 2nd ed., Norwood, MA: Artech House, 2011.

[65] Adamy, D. L., *EW 104: Electronic Warfare Against a New Generation of Threats*, Norwood, MA: Artech House, 2015.

[66] Tsui, J. B. Y., and C. H. Cheng, *Digital Techniques for Wideband Receivers*, 3rd ed., Raleigh, NC: SciTech Publishing, 2016.

[67] Wheeler, H. A., "The Interpretation of Amplitude and Phase Distortion in Terms of Paired Echoes," *Proc. IRE*, Vol. 27, No. 6, 1939, pp. 359–384.

[68] Di Toro, M. J., "Phase and Amplitude Distortion in Linear Networks," *Proc. IRE*, Vol. 36, No. 1, 1948, pp. 24–36.

[69] Franco, J. V., and W. L. Rubin, "Analysis of Signal Processing Distortion in Radar Systems," *IRE Trans.*, Vol. MIL-6, No. 2, 1962, pp. 219–227.

[70] Khazan, V. L., Y. S. Poberezhskiy, and N. P. Khmyrova, "Influence of the Linear Two-Port Network Parameters on the Spread Spectrum Signal Correlation Function" (in Russian), *Proc. Conf. Problems of Optimal Filtering*, Vol. 2, Moscow, Russia, 1968, pp. 53–62.

[71] Poberezhskiy, Y. S., "Statistical Estimate of Linear Distortion in the Narrowband Interference Suppressor of the Oblique Sounding System Receiver" (in Russian), *Problems of Radio-Electronics*, TRC, No. 7, 1970, pp. 32–39.

[72] Poberezhskiy, Y. S., "Derivation of the Optimum Transfer Function of a Narrowband Interference Suppressor in an Oblique Sounding System" (in Russian), *Problems of Radio-Electronics*, TRC, No. 9, 1969, pp. 3–11.

[73] Poberezhskiy, Y. S., "Optimum Transfer Function of a Narrowband Interference Suppressor for Communication Receivers of Spread Spectrum Signals in Channels with Slow Fading" (in Russian), *Problems of Radio-Electronics*, TRC, No. 8, 1970, pp. 104–110.

[74] Poberezhskiy, Y. S., "Optimum Filtering of Sounding Signals in Non-White Noise," *Telecommun. and Radio Engineering*, Vol. 31/32, No. 5, 1977, pp. 123–125.

[75] Devarajan, S., et al., "A 12-Bit 10-GS/s Interleaved Pipeline ADC in 28-nm CMOS Technology," *IEEE J. Solid-State Circuits*, Vol. 52, No. 12, 2017, pp. 3204–3218.

CHAPTER 3
Digital Transmitters

3.1 Overview

As explained in Chapter 2, the use of digital communication radios as a case study for examining and developing D&R circuits and algorithms allows applying the obtained results to virtually all other technical fields because these radios impose very high and diverse requirements on these procedures. Information on digital communication radios presented in Chapters 1 and 2 is an introduction to their analysis in this and next chapters. The study of digital Txs below differs from those in other publications (see, for instance, [1–18]) by its focus on D&R procedures. All signal processing operations in Txs are examined from the standpoint of their relation to the D&R. Digitization of Txs' input signals and reconstruction of their output signals are described. Connection between power utilization and complexity of reconstruction in Txs as well as the approaches to improving power utilization and easing the requirements for Txs' reconstruction circuits are discussed.

Section 3.2 shows that modern IC and DSP technologies support the development of not only multipurpose and/or multistandard software-defined radios (SDRs) and cognitive radios (CRs), but also inexpensive, low-power, single-purpose digital radios. The differences between their Txs are discussed. Despite the differences, many functions are common for most digital Txs. A typical architecture of a multipurpose digital Tx is presented. The influence of the operations performed in the TDP on the reconstruction circuits is explained. Nonrecursive direct digital synthesizer (DDS) is described, largely due to the similarity of its algorithms to those of some digital weight function generators (WFGs) used for D&R based on the direct and hybrid interpretations of the sampling theorem (See Chapter 6).

Section 3.3 describes and analyzes D&R in digital Txs. Digitization of analog input signals is considered in Section 3.3.1. Since these signals are baseband, general information on digitization of such signals is also provided. Reconstruction of Txs' output signals is considered in Section 3.3.2, and both baseband and bandpass reconstruction techniques are discussed in detail. Comparison of these techniques is performed in Section 3.3.3. Architectures of a conversion block, which completes the signal reconstruction and translates the reconstructed signals to the format required for the Tx power amplifier (PA), are also analyzed there.

Approaches to improving Tx power utilization and simplifying reconstruction of its output signals are discussed in Sections 3.4. It is shown that these approaches are different for energy-efficient and bandwidth-efficient signals, and the power utilization improvement does not always reduce the reconstruction complexity.

3.2 Digital Tx Basics

3.2.1 Txs of Different Categories of Digital Radios

All definitions related to digital radios are conditional and somewhat vague. Even the term "digital radio" sometimes refers to digital broadcasting, especially to digital audio broadcasting. In most cases and in this book, however, it is used for radios where main signal processing operations, such as channel encoding/decoding, modulation/demodulation, spreading/despreading, and most filtering, are performed in TDPs and RDPs. Many other functions, for example, frequency synthesis, synchronization, source coding, automatic control operations, and interference rejection, are fully or partially executed in the digital domain. Still, even the most advanced digital radios contain analog and mixed-signal portions. This is especially true for digital Txs where the most energy-consuming units, PAs, remain analog. Thus, the point at which a radio can be called digital is conditional. In SDRs the minimum number and importance of functions that should be software-defined are not specified. CRs should assess the RF environment and adapt their operations to it and to the user needs by learning and memorizing the outcomes of prior decisions. Some of the first such radios (without using the term "CRs") were developed in Russia to cope with intentional and unintentional jamming in the HF band in the early 1970s (see, for example, [19, 20]). Those radios could hardly be considered digital, but they performed many cognitive functions.

Despite these ambiguities, it is clear that all multipurpose and/or multistandard digital radios are currently evolving as software-defined, and many SDRs are getting cognitive capabilities. CRs were initially focused on spectrum sensing to use its best available parts without interfering with other users and on power control in spectrum-sharing systems. Now they also utilize spatial signal characteristics using antenna arrays. Cognitive networks allow sharing the sensing results among CRs and optimize the collective use of spectral, space, and energy resources. The rising density of radio stations, spectral and spatial sensing, dynamic access, and growing data rates require increasing the dynamic ranges, bandwidths, and flexibility of SDRs and CRs. These properties of digital Txs and Rxs directly affect the requirements for their D&R circuits. SDRs and CRs form the first major category of digital radios.

Modern IC and DSP technologies support not only the progress of SDRs and CRs but also fast evolution and proliferation of inexpensive, low-power, single-purpose digital radios for many applications, including personal area and sensor networks. The latter radios form the second major category of digital radios. They minimally use versatility and flexibility of DSP but still utilize its other advantages: high and independent of destabilizing factors signal processing accuracy, possibility of regenerating information from digital signals distorted during their transmission, storage, or processing, as well as large scale of integration and low production cost. Such radios are ubiquitous, small, more specialized, and inexpensive devices emitting low-power signals due to short ranges and/or low rates of transmissions. The power of transmitted signals may be so low that Txs can consume less power than Rxs where processing is more complex. Some of such Rxs can be powered by the RF radiation of various sources intercepted by their antennas. Normally, however, they are powered by small batteries capable of supporting their operation for months or years without recharging or replacing. Digital radios of the second category are implemented mostly as ASICs. Limited battery capacity necessitates the use of energy-efficient signals. Low power of transmitted signals allows placing a Tx drive and a PA on the same chip. The radios' small size and low power consumption limit the complexity of their processing algorithms and stability of their local oscillators (LOs). Since small antenna sizes limit the directivity, some networks of densely located sensor radios form phased antenna arrays to communicate with remote radios or networks.

Although multipurpose and/or multistandard SDRs and CRs can have different size, weight, cost, and power consumption, they are usually larger, heavier, and more expensive and power-consuming than single-purpose radios. They transmit larger variety of information over longer distances with superior throughput. They are powered by self-contained or external batteries, or stationary power sources. Since high power of transmitted signals makes the PA technology different from that of Tx drives, PAs are usually separate blocks or separate chips in the Txs. The complexity of DSP is minimally limited in these radios. Therefore, they are adaptive and employ the most effective encryption, modulation, and coding algorithms. Their TDPs and RDPs use various hardware platforms. ASIC implementation of these portions (with embedded general-purpose processor (GPP) cores) is the best for mass-produced SDRs and CRs. Field programmable gate arrays (FPGAs) are preferable platforms for the TDPs and RDPs in radios produced in moderate volumes. Digital signal processors (DSPs) are employed in relatively small SDRs. Standalone GPPs are used mostly for rapid prototyping and testing of signals and processing algorithms in laboratories where they may not operate in real time and their size, weight, and power consumption are unimportant. Joint operation of GPPs and DSPs or specialized processing units (SPUs), such as graphics processing units (GPUs), increases the throughput of TDPs and RDPs and preserves their high versatility. In these cases, the most sophisticated but low-speed processing is performed by GPPs, whereas DSPs handle signals at the D/A inputs in TDPs, the A/D outputs in RDPs, and other high-speed processing stages in TDPs and RDPs. The most complex processing in RDPs and TDPs may require all the types of devices mentioned above.

The improvement of power utilization in Txs of different categories requires different approaches. Since the second-category radios employ energy-efficient signals, the signals' crest factor reduction improves the Tx power utilization and simultaneously simplifies reconstruction. This approach is insufficient for the first-category radios that typically use not only energy-efficient but also bandwidth-efficient signals whose crest factors cannot be significantly reduced. The ways to improve power utilization in Txs of both categories are discussed in Section 3.4.

3.2.2 Architecture of a Digital Tx

Although most processing in digital Txs is performed in their TDPs, analog bandpass real-valued signals have to be reconstructed at the end. As technology progresses, TDPs execute more functions, and reconstruction moves closer to the antennas. Still, the final interpolating filtering and amplification, as well as antenna coupling, remain analog. Txs with substantial transmit power comprise two functionally dissimilar parts: a Tx drive (or exciter) and a PA. The Tx drive performs "intelligent" functions, such as D&R, source and channel encoding, modulation, spreading, multiplexing, frequency synthesis and translations, some amplification, most of the filtering, and control functions, for example, automatic level control (ALC). Thus, it can be considered the Tx's "brains," whereas the PA, which performs energy-consuming final amplification, some filtering, and antenna coupling, can be considered its "muscles."

From the D&R standpoint, the division of a Tx into a Tx drive and a PA is less important than the separation among a digitizer of input analog signals, a TDP, an AMB, and a part containing the analog and mixed-signal blocks not included in the digitizer or AMB. In the block diagram in Figure 3.1, which is a more detailed version of those in Figures 1.19(a) and 2.2(a), such blocks are the primary frequency synthesizer and MFS. Some Tx input signals (e.g., data, measurement results, speech, or music) are originally functions of time; others, such as plots and pictures, are converted into those by scanning. The input signals undergo source encoding to reduce the redundancy unused for improving the transmission reliability. For analog signals, a part of this encoding can be performed during their digitization. Although such an option is not discussed in this chapter, it is examined in Chapter 7.

While encryption usually does not influence signal reconstruction in Txs, the redundancy introduced by channel encoding complicates it due to either widened signals' bandwidths or (like trellis coding) increased sensitivity of signal constellations to distortions in Txs and Rxs. Recall that distortions of signal constellations are characterized by error vector magnitude (EVM). Modulation influences the reconstruction complexity through the bandwidths, crest factors, and EVM requirements. The best bandwidth-efficient signals (e.g., QAM) have high crest factors, and only some less efficient ones (e.g., M-ary PSK) have low crest factors. Large constellations of bandwidth-efficient signals are sensitive to distortions and require low EVM. Energy-efficient signals typically allow minimization of their crest factors and their constellations are more tolerant to distortions, but their bandwidths are much wider than those of bandwidth-efficient signals for the same bit rates. For these signals, EVM is determined by the needed accuracy of symbol-shaping and interpolating filtering. Normally, the reconstruction of bandwidth-efficient signals is more complex than that of energy-efficient ones.

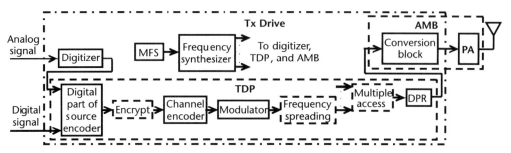

Figure 3.1 Block diagram of a multipurpose digital Tx.

Frequency spreading affects reconstruction mostly by increasing signals' bandwidths, but crest factors of SS signals are also important (see Section 3.4). Multiple access techniques differently influence reconstruction. TDMA increases the signals' bandwidths, whereas CDMA and FDMA increase the signals' bandwidths and crest factors, as well as complexity of their constellations (the constellations produced by CDMA are similar to those of QAM). Symbol-shaping filtering prior to reconstruction and digital interpolating filtering during it also affect the reconstruction complexity. Since only the combined result of all these operations is important, the effect produced by one of them can be compensated or absorbed by other operation(s). For instance, in signals produced by a sequence of channel encoding, modulation, and spreading, a bandwidth increase caused by channel encoding and/or modulation cannot influence reconstruction if it does not change the SS signal bandwidth.

As noted in Sections 1.4.1 and 2.3.2, signals in TDPs are usually represented by their digital baseband complex-valued equivalents, and reconstruction of analog bandpass signals can be baseband or bandpass. The DPR performs digital operations for any type of reconstruction. Although these operations differ for baseband and bandpass reconstruction, they both include upsampling with digital interpolating filtering. The interpolating filter passband should be wider than that of the symbol-shaping filter if they are not combined. Besides the upsampling, the DPR can also predistort signals to compensate their distortion in the subsequent mixed-signal and/or analog circuits. The conversion block completes this reconstruction and carries out all other analog operations needed prior to the Tx PA. Reconstruction of TDP output signals is considered in the next section.

In Txs and Rxs, the MFSs are sources of accurate and stable frequencies. All other frequencies are synthesized from them. Currently, most MFSs are high-precision crystal oscillators, but they encounter competition from MEMS-based oscillators, especially in inexpensive radios. For the frequency stability better than 10^{-9} ppm, atomic standards are used as MFSs. Various techniques, such as frequency multiplication, division, mixing, DDS, and phase-locked loops (PLLs), are used in frequency synthesizers. The primary frequency synthesizers of most digital radios initially generate the required frequencies digitally. Since digitally-generated frequencies may have unacceptable level of spurious components, they are not sent to other blocks but applied to an analog voltage-controlled oscillator (VCO) within a PLL that filters out spurious components and phase noise. Therefore, the VCO output frequency is "clean" and has accuracy and stability determined by the MFS. The PLL-based synthesizers are widely used because they can inexpensively generate accurate, stable, and pure frequencies within broad ranges. The generated frequencies are sent to the digitizer, TDP, and conversion block of the AMB. In the TDP, these frequencies are used not only directly but also as references for generating other needed frequencies. Analysis of frequency synthesizers is out of this book's scope. However, nonrecursive DDSs are discussed below due to the similarity of their algorithms to those of WFGs used for the D&R based on the direct and hybrid interpretations of the sampling theorem (see Chapter 6). Note that the aforementioned phase noise represents short-term random fluctuations of signal phase in the frequency domain. In the time-domain analysis of D&R circuits, these fluctuations are referred to as jitter.

3.2.3 Direct Digital Synthesis

The major blocks of nonrecursive DDSs, suggested in [21] and considered in many publications, including [2, 9, 21–24], are the phase accumulator and the digital functional converter (DFC) that converts the digital words sent from the phase accumulator into the digital words representing the sine wave values (see Figure 3.2(a)). Typically, DFCs simultaneously generate sine and cosine waves. Since their generation is similar, only the sine wave generation is shown in Figure 3.2 for conciseness. The D/A and subsequent analog interpolating LPF depicted in Figure 3.2(a) are optional: they are needed only if the output sine wave is used in the analog domain.

The phase accumulator contains an N-bit register and adds a frequency control code word k to that in the register on every clock pulse. It operates as a modulo-A counter with increment k where A can, in principle, be any integer within interval $[k, 2^N]$. The code word at the accumulator output represents the phase of the sine wave generated by the DDS. The larger k, the earlier the phase accumulator overflows, the faster the DFC completes a sine wave cycle, and the higher the DDS output frequency is. Only $M < N$ most significant bits (MSBs) are sent to the DFC. The maximum accuracy of frequency tuning is determined by N, but the phase quantization error is determined by M. This error is one of the primary sources of spurious products. Another source is the quantization error caused by a finite length of the words representing the sine wave.

The arrangement of M bits sent to the DFC is illustrated by the timing diagrams in Figure 3.2(b). Two MSBs determine the current sine wave quadrant. The first MSB is designated as a sign bit because it controls the signs of the sine wave half-cycles at the DDS output. The second MSB is designated as a quadrant bit because it, together with the sign bit, indicates the quadrant. The other $M - 2$ bits sent to the DFC represent the phase values within each quadrant. When $A = 2^N$ and the phase word is modulo-2^N incremented by the frequency control word k at the clock rate f_c, the DDS output sine wave frequency is

$$f_{out} = \frac{k}{2^N} f_c \tag{3.1}$$

The minimum output frequency f_{min} and the smallest frequency increment f_{inc} correspond to $k = 1$. From (3.1),

$$f_{min} = f_{inc} = \frac{1}{2^N} f_c \tag{3.2}$$

To provide sufficient analog interpolating filtering at the D/A output when an analog sine wave is needed, the maximum DDS frequency should meet the condition:

$$f_{max} \leq 0.25 f_c \tag{3.3}$$

Recall that a sine wave (as well as a cosine wave) is unique only within a quarter-cycle. Therefore, actual conversion of the phase values into the sine wave values is performed by the single-quadrant converter. The other DFC blocks (i.e., the phase code complement and sign control) extend the results of the single-quadrant conversion to the full sine wave cycle. The phase code complement, controlled by the quadrant bit, converts the modulo-0.5π phase values into their complements to 0.5π if the full sine wave phase is in the second or fourth quadrant, as shown in Figure 3.2(c). When the phase values are represented by binary code, these complements are obtained by inverting all $M - 2$ MSBs of the phase word, as shown in the first four columns of Table 3.1. In the phase accumulators of some DDSs, however, the employment of biased binary-decimal codes is more convenient. The last four columns of Table 3.1 demonstrate that complementing of code words in one of such codes, namely, excess-3 binary-decimal (XS-3) code, also can be achieved by inverting all their bits. Although the arithmetic of biased binary-decimal codes is somewhat more complex than that of nonbiased binary-decimal codes, their self-complementary property justifies their use in DDSs. When a biased binary-decimal code is used, the role of the two MSBs of the phase accumulator output word remains the same.

In the single-quadrant converter, the phase values are converted into the corresponding values of sine using one of three techniques: (1) lookup table when sine values are stored in a read-only memory (ROM) or a programmable ROM (PROM) and phase values are used as addresses, (2) calculation of sine values from phase values using, for instance, Taylor power series expansion (in some cases, CORDIC algorithm can be used instead), and (3) a hybrid technique that combines the use

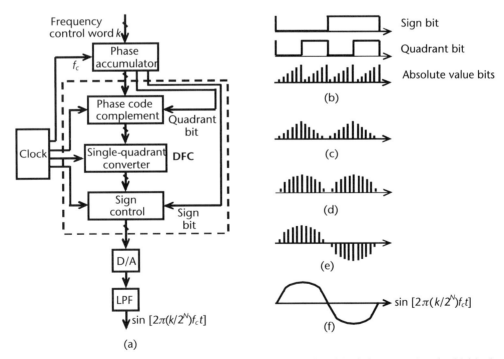

Figure 3.2 Block diagram of a DDS and timing diagrams of its blocks' output signals: (a) block diagram, (b) phase accumulator output, (c) phase code complement output, (d) single-quadrant converter output, (e) sign control output, and (f) LPF output.

of a lookup table and calculations. When speed is a primary requirement, the first technique is preferable. As shown in [25, 26], considering it a special case of a more general Boolean technique allows increasing the speed and reducing the size and power consumption of DDSs. This technique is explained below due to its importance for both DDSs and WFGs. Independently of the single-quadrant converter technique, its output sine values (see Figure 3.2(d)) are sent to the sign control directed by the sign bit. The digital values of the sine wave, shown in Figure 3.2(e), enter the D/A if this wave should also be formed in the analog domain. The discrete-time signal from the D/A after analog interpolating filtering by the LPF becomes the analog sine wave as depicted in Figure 3.2(f). Adding cosine wave synthesis is very easy because cosine is the same waveform shifted by a quarter-cycle.

As to the Boolean technique, it regards the single-quadrant converter a logical structure that can be optimized. To that end, the bits q_1, q_2, q_3, ... of a code word representing a sine value at the converter output are expressed as logical functions of the bits d_1, d_2, d_3, ... of an input phase word (here, the indexes of q_i and d_j start from the MSB). It is clear that a regular DDS lookup table has large redundancy because the values of phase and sine are highly correlated within a quadrant. The code words representing the phase and sine values often have the same length. Below, both lengths are identical and short ($M - 2 = 4$) to simplify the explanation.

To achieve maximum speed, the logical structure should have minimal depth, provided by the conjunctive or disjunctive normal form (CNF or DNF) of the logical

Table 3.1 Complementing in Binary and Excess-3 Binary-Decimal Systems

Binary Coding				Excess-3 Binary-Decimal (XS-3) Coding			
Binary	Decimal	Binary, Bits Inverted	Decimal	XS-3	Decimal	XS-3, Bits Inverted	Decimal
0000	0	1111	15	—	—	—	—
0001	1	1110	14	—	—	—	—
0010	2	1101	13	—	—	—	—
0011	3	1100	12	0011	0	1100	9
0100	4	1011	11	0100	1	1011	8
0101	5	1010	10	0101	2	1010	7
0110	6	1001	9	0110	3	1001	6
0111	7	1000	8	0111	4	1000	5
1000	8	0111	7	1000	5	0111	4
1001	9	0110	6	1001	6	0110	3
1010	10	0101	5	1010	7	0101	2
1011	11	0100	4	1011	8	0100	1
1100	12	0011	3	1100	9	0011	0
1101	13	0010	2	—	—	—	—
1110	14	0001	1	—	—	—	—
1111	15	0000	0	—	—	—	—

functions q_1, q_2, q_3, and q_4. of the independent variables d_1, d_2, d_3, and d_4. These forms also guarantee identical delays for all q_i. The minimum number of independent variables in these functions' equations corresponds to the minimum memory size and power consumption. The number of independent variables in each of the logical functions can be minimized using Karnaugh mapping when $M - 2$ is small or using computerized Quine-McCluskey algorithm when $M - 2 > 4$. If the phase values at the input of a single-quadrant phase-sine converter are binary coded, the minimal DNFs are

$$\begin{cases} q_1 = d_1 \vee d_2 d_3 \vee d_2 d_4, \\ q_2 = d_1 d_3 \vee d_1 d_4 \vee d_2 \neg d_3 \neg d_4 \vee \neg d_2 d_3, \\ q_3 = d_1 d_2 \vee d_1 d_3 d_4 \vee d_1 \neg d_3 \neg d_4 \vee \neg d_1 \neg d_2 \neg d_3 d_4 \vee d_2 d_3 d_4 \vee d_2 \neg d_3 \neg d_4, \\ q_4 = d_1 d_2 d_4 \vee d_1 d_3 \neg d_4 \vee \neg d_1 \neg d_2 d_3 d_4 \vee d_2 d_3 \neg d_4 \vee \neg d_2 \neg d_3 \neg d_4. \end{cases} \quad (3.4)$$

Equations (3.4) reflect the minimal DNFs for each output bit of the converter, but only the joint minimization of the total number L of the independent variables in all unique conjunctive clauses of the equation system guarantees the minimum size of the converter memory. Indeed, only one logical circuit is needed to determine all identical conjunctive clauses in the system. Consequently, the joint minimization should reduce the number of unique conjunctive clauses at the expense of an increased number of identical ones.

An effective heuristic procedure of such minimization suggested in [25] is as follows. First, the system of $M - 2$ minimal DNFs such as (3.4) should be formed. Second, several conjunctive clauses C_j containing a large number of the same independent variables should be identified in $m \leq M - 2$ minimal DNFs q_j of this system, and each C_j should be replaced with the conjunctive clause

$$C = \bigwedge_{j=1}^{m} C_j \quad (3.5)$$

if the logical functions of all q_j remain the same and L is reduced after the replacement, that is,

$$\sum_{j=1}^{m} r_j > r \quad (3.6)$$

where r_j and r are the numbers of the independent variables in C_j and C, respectively. This replacement should be iterated while possible. Since the final logical structure depends on the initial system of minimal DNFs and the replacement sequence, this procedure should be performed several times, and the equation system with the minimum L should be selected. This is a system of jointly minimal DNFs or very close to it.

The joint minimization of (3.4) results in the replacement of $d_2 d_3$ in q_1 with $d_2 d_3 \neg d_4$ from q_4 and $d_1 d_3$ in q_2 with $d_1 d_3 \neg d_4$ from q_4. This allows rewriting (3.4) as

$$\begin{cases} q_1 = d_1 \vee \underline{d_2 d_3 \neg d_4} \vee d_2 d_4, \\ q_2 = \underline{\underline{d_1 d_3 \neg d_4}} \vee d_1 d_4 \vee \underline{\underline{d_2 \neg d_3 \neg d_4}} \vee \neg d_2 d_3, \\ q_3 = d_1 d_2 \vee d_1 d_3 d_4 \vee d_1 \neg d_3 \neg d_4 \vee \neg d_1 \neg d_2 \neg d_3 d_4 \vee d_2 d_3 d_4 \vee \underline{\underline{d_2 \neg d_3 \neg d_4}}, \\ q_4 = d_1 d_2 d_4 \vee \underline{d_1 d_3 \neg d_4} \vee \neg d_1 \neg d_2 d_3 d_4 \vee \underline{d_2 d_3 \neg d_4} \vee \neg d_2 \neg d_3 \neg d_4. \end{cases} \quad (3.7)$$

In (3.7), identical conjunctive clauses are underlined. System (3.7) represents jointly minimal DNFs where L is about 10% smaller than in (3.4). The number of gates in the converter corresponding to (3.7) is approximately three times smaller than in a nonoptimized single-quadrant lookup table.

If excess-3 binary-decimal code is used and $M - 2 = 4$, the system of jointly minimal DNFs is

$$\begin{cases} q_1 = d_1 \vee d_2 d_3, \\ q_2 = \underline{d_1 d_3} \vee d_1 d_4 \vee d_2 \neg d_3, \\ q_3 = \underline{d_1 d_3} \vee \underline{\underline{d_1 \neg d_3 \neg d_4}} \vee d_2 d_4, \\ q_4 = \underline{\underline{d_1 \neg d_3 \neg d_4}} \vee \neg d_2 d_4. \end{cases} \quad (3.8)$$

In (3.8), as in (3.7), identical conjunctive clauses are underlined. The number of gates in the converter corresponding to (3.8) is about five times smaller than in a nonoptimized single-quadrant lookup table.

A similar technique allows forming a system of jointly minimal CNFs. Thus, the joint minimization described above is independent of coding used in a DDS. It is equally applicable to minimal DNFs and CNFs. It can also be used when the number of phase word bits at a single-quadrant converter input differs from the number of bits representing the sine wave at its output, as shown in [25].

The logical structures obtained as a result of the minimization described above are based on AND and OR gates. In most cases, however, these gates have longer time delays and higher power consumption than NAND and NOR gates. The expressions below allow transitioning from jointly minimal systems of DNFs or CNFs to the jointly minimal systems based on NAND or NOR gates [26]:

$$\bigvee_i C_i = \neg \bigwedge_i \neg C_i \quad (3.9)$$

$$\bigwedge_i D_i = \neg \bigvee_i \neg D_i \quad (3.10)$$

where C_i is the ith conjunctive clause in a DNF and D_i is the ith disjunctive clause in a CNF. Identities (3.9) and (3.10) follow from De Morgan's laws with the allowance for the fact that double negation is assertion. The systems of equations below are deduced by applying (3.9) to jointly minimal DNFs (3.7) and (3.8), respectively:

3.2 Digital Tx Basics

$$\begin{cases} \neg q_1 = \neg d_1 \wedge \neg(d_2 d_3 \neg d_4) \wedge \neg(d_2 d_4), \\ \neg q_2 = \neg(d_1 d_3 \neg d_4) \wedge \neg(d_1 d_4) \wedge \neg(d_2 \neg d_3 \neg d_4) \wedge \neg(\neg d_2 d_3), \\ \neg q_3 = \neg(d_1 d_2) \wedge \neg(d_1 d_3 d_4) \wedge \neg(d_1 \neg d_3 \neg d_4) \wedge \neg(\neg d_1 \neg d_2 \neg d_3 d_4) \wedge \neg(d_2 d_3 d_4) \wedge \neg(d_2 \neg d_3 \neg d_4), \\ \neg q_4 = \neg(d_1 d_2 d_4) \wedge \neg(d_1 d_3 \neg d_4) \wedge \neg(\neg d_1 \neg d_2 d_3 d_4) \wedge \neg(d_2 d_3 \neg d_4) \wedge \neg(\neg d_2 \neg d_3 \neg d_4) \end{cases} \quad (3.11)$$

$$\begin{cases} \neg q_1 = \neg d_1 \wedge \neg(d_2 d_3), \\ \neg q_2 = \neg(d_1 d_3) \wedge \neg(d_1 d_4) \wedge \neg(d_2 \neg d_3), \\ \neg q_3 = \neg(d_1 d_3) \wedge \neg(d_1 \neg d_3 \neg d_4) \wedge \neg(d_2 d_4), \\ \neg q_4 = \neg(d_1 \neg d_3 \neg d_4) \wedge \neg(\neg d_2 d_4). \end{cases} \quad (3.12)$$

The logical structures corresponding to (3.11) and (3.12) are shown in Figure 3.3(a, b), respectively.

A comparison of (3.7) and (3.8) to (3.11) and (3.12) shows that the transformations according to (3.9) and (3.10) change neither the depth of the logical structures nor L. Simultaneously, the replacement of AND and OR gates by NAND and NOR gates reduces the converter delay and power consumption. The reduction factors depend on the DDS technology. In some cases, the delay can be reduced by a factor of 1.5, and the power consumption by a factor of 2. An additional advantage of the logical structures corresponding to (3.11) and (3.12) is that they require the same type of gates for both stages, unlike the structures corresponding to (3.7) and (3.8).

The DFC optimization described above is focused on increasing its speed and minimizing the required memory (i.e., minimizing DFC size, power consumption, weight, and cost). These DFC parameters, being important for DDSs, are even more essential for WFGs. Therefore, it is reasonable to summarize the optimization procedure in more general terms, taking into account both DDSs and WFGs. It consists of six steps.

The unique repeating parts of the DFC output signal that fully characterize it should be identified first. In phase-sine converters, the sine values are unique only within a quarter-cycle, but they allow generating the entire output signal. In WFGs, the extent of the DFC simplification depends on a specific weight function, but the duration of a weight function's unique part cannot exceed its half-length because all weight functions are symmetric about their midpoints. At the second step, the weight function values should be calculated for all predetermined time instants within the weight function unique parts. At the third step, the bits of code words representing the weight function values should be expressed as CNFs or DNFs of the logical functions of the bits representing the corresponding time instants. At the fourth step, these CNFs or DNFs should be minimized independently of each other, using Karnaugh mapping or computerized Quine-McCluskey algorithm. At the fifth step, the obtained minimal CNFs or DNFs should undergo the joint minimization procedure described above. Finally, at the sixth step, (3.9) or (3.10) should

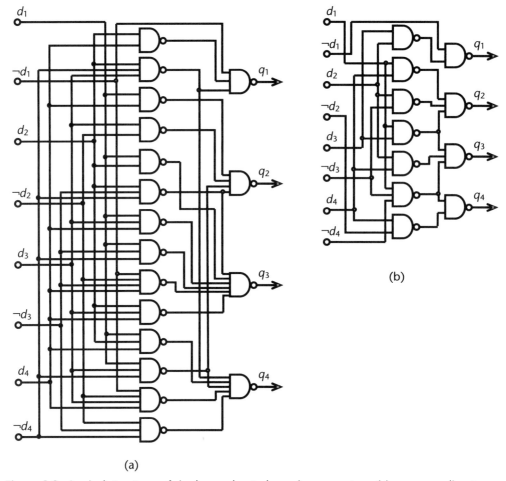

Figure 3.3 Logical structures of single-quadrant phase-sine converters: (a) corresponding to (3.11) and (b) corresponding to (3.12).

be used for transforming the jointly minimal CNFs or DNFs, to replace AND and OR gates with NAND or NOR gates.

3.3 D&R in a Digital Tx

3.3.1 Digitization of TDP Input Signals

As mentioned in Section 3.2.2, the digitization, discussed below and reflected by the block diagram in Figure 3.4, is not combined with source encoding. It uses uniform sampling and quantization, that is, samples are formed at a constant rate, and quantization steps are identical for all signal levels. This approach has minimum sensitivity to the input signal statistics and imposes minimum restrictions on the subsequent DSP. Yet it requires higher speed and resolution of quantization and processing than the digitization combined with the initial part of source encoding.

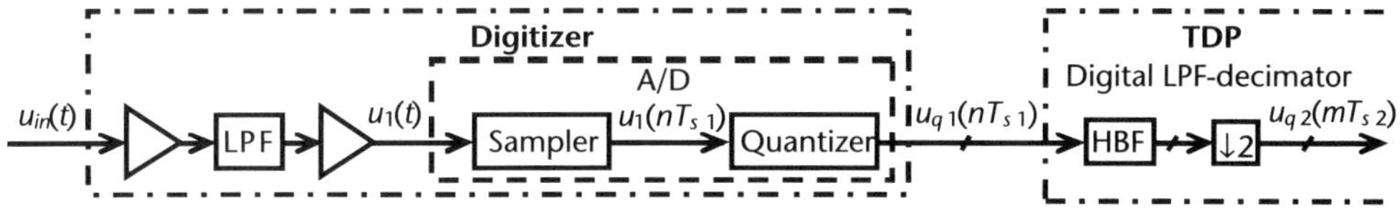

3.3 D&R in a Digital Tx

Figure 3.4 Digitization of baseband signals.

The block diagram in Figure 3.4 shows that digitization comprises analog, mixed-signal, and digital operations.

Uniform sampling based on the classical sampling theorem is applicable only to bandlimited signals. Therefore, it requires antialiasing filtering prior to or simultaneously with sample generation. Since the Tx input signals are usually baseband, this filtering is performed by an LPF whose passband should ideally be equal to the one-sided bandwidth B of the desired signal $u(t)$. In practice, however, it can be wider than B, especially if input signals have different bandwidths. If the input signal $u_{in}(t)$ bandwidth is wider than B, its spectrum $S_{in}(f)$ may contain the spectra of interfering signals (ISs) besides the spectrum $S(f)$ of $u(t)$. The LPF usually has input and output buffer amplifiers (BAs). Samples are generated by the sampler at the rate f_{s1} with period $T_{s1} = 1/f_{s1}$. Currently, THAs are usually employed as samplers, and they are placed together with quantizers in the same A/D packages.

Although quantized signals are already represented digitally at the TDP input, they undergo several additional digitization-related operations that typically include downsampling with digital decimating filtering because f_{s1} is usually selected relatively high to lower the requirements for the antialiasing filter's transition band and accommodate its excessive bandwidth. The downsampling increases the subsequent DSP efficiency. As noted in Appendix B, the use of digital LPFs with finite impulse responses (FIRs) as decimating filters simplifies the design of digital radios due to the ease of combining the steps of downsampling or upsampling and achieving perfectly linear PFRs, as well as the absence of round-off error accumulation inherent in infinite impulse response (IIR) filters. Among FIR filters, half-band filters (HBFs) (see Section B.4) or their cascade structures minimize the computational intensity of downsampling if the decimation factor is a power of 2 [27–35]. If needed, the TDP input stages can also correct some distortions caused by the preceding analog and mixed-signal circuits.

The spectral diagrams in Figure 3.5 illustrate the digitization. Here and below, like in Figures 1.16 and 1.21 through 1.23, the triangular shape of the spectra was selected for its capability to indicate possible spectral inversions. The AFRs of the analog antialiasing LPF $|H_{a.f}(f)|$ and digital decimating LPF $|H_{d.f}(f)|$ are also shown in this figure. In this and subsequent chapters, in contrast with Appendix B, the AFRs of both decimating and interpolating filters have the same subscripts d, meaning digital. In Figure 3.5(a), the input signal $u_{in}(t)$ with the spectrum $S_{in}(f)$ contains two ISs $u_{i1}(t)$ and $u_{i2}(t)$ with the spectra $S_{i1}(f)$ and $S_{i2}(f)$, respectively, in addition to $u(t)$ with the spectrum $S(f)$. Sampling causes proliferation of the spectrum $S_1(f)$ of the sampler input signal $u_1(t)$. Therefore, the spectrum $S_{d1}(f)$ of the discrete-time signal $u_1(nT_{s1})$ at the sampler output is

$$S_{d1}(f) = \frac{1}{T_{s1}} \sum_{k=-\infty}^{\infty} S_1(f - kf_{s1}) \qquad (3.13)$$

Thus, $|S_{d1}(f)|$ is a periodic function of frequency with the period f_{s1} comprising the spectral replicas $|S_1(f)|$ centered at kf_{s1} where k is any integer, and the replicas of $|S(f)|$ occupy intervals:

$$\left[kf_{s1} - B, \; kf_{s1} + B\right] \qquad (3.14)$$

If $u_1(nT_{s1})$ is uniformly quantized with high accuracy, the spectrum $S_{q1}(f)$ of the quantized (i.e., digital) signal $u_{q1}(nT_{s1})$ is virtually identical to $S_{d1}(f)$. Therefore, only $S_{q1}(f)$ is presented in Figure 3.5(b). In Figure 3.5(a), $|H_{a.f}(f)|$ shows that the antialiasing LPF has to suppress the $u_{in}(t)$ spectral components within intervals (3.14) where the $S(f)$ replicas appear after sampling, while its spectral components within the gaps between these intervals do not have to be suppressed by this LPF because they can be rejected later by digital filtering in the TDP. Therefore, these gaps are often called "don't care" bands. Still, some weakening of ISs within don't care bands by antialiasing filtering may lower the required resolution of the quantizer and subsequent DSP. Traditional analog filtering cannot utilize the existence of don't care bands, but these bands allow increasing the efficiency of antialiasing and interpolating filtering based on the direct and hybrid interpretations of the sampling theorem (see Chapters 5 and 6). Figure 3.5(a, b) shows that the antialiasing filter rejects $u_{i2}(t)$ but only slightly weakens $u_{i1}(t)$ because $S_{i1}(f)$ is located within its

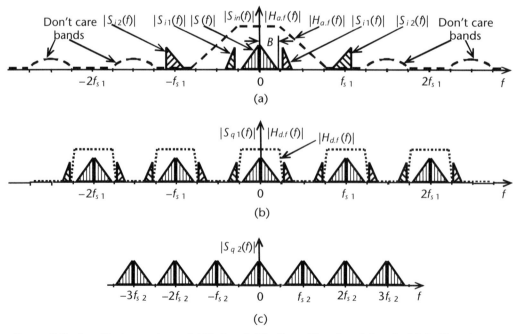

Figure 3.5 Amplitude spectra and AFRs for digitization of baseband signals: (a) $S_{in}(f)$ and $|H_{a.f}(f)|$ (dashed line), (b) $S_{q1}(f)$ and $|H_{d.f}(f)|$ (dotted line), and (c) $|S_{q2}(f)|$.

transition band. Ultimately, $u_{i1}(t)$ is rejected by the digital decimating LPF during downsampling that halves f_{s1}, reducing the required speed of the subsequent DSP. The amplitude spectrum $|S_{q2}(f)|$ of $u_{q2}(mT_{s2})$ is shown in Figure 3.5(c).

3.3.2 Reconstruction of TDP Output Signals

Both baseband and bandpass reconstructions of analog bandpass signals are considered below. The block diagram in Figure 3.6 reflects baseband reconstruction that comprises converting the digital baseband complex-valued equivalent $Z_{q1}(nT_{s1})$ into the analog domain with obtaining the analog baseband complex-valued equivalent $Z(t)$, and forming the analog bandpass real-valued signal u_{out} from $Z(t)$. Prior to entering the D/As, $Z_{q1}(nT_{s1})$, represented by $I_{q1}(nT_{s1})$ and $Q_{q1}(nT_{s1})$, usually undergoes upsampling with digital interpolating filtering (see Appendix B). This upsampling is needed because most of the TDP signal processing is performed at the minimum possible sampling rate to efficiently utilize the digital hardware, but wide transition bands of the analog interpolating LPFs require increasing this rate at the D/As' inputs.

Since an even AFR and a linear PFR of the digital interpolating filter in the DPR are desirable, this complex-valued filter is reduced to two identical real-valued LPFs because its coefficients become real-valued. The LPFs are usually HBFs or cascade structures of HBFs (see Section B.4) if the upsampling factor is a power of 2, as shown in Figure 3.6. When the digital interpolating filter predistorts signals to compensate their linear distortion in the Tx mixed-signal and/or analog circuits, its coefficients are often complex-valued, and the interpolating filter consists of four real-valued LPFs. Due to the upsampling, the sampling rate f_{s2} of $I_{q2}(mT_{s2})$ and $Q_{q2}(mT_{s2})$ is higher than the rate f_{s1} of $I_{q1}(nT_{s1})$ and $Q_{q1}(nT_{s1})$.

The transitions between adjacent analog samples at the outputs of the D/As contain glitches caused by switching time disparities among the D/A bits and between on and off switching. The pulse shapers (PSs), controlled by the gating pulse generator (GPG), select the undistorted segments of the D/As' output samples as illustrated by the timing diagrams in Figure 3.7. Here Δt_s is the gating pulse length, whereas Δt_d is the time delay of gating pulses relative to the fronts of the D/A output samples. The time delay must be equal to or longer than the length of the sample's distorted portion. The selected parts of the samples are amplified by the BAs and interpolated by the analog LPFs. This interpolation transforms discrete-time signals $I(mT_{s2})$ and

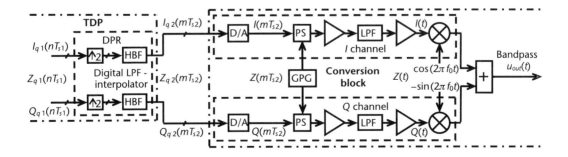

Figure 3.6 Baseband reconstruction of bandpass signals.

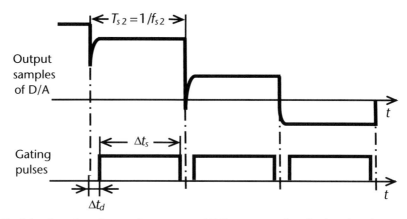

Figure 3.7 Selection of undistorted segments of D/A output pulses for baseband reconstruction.

$Q(mT_{s2})$ into analog signals $I(t)$ and $Q(t)$ that are the I and Q components of the analog baseband complex-valued equivalent $Z(t)$, which is then converted into the output bandpass real-valued signal $u_{out}(t)$ as shown in Figure 3.6.

Figure 3.8 shows the signal spectrum transformations during this reconstruction and the required AFRs of the interpolating filters. The amplitude spectrum $|S_{q1}(f)|$ of $Z_{q1}(nT_{s1})$ and AFR $|H_{d,f}(f)|$ of the digital interpolating LPF in the DPR are depicted in Figure 3.8(a). Spectrum $S_{q1}(f)$ comprises the replicas of the spectrum $S_Z(f)$ of $Z(t)$ centered at kf_{s1} where k is any integer.

Figure 3.8(b) shows the amplitude spectrum $|S_{q2}(f)|$ of $Z_{q2}(mT_{s2})$, which is virtually identical to the amplitude spectrum $|S_{d2}(f)|$ of the discrete-time complex-valued signal $Z(mT_{s2})$ when the D/As are accurate. The required AFR $|H_{a,f}(f)|$ of the analog interpolating filter is also displayed in Figure 3.8(b). The upsampling reflected by the spectral diagrams in Figure 3.8(a, b) doubles the sampling rate. The analog interpolating filter reconstructs analog $Z(t)$ by rejecting all the replicas of $S_Z(f)$ in $S_{d2}(f)$ except the baseband one. As in the case of the antialiasing filter with the AFR in Figure 3.5(a), the don't care bands of the analog interpolating filter with the AFR depicted in Figure 3.8(b) are not utilized by traditional filtering techniques, but they allow increasing the efficiency of interpolating filters based on the direct and hybrid interpretations of the sampling theorem. Figure 3.8(c, d) displays the amplitude spectra $|S_Z(f)|$ of $Z(t)$ and $|S_{out}(f)|$ of $u_{out}(t)$, respectively. Note that $B = 2B_z$.

The block diagram in Figure 3.9 reflects bandpass reconstruction that comprises forming a digital bandpass real-valued signal $u_q(lT_{s3})$ from its digital baseband complex-valued equivalent $Z_{q1}(nT_{s1})$, converting $u_q(lT_{s3})$ into the analog domain, and translating the obtained $u_{out}(t)$ to the Tx RF (if needed). At the DPR input, $Z_{q1}(nT_{s1})$, represented by $I_{q1}(nT_{s1})$ and $Q_{q1}(nT_{s1})$, has the lowest sampling rate f_{s1}. The upsampling with digital interpolating filtering in the DPR (see Appendix B) increases the sampling rate not only to make it adequate to the transition bands of the analog interpolating filter, but also to create room for digital upconversion required for the bandpass reconstruction. This upsampling is usually performed in several stages.

3.3 D&R in a Digital Tx

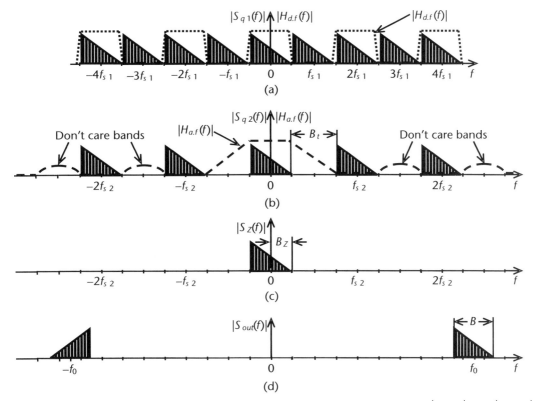

Figure 3.8 Amplitude spectra and AFRs for baseband reconstruction: (a) $|S_{q1}(f)|$ and $|H_{d.f}(f)|$ (dotted line), (b) $|S_{q2}(f)|$ and $|H_{a.f}(f)|$ (dashed line), (c) $|S_Z(f)|$, and (d) $|S_{out}(f)|$.

In Figure 3.9, it is executed in two stages. Each stage doubles the sampling rate and contains two identical HBFs. The first stage transforms $Z_{q1}(nT_{s1})$ into $Z_{q2}(mT_{s2})$, and the second stage transforms $Z_{q2}(mT_{s2})$ into $Z_{q3}(lT_{s3})$. The digital baseband complex-valued equivalent $Z_{q3}(lT_{s3})$ is converted into the digital bandpass real-valued signal $u_q(lT_{s3})$ sent to the D/A. The conversion of $Z_{q3}(lT_{s3})$ into $u_q(lT_{s3})$ is performed as shown in Figure 1.22 (see Section 1.4.2). Besides the upsampling and conversion of $Z_{q3}(lT_{s3})$ into $u_q(lT_{s3})$, the DPR may also predistort signals to compensate their

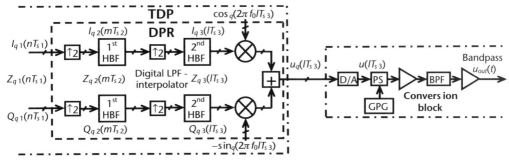

Figure 3.9 Bandpass reconstruction of bandpass signals.

distortion in the subsequent mixed-signal and/or analog circuits. It is usually better to predistort them at the stage with the lowest sampling rate.

As shown in Figure 3.10, the PS, controlled by the GPG, selects only a short segment of every D/A output sample, thus losing most of its energy, because Δt_s should meet the condition:

$$\Delta t_s \leq 0.5 T_0 = \frac{0.5}{f_0} \qquad (3.15)$$

to increase the signal energy within the analog interpolating BPF passband. In (3.15), f_0 is the center frequency of $u_{out}(t)$, and $T_0 = 1/f_0$ (compare the gating pulses' lengths in Figures 3.7 and 3.10). Note that a rectangular shape of gating pulses is not necessarily optimal, and more efficient methods of increasing the signal energy within the BPF passband are described in Chapter 6. The selected segments of the D/A output samples are amplified by the BA and interpolated by the analog BPF, which transforms the discrete-time $u(lT_{s3})$ into analog $u_{out}(t)$.

Spectral diagrams in Figure 3.11 illustrate bandpass reconstruction. The amplitude spectrum $|S_{q1}(f)|$ of $Z_{q1}(nT_{s1})$ and AFR $|H_{d.f1}(f)|$ of the first-stage HBF are depicted in Figure 3.11(a), whereas the amplitude spectrum $|S_{q2}(f)|$ of $Z_{q2}(mT_{s2})$ and AFR $|H_{d.f2}(f)|$ of the second-stage HBF are displayed in Figure 3.11(b). Figure 3.11(c) presents the amplitude spectrum $|S_{q3}(f)|$ of $Z_{q3}(lT_{s3})$ obtained after upsampling. In Figure 3.11(d), the amplitude spectrum $|S_{q3BP}(f)|$ of $u_q(lT_{s3})$ also represents the amplitude spectrum $|S_{d3BP}(f)|$ of the discrete-time bandpass real-valued signal $u(lT_{s3})$ because of the presumably accurate D/A conversion.

The AFR $|H_{a.f}(f)|$ of the analog interpolating BPF is shown in Figure 3.11(d) as well. The prior remarks on the don't care bands utilization in antialiasing and interpolating filters are also applicable to $|H_{a.f}(f)|$ in this figure. The amplitude spectrum $|S_{out}(f)|$ of $u_{out}(t)$ selected by the BPF is depicted in Figure 3.11(e). Comparison of the spectral diagrams in Figure 3.11(a, e) shows that $S_{out}(f)$ is inverted relative to $S_{q1}(f)$. Such an inversion can be needed to match, for instance, the output of the Tx modulator and the input of the Rx demodulator, or it can be made to simplify the signal transformations in a Tx. In the latter case, the inversion can be

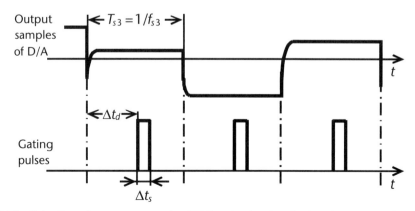

Figure 3.10 Selection of proper segments of D/A output pulses for bandpass reconstruction.

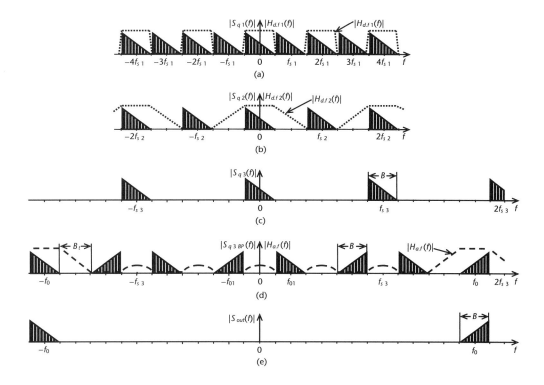

Figure 3.11 Amplitude spectra and AFRs for bandpass reconstruction: (a) $|S_{q1}(f)|$ and $|H_{d.f1}(f)|$ (dotted line), (b) $|S_{q2}(f)|$ and $|H_{d.f2}(f)|$ (dotted line), (c) $|S_{q3}(f)|$, (d) $|S_{q3BP}(f)|$ and $|H_{a.f}(f)|$ (dashed line), and (e) $|S_{out}(f)|$.

easily corrected by changing the sign of the Q component of the signal's baseband complex-valued equivalent. The signal spectra in Figure 3.8(a, b) and in Figure 3.11(a, b) are identical, but the subsequent spectra in Figures 3.8 and 3.11 diverge, demonstrating the difference between the baseband and bandpass reconstructions and possible inversions of the signal spectrum.

Bandpass reconstruction imposes certain restrictions on the relations among f_0, f_s, B, and transition band B_t. First, let us show why the relation

$$f_s = \frac{f_0}{|k \pm 0.25|} \qquad (3.16)$$

between f_0 and f_s is considered optimal when k is an integer. The spectral diagram in Figure 3.11(d) where f_0 and f_s meet (3.16) shows that (3.16) provides the equal distances between all the neighboring replicas of $|S_{out}(f)|$ in $|S_{q3BP}(f)|$. Such distances maximize the permissible transition bands of analog interpolating BPFs for given f_s and B, assuming that these bands are identical and equal to B_t. In practice, they can differ, but usually insignificantly. The increased B_t simplifies the realization and reduces the cost of the BPFs. The equidistant positions of the spectral replicas also minimize the number and power of even-order intermodulation products (IMPs) within $S_{out}(f)$. Specifically, they prevent the appearance of second-order IMPs within $|S_{out}(f)|$ when $f_s/B \geq 6$. In addition, they simplify the conversion of

$Z_{q3}(lT_{s3})$ into $u_q(lT_{s3})$. Indeed, if (3.16) is true, the lowest-frequency replicas in the spectra of digital $\cos_q(2\pi f_0 lT_s)$ and $\sin_q(2\pi f_0 lT_s)$ are located at

$$f_{01} = 0.25 f_s \qquad (3.17)$$

In this case, the cosine and sine values are

$$\cos_q\left(2\pi f_0 lT_s\right) = \cos_q(0.5\pi l) \text{ and } \sin_q\left(2\pi f_0 lT_s\right) = \sin_q(0.5\pi l) \qquad (3.18)$$

where l is an integer. Consequently, they can be equal only to +1, 0, and −1 as shown in Figure 3.12.

For the DPR depicted in Figure 3.9, this means that multiplying $I_{q3}(lT_{s3})$ by $\cos_q(2\pi f_0 lT_{s3})$ is reduced to zeroing the odd samples of $I_{q3}(lT_{s3})$ and alternating the signs of its even samples. Likewise, multiplying $Q_{q3}(lT_{s3})$ by $-\sin_q(2\pi f_0 lT_{s3})$ is reduced to zeroing the even samples of $Q_{q3}(lT_{s3})$ and alternating the signs of its odd samples. The samples that will be zeroed should not be calculated. Thus, (3.16) simplifies the conversion of digital baseband complex-valued equivalents into the corresponding digital bandpass real-valued signals. Due to these advantages, f_s that satisfies (3.16) is called optimal, although it has a drawback: it maximizes the power of odd-order IMPs within the signal spectrum. Note that, according to (3.16), the same f_s is optimal for all signal center frequencies f_0 that are equal to

$$0.25 f_s, 0.75 f_s, 1.25 f_s, 1.75 f_s, 2.25 f_s, 2.75 f_s, 3.25 f_s, \ldots \qquad (3.19)$$

that is, f_s is optimal for all f_0 located in the middles of Nyquist zones. These zones on the positive frequencies are located between $0.5(m-1)f_s$ and $0.5m f_s$ where m

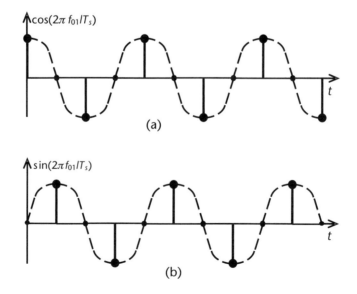

Figure 3.12 Samples of cosine and sine signals for $f_{01} = 0.25 f_s$: (a) $\cos(2\pi f_{01} lT_s)$ and (b) $\sin(2\pi f_{01} lT_s)$.

is any positive integer. Thus, the first Nyquist zone is from dc to $0.5f_s$, the second Nyquist zone is from $0.5f_s$ to f_s, and so on (see Figure 3.13). In Figure 3.11(d), each spectral replica in $S_{q3BP}(f)$ and in $S_{d3BP}(f)$ is located within a separate Nyquist zone and B_t is selected maximally wide to simplify the BPF realization. In this case, $B + B_t = 0.5f_s$, and any change of f_0 within a Nyquist zone makes the bandpass reconstruction impossible. Changing f_0 within a certain frequency interval (and deviating from (3.16)) without losing the reconstruction capability is possible only if

$$B + B_t < 0.5f_s \tag{3.20}$$

Let us determine the boundaries of this interval, assuming initially that both BPF transition bands are equal to B_t. In Figure 3.13(a), the desired spectral replica of an analog bandpass signal in the spectrum $S_{dBP}(f)$ of a discrete-time signal is centered about f_0. It occupies the leftmost position within the second Nyquist zone that allows the analog interpolating BPF with the AFR $|H_{a.f}(f)|$ (dashed line) to select it and reject all other replicas. In this case, $f_0 = 0.5(f_s + B + B_t)$. Therefore, the leftmost f_0 within the mth Nyquist zone is

$$f_0 = 0.5\left[(m-1)f_s + B + B_t\right] \tag{3.21}$$

In Figure 3.13(b), the replica occupies the rightmost position within this Nyquist zone, which allows the BPF to perform its functions. In that case, $f_0 = f_s - 0.5(B + B_t)$, and the rightmost f_0 within the mth Nyquist zone is

$$f_0 = 0.5\left[mf_s - (B + B_t)\right] \tag{3.22}$$

Consequently, the analog interpolating BPF can perform its functions if and only if

$$0.5(B + B_t) \le f_0 \bmod(0.5f_s) \le 0.5\left[f_s - (B + B_t)\right] \tag{3.23}$$

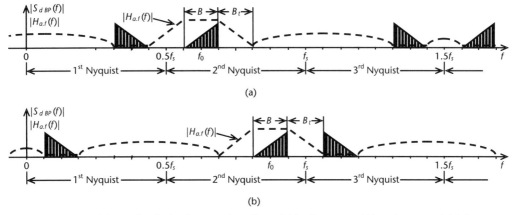

Figure 3.13 Positions of a desired spectral replica within the second Nyquist zone: (a) leftmost position and (b) rightmost position.

where $f_0 \mod(0.5f_s)$ is a remainder of the division of f_0 by $0.5f_s$. From (3.23) and Figure 3.13, the interpolating BPF with lower and upper transition bands $B_{t.l}$ and $B_{t.u}$, respectively, can perform its functions if and only if

$$0.5(B + B_{t.l}) \leq f_0 \mod(0.5f_s) \leq 0.5[f_s - (B + B_{t.u})] \quad (3.24)$$

Inequalities (3.23) and (3.24) determine the boundaries $f_{\min}$ and $f_{\max}$ of digital frequency tuning. In most cases, $|B_{t.l} - B_{t.u}| \ll 0.5(B_{t.l} + B_{t.u})$. Therefore, (3.23) is sufficiently accurate for theoretical analysis, whereas (3.24), being filter-specific, is more appropriate for a particular design. As follows from (3.23), f_0 can be tuned within a wide frequency range in each Nyquist zone if

$$B + B_t \ll f_s \quad (3.25)$$

Thus, a high f_s allows a large tuning range, but it raises the required speed of the DPR and D/A. Selecting different f_s for different parts of this range reduces the maximum f_s. For a given maximum f_s, the number of different f_s needed for tuning depends not only on the range width $f_{\max} - f_{\min}$ but also on its position on the frequency axis.

3.3.3 Comparison of Reconstruction Techniques and Conversion Block Architectures

Since both baseband and bandpass reconstruction techniques are used for the same purpose, it is important to compare them. Spectral diagrams in Figures 3.8 and 3.11 show that wider bandwidths of the D/As' input signals and transition bands B_t of the subsequent analog interpolating filters require higher speed of the D/As. This speed does not depend on the input signals' center frequencies (but the dynamic range does). Since the bandwidth B of a bandpass real-valued signal is two times wider than the bandwidth B_Z of its baseband complex-valued equivalent, the speed of a D/A used for bandpass reconstruction should be two times higher than that of each of two D/As needed for baseband reconstruction when the analog interpolating filters' B_t are the same (compare the spectral diagrams in Figure 3.8(b, c) to those in Figure 3.11(c, d)).

The timing diagrams in Figures 3.7 and 3.10 show that conventional PSs utilize the energy of D/As' output samples more efficiently for baseband reconstruction than for bandpass one. Moreover, the utilization efficiency of the bandpass PSs declines as the signal center frequency f_0 increases. IC implementation of analog interpolating LPFs is simpler than that of analog interpolating BPFs, and the LPFs are easier to make adjustable. The most effective BPFs, for example, crystal, electromechanical, ceramic, surface acoustic wave (SAW), and bulk acoustic wave (BAW) filters, are nonadjustable and incompatible with the IC technology. These facts reflect the advantages of the baseband reconstruction over the bandpass one.

The spectral diagrams in Figures 3.8 and 3.11 as well as the block diagrams in Figures 3.6 and 3.9 also demonstrate the advantages of bandpass reconstruction over baseband one. As shown in Figure 3.8(c), the spectrum $S_Z(f)$ of the reconstructed analog complex-valued equivalent $Z(t)$ has baseband location. At this location dc offset, flicker noise, and the largest number and power of even-order IMPs appear within the spectrum $S_{out}(f)$ of $u_{out}(t)$. In the case of bandpass reconstruction (see Figure 3.11(d, e)), the spectra of the discrete-time and analog signals do not contain baseband components, and, therefore, $S_{out}(f)$ is not influenced by dc offset and flicker noise. For the same reason, the number of even-order IMPs within $S_{out}(f)$ caused by the same nonlinearity can be minimized by selecting an optimal f_s. The block diagrams in Figures 3.6 and 3.9 show that baseband reconstruction requires separate I and Q channels in the discrete-time and analog domains, whereas bandpass reconstruction requires a single channel in these domains. Amplitude and phase imbalances between the I and Q channels (commonly referred to as IQ imbalance) are a problem only in the discrete-time and analog circuits. In the digital domain, they can be easily made negligible. Therefore, just baseband reconstruction suffers from this problem.

The IQ imbalance, dc offset, flicker noise, and IMPs limit the Tx dynamic range and accuracy of modulation. In particular, dc offset produces or changes the carrier in a transmitted signal. For these reasons, bandpass reconstruction is preferable when high modulation accuracy is required. An emergence of a carrier caused by dc offset is undesirable in military applications. The level and spectral distribution of flicker noise depends on the employed semiconductor technology. In principle, the IMPs can be reduced by predistortion in the digital domain, while the dc offset and IQ imbalance allow adaptive digital precompensation. However, these measures increase the complexity and cost of Txs.

As to the described above advantages of baseband reconstruction over bandpass one, the required speed of D/As is halved at the expense of doubling their number, and the other advantages are not fundamental but provisional. They are caused partly by technological limitations but mostly by incomplete understanding of technical opportunities provided by the sampling theory. Indeed, the root causes of the drawbacks (i.e., inefficient utilization of the energy of D/A output samples, inflexibility of traditional analog BPFs, and their incompatibility with the IC technology) are to a certain extent or completely overcome in reconstruction techniques based on the hybrid or direct interpretations of the sampling theorem, as shown in Chapters 5 and 6. For these reasons, baseband reconstruction, currently used in many inexpensive Txs, will probably be replaced by bandpass one even there.

Although conversion blocks are not always needed in digital Txs, in the Txs similar to that shown in Figure 3.1, they complete reconstruction and carry out other operations needed prior to the PAs. The latter operations may include additional analog filtering, amplification, and translation of the analog bandpass signals, reconstructed at an IF, to the RF. Therefore, possible conversion block architectures depend on, but are not completely determined by, the Tx reconstruction technique. The direct upconversion (Figure 3.14(a)), offset upconversion with baseband reconstruction (Figure 3.14(b)), offset upconversion with bandpass reconstruction (Figure 3.14(c)), and direct RF reconstruction (Figure 3.14(d)) architectures reflect four different approaches to the block design. The low-IF architecture is not discussed here because it does not provide notable advantages for digital Txs. The block diagrams in Figure 3.14 are simplified. For instance, the Tx level control circuits are represented only by a variable-gain amplifier (VGA) in the RF section. In practice, this control is usually more complex.

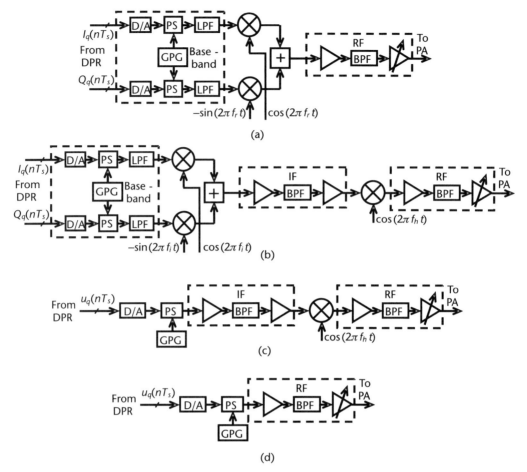

Figure 3.14 Conversion block architectures: (a) direct upconversion, (b) offset upconversion with baseband reconstruction, (c) offset upconversion with bandpass reconstruction, and (d) direct RF reconstruction.

The direct upconversion architecture in Figure 3.14(a), known for analog Txs since the 1920s, became technically sound and economically beneficial only in the 1980s. Its main advantages are better compatibility with the IC technology allowing on-chip implementation of Tx drives, higher flexibility, and lower cost compared to other architectures. However, it not only inherits all the baseband reconstruction drawbacks (i.e., IQ imbalance, dc offset, flicker noise, and the largest number of even-order IMPs), but also worsens many of them. For example, minimizing the IQ imbalance within the Tx frequency range is more difficult than at a single frequency. The imbalance and feedthrough cause an LO spur and an inverted signal image that falls within the transmitted signal spectrum. An additional problem is that the Tx VCO is susceptible to pulling from the PA output. All attempts to compensate or reduce these phenomena complicate the Txs and make them more expensive. The direct upconversion architecture also increases the Tx power consumption because gain is provided mostly by the RF amplifiers that are more power-hungry than IF amplifiers. A high gain of the RF stages increases noise emission as well.

The offset upconversion architecture with baseband reconstruction in Figure 3.14(b) radically reduces the feedthrough and the possibility of the VCO pulling because the Tx output frequencies differ from the Tx VCO frequency. It provides better reconstruction filtering than the previous architecture due to the use of effective BPFs (crystal, electromechanical, ceramic, SAW, BAW, etc.). However, these improvements are achieved at the expense of lower flexibility and reduced scale of integration because the effective IF BPFs cannot change their parameters and are incompatible with the IC technology. This architecture slightly improves modulation accuracy compared to that of the previous architecture because bandpass signals are formed at a constant IF. It can somewhat reduce the Tx power consumption compared to the direct upconversion architecture due to performing a significant part of amplification at the IF, instead of the RF. However, this effect is weakened by the increased Tx parts count, caused by the lower compatibility with the IC technology.

The architecture in Figure 3.14(c) retains all the advantages provided by the offset upconversion. In addition, it reduces the number and power of even-order IMPs within the signal spectrum and excludes the dc offset, flicker noise, and *IQ* imbalance due to the bandpass reconstruction. As a result, it increases the modulation accuracy and dynamic range compared to the previous architectures. It allows selection of an optimal f_s that, in addition to minimizing the number of even-order IMPs within the signal spectrum and the requirements for an analog interpolating BPF, simplifies the digital interpolating filtering and forming of digital bandpass signals in the DPR. Therefore, various versions of this architecture are used in high-quality digital Txs. Among them, the version with two frequency translations is often employed. In that version, the first IF is usually selected low enough to simplify the reconstruction and related analog interpolating filtering, whereas the second IF is chosen high enough for better image rejection and smaller number of spurious responses. Still, this architecture suffers from the inflexibility of analog interpolating BPFs, their incompatibility with the IC technology, and inefficient utilization of the D/A output samples' energy when conventional techniques are used. These problems can be solved based on the direct or hybrid interpretations of the sampling theorem. Therefore, it can be expected that the application area of this architecture will become wider in the future.

In the direct RF reconstruction architecture in Figure 3.14(d), one of the spectral replicas constituting the discrete-time bandpass signal located at the Tx RF is selected by the analog interpolating BPF after the signal D/A conversion. This can be illustrated by the spectral diagram in Figure 3.11(d), assuming that f_0 there corresponds to the Tx RF. The obvious advantage of this architecture is that almost all signal processing is executed in the digital domain, providing the highest accuracy and flexibility. However, it significantly raises the requirements for the DPR, D/A, PS, and analog interpolating BPF compared to the architecture in Figure 3.14(c). The requirements for the DPR and D/A are increased due to the necessity to perform the Tx frequency tuning in the digital domain and a wider (sometimes significantly) transition band B_t of the tunable analog interpolating RF BPF. In addition, a high f_0 of the spectral replica selected by the analog BPF reduces the energy of each signal sample. The design of a sufficiently small and technologically sound tunable analog BPF at the Tx RF with adequate suppression in the stopbands and reasonably narrow B_t is difficult. When the middle of the Tx frequency range is substantially larger than the range width, that is,

$$\frac{0.5(f_{\min} + f_{\max})}{f_{\max} - f_{\min}} \gg 1, \quad (3.26)$$

the analog interpolating BPF can be designed with a constant bandwidth that is slightly wider than the Tx frequency range. This simplifies the BPF design and the Tx frequency tuning, but may significantly increase f_s, complicating the DPR and D/A. In general, the major trends of the technological progress favor the direct RF architecture, and the direct and hybrid interpretations of the sampling theorem also allow solving many of its problems. Despite all these factors, the architecture in Figure 3.14(d) will not be able to compete with that in Figure 3.14(c) in some cases, including Txs of the highest frequency bands, and Txs operating in multiple frequency bands.

3.4 Power Utilization Improvement in Txs

3.4.1 Power Utilization in Txs with Energy-Efficient Modulation

It was explained in Section 2.3.3 that (a) the highest energy efficiency of signals requires maximum utilization of Tx power and minimum energy per bit E_b for a given bit error rate P_b, (b) modulation allowing the Tx to operate in saturation mode maximizes the Tx power utilization, and (c) many energy-efficient signals cannot operate in this mode. As shown below, the ways to maximize Tx power utilization for most energy-efficient signals can be reduced to minimizing the signals' crest factors without regeneration of their spectrum sidelobes. An energy-efficient modulation named alternating quadratures DBPSK (AQ-DBPSK) is analyzed in the next section. The Tx power utilization improvement for bandwidth-efficient modulation techniques is discussed in Section 3.4.3.

Digital data processing in a TDP, illustrated by the block diagram in Figure 3.15, is used to analyze the Tx power utilization by energy-efficient signals. Here, the bit stream from the TDP source encoder is distributed between the I and Q channels. In each channel, the partial bit stream undergoes channel encoding, modulation, and DS spreading. For simplicity, channel encoding is not considered, only binary and quaternary modulation and spreading techniques are analyzed, and signals in the I and Q channels are modulated independently.

The bit rates at the outputs of channel encoders are higher than those at their inputs due to the code redundancy. As mentioned in Section 2.3.3, BPSK is the most noise-resistant type of binary modulation in AWGN channels because its signals are antipodal and therefore have the maximum Euclidean distance [36–38]. For instance, it requires two times smaller E_b than BFSK for a given P_b. Both BPSK and BFSK demodulators are amplitude-invariant, and this property is important in channels with fast fading. The major problem of BPSK, the initial phase ambiguity, was solved by N. T. Petrovich who proposed differential BPSK (DBPSK) in the early 1950s [36]. In contrast with BPSK where data are carried by the absolute phase values (e.g., 0°

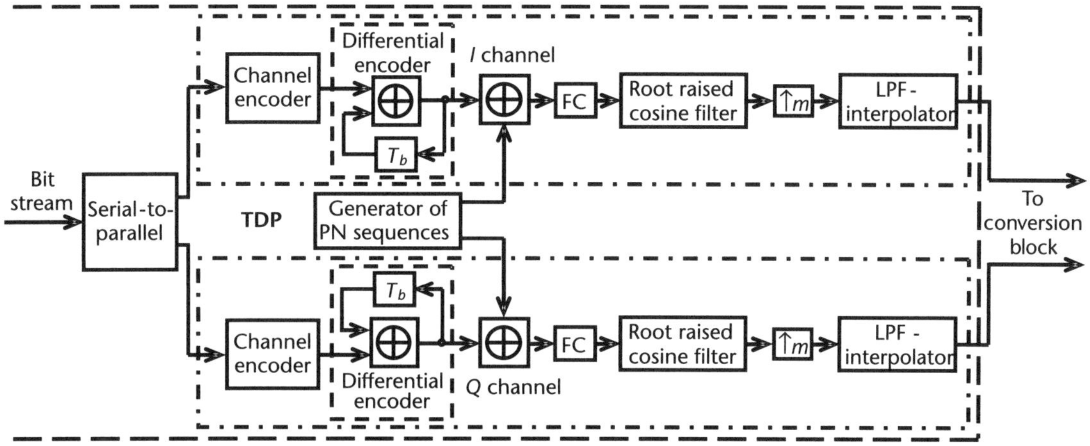

Figure 3.15 Dual independent DBPSK with DS spreading in a TDP.

and 180°), DBPSK transmits data by the phase differences between neighboring symbols. This eliminates the phase ambiguity problem. Despite the emergence of other methods of coping with phase ambiguity, DBPSK is still widely used due to its simplicity and insignificance of energy loss compared to BPSK. Indeed, noncoherent demodulation of DBPSK requires less than 1 dB increase in E_b compared to coherent BPSK demodulation when $P_b \leq 10^{-4}$. Note that differential QPSK (DQPSK) is less common because of its complexity and higher energy loss compared to QPSK.

BPSK, DBPSK, QPSK, and many other energy-efficient signals, such as M-ary orthogonal and biorthogonal signals, have constant envelopes. However, their high spectral sidelobes, interfering with neighboring channels, should be suppressed by filtering that increases the signals' crest factors, mainly due to 180° phase transitions between adjacent symbols (recall that these transitions provide the lowest E_b for a given P_b). The increased crest factors reduce the Tx power utilization. Hard-limiting of the filtered signals restores their near-constant envelopes at the cost of the sidelobe regeneration. An effective solution to this problem for QPSK is a half-symbol offset between its I and Q components, that is, offset QPSK (OQPSK) modulation, which limits the absolute values of phase transitions to no more than 90° at a time, significantly reducing crest factors and improving the Tx power utilization. The same effect is produced by half-symbol offset between the independent BPSK or DBPSK signals in I and Q channels. An even smaller crest factor is provided by minimum shift keying (MSK) that is also widely used and can be considered either a form of OQPSK with cosine symbol weighting or a form of BFSK [37]. The latter MSK interpretation allows its noncoherent demodulation, although with a substantial energy loss. The disadvantages of MSK, compared to OQPSK, are wider spectrum and simplicity of determining the symbol rate by unauthorized Rxs.

However, when modulation is followed by DS spreading (like in Figure 3.15), the Tx output signals' crest factors are determined not by modulation but by spreading, and a half-chip offset between the signals in I and Q channels reduces the crest factors of many signals, including BPSK, DBPSK, as well as orthogonal and biorthogonal signals based on Walsh functions. This offset is not reflected in Figure 3.15. Here, bits from the channel encoders are differentially encoded in both channels. Each differential encoder comprises a memory element with the delay equal to the bit duration T_b and an exclusive-or (XOR) gate that operates as follows:

$$0 \oplus 0 = 0, 0 \oplus 1 = 1, 1 \oplus 0 = 1, 1 \oplus 1 = 0 \quad (3.27)$$

Differential encoding transforms a sequence $\{a_k\}$ of bits $a_1, a_2, a_3, ..., a_k, ...$ from the channel encoder into a sequence $\{b_k\}$ of bits $b_1, b_2, b_3, ..., b_k, ...$ according to the rule:

$$b_1 = a_1, \text{ and } b_k = a_k \oplus b_{k-1} \text{ for } k \geq 2 \quad (3.28)$$

The differential encoders' output bits modulate the spreading PN sequences using XOR operations in both I and Q channels. The chip rate of the PN sequences is an integer multiple of the bit rate, and their ratio determines the spreading processing gain G_{ps}. Phase modulation and spreading are performed jointly by XOR gates at the differential encoders' outputs in the I and Q channels. When BPSK spreading is used, the same PN sequence is sent to both I and Q channels. For QPSK spreading, different PN sequences are generated for these channels. Note that dual independent DBPSK without spreading requires coherent demodulation. DS spreading with orthogonal PN sequences in I and Q channels allows noncoherent demodulation. Moreover, even a half-chip offset between the I and Q signals, which reduces the Tx output signals' crest factors, does not prevent noncoherent demodulation when $G_{ps} \geq 16$ because the signals still remain quasi-orthogonal.

The representation of binary symbols by ones and zeroes allowed executing modulation and spreading using XOR gates, but this format is unacceptable for digital filters. Therefore, the format converters (FCs) convert 1 to −1 and 0 to +1 prior to filtering in each channel. In the absence of spreading, symbol-shaping filtering is performed at the modulator output to limit the signal bandwidth while minimizing ISI. This ISI is avoided if at the sampling instant of each symbol at the Rx demodulator output all other symbols have near-zero levels (see Appendix B). There are several types of symbol-shaping filters (e.g., raised cosine, Gaussian, and root raised cosine). Filters of the last type should be used jointly in the Tx and Rx of a communication system to create the effect of a raised cosine filter. In the case of spreading, chip-shaping filtering is also useful. Symbol- or chip-shaping filtering is performed prior to or jointly with digital interpolating filtering.

Thus, a half-symbol or a half-chip offset is an effective way to reduce crest factors of energy-efficient signals. This reduction not only improves the Tx power utilization but also simplifies signal reconstruction. When crest factors of energy-efficient signals are sufficiently small, the required accuracy of symbol- or chip-shaping and interpolating filtering determine the reconstruction complexity for a given signal bandwidth.

3.4.2 AQ-DBPSK Modulation

Small single-purpose digital radios, powered by miniature batteries and intended for communications over short distances with a low throughput, are widely used in various sensor networks (see, for instance, [39]). To be energy-efficient, modulation techniques in these radios should not only minimize E_b for a given P_b and maximize the Tx power utilization, but also meet three additional requirements. The first requirement is simplicity of Tx and especially Rx circuits, needed because the transmitted signals' power is comparable with the Rx power consumption. The second one is tolerance to frequency offsets between Rx and corresponding Tx, needed due to the difficulty of achieving high frequency stability in these radios. The third one is fast synchronization that saves more energy for the payload data transmission, especially if the data are transmitted in bursts.

None of the energy-efficient modulations discussed in the previous section meets all these requirements. Therefore, a modulation named alternating quadratures DBPSK (AQ-DBPSK), which preserves the DBPSK advantages (such as low E_b for a given P_b and simplicity of modulation and demodulation), while mitigating its drawbacks (such as poor utilization of Tx power and insufficient tolerance to the frequency offset between the Rx and corresponding Tx, although this tolerance is higher than that of other phase modulation techniques) was proposed in [40–42]. AQ-DBPSK sends odd symbols in quadrature with even ones, reducing the phase transitions between all adjacent symbols to ±90°. Therefore, the crest factor of AQ-DBPSK is similar to that of OQPSK, but AQ-DBPSK allows noncoherent demodulation. At the same time, the data are transmitted by the phase differences equal to 0° or 180° between the same-parity symbols. Therefore, AQ-DBPSK and DBPSK have the same E_b for a given P_b in AWGN channels. Consequently, AQ-DBPSK has higher energy efficiency than DBPSK because it better utilizes the Tx power. AQ-DBPSK allows various demodulation techniques. Two of them, demodulation optimal in AWGN channels and frequency-invariant demodulation, are described in Section 4.5.2, where the AQ-DBPSK advantages are fully explained. Two different embodiments of AQ-DBPSK modulation are described below. The first one performs separate differential encoding of the same-parity symbols in I and Q channels, whereas the second embodiment carries out joint differential encoding of these symbols.

The AQ-DBPSK modulator's first embodiment is illustrated by its block diagram in Figure 3.16(a), signal constellations in Figure 3.16(b), and timing diagrams in Figure 3.16(c) (digital numbers in the timing diagrams are depicted by their analog equivalents). In the FC, the modulator input bits of length T_b from a channel encoder are mapped into 2-bit digital symbols of the same length according to the rule: 1 to –1 and 0 to +1 (see the first two timing diagrams of Figure 3.16(c)). The 2-bit symbols represented by sign and magnitude (–1 by 11 and +1 by 01) are sent to the serial-to-parallel converter. Here, switches S_1 and S_2 controlled by the direct and inverse outputs O and $\bar{O}$ of a modulo-2 counter (MTC), distribute the odd and even symbols between different channels. The binary sequence at the MTC output O and the stream of the two-bit symbols from S_1 (I channel) are shown in the third and fourth timing diagrams of Figure 3.16(c), respectively. In the latter diagram, the spaces correspond to the positions of even 2-bit symbols removed by S_1. The binary sequence at the MTC output $\bar{O}$ and the stream of the 2-bit symbols from S_2 (Q channel) are shown in the sixth and seventh timing diagrams of Figure 3.16(c), respectively. In the seventh diagram, the spaces correspond to the positions of odd 2-bit symbols removed by S_2.

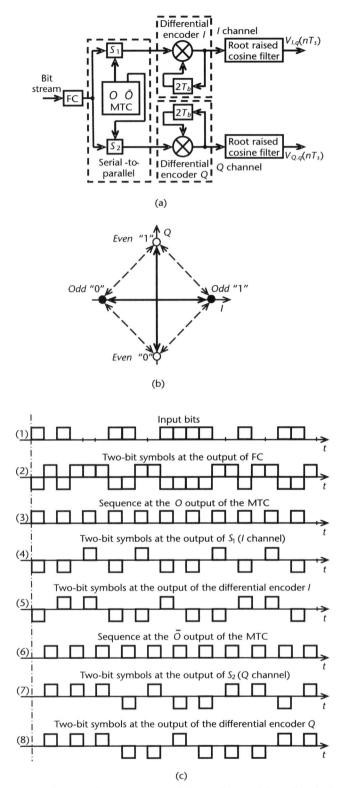

Figure 3.16 First embodiment of AQ-DBPSK modulator: (a) modulator block diagram, (b) signal constellation, and (c) timing diagrams of signals at modulator blocks' outputs.

In the channels, 2-bit symbols separately undergo differential encoding of the same-parity symbols. Each differential encoder, comprising a digital memory with the $2T_b$ delay and a digital multiplier, transforms an input sequence $\{a_k\}$ of 2-bit symbols $a_1, a_2, a_3, ..., a_k, ...$ into a sequence $\{b_k\}$ of 2-bit symbols $b_1, b_2, b_3, ..., b_k, ...$ according to the rule:

$$b_1 = a_1, \; b_2 = a_2, \; \text{and} \; b_k = a_k \times b_{k-2} \; \text{for} \; k \geq 3 \tag{3.29}$$

Two ones should be written to the memory of each encoder upon initialization.

The differential encoders' output symbols in the I and Q channels are shown in the fifth and eighth timing diagrams of Figure 3.16(c), respectively. Since the odd and even symbols are transmitted over different channels, the phase transitions between adjacent symbols can be equal only to $\pm 90°$. In both channels, the differential encoders' output symbols undergo digital symbol-shaping filtering by root raised cosine (or raised cosine) filters. The multibit digital signals $V_{Iq}(nT_s)$ and $V_{Qq}(nT_s)$ at the outputs of the symbol-shaping filters are, respectively, I and Q components of the digital complex-valued baseband output signal of the AQ-DBPSK modulator. The symbol-shaping filters and the subsequent digital and analog filters of a Tx suppress the spectral sidelobes of the analog bandpass modulated signal $u_{out}(t)$ reconstructed from $V_I(t)$ and $V_Q(t)$ according to the equation

$$u_{out}(t) = V_I(t)\cos(2\pi f_0 t) - V_Q(t)\sin(2\pi f_0 t) \tag{3.30}$$

where f_0 is the center frequency of $u_{out}(t)$. Since the absolute values of the phase transitions between adjacent symbols cannot exceed 90°, the spectral sidelobe suppression causes insignificant amplitude fluctuations of $u_{out}(t)$. At the same time, the phase shifts of only 0° to 180° between the same-parity symbols, used for the data transmission (see Figure 3.16(b)), provide high noise immunity of communications.

The second embodiment of the AQ-DBPSK modulator is illustrated by its block diagram in Figure 3.17(a), signal constellations in Figure 3.17(b), and timing diagrams in Figure 3.17(c). All input 1-bit symbols undergo joint differential encoding of the same-parity symbols performed by one differential encoder that comprises a memory element with the $2T_b$ delay and an XOR gate. When the differential encoder's input and output symbols are denoted c_k and d_k, respectively, the encoding rule is:

$$d_1 = c_1, \; d_2 = c_2, \; \text{and} \; d_k = c_k \oplus d_{k-2} \; \text{for} \; k \geq 3 \tag{3.31}$$

Two zeros should be written to the encoder memory upon initialization. The modulator input one-bit symbols and differentially encoded 1-bit symbols are shown in the first and second timing diagrams of Figure 3.17(c), respectively. The encoded symbols enter both I and Q channels. In the I channel, these symbols are XOR-ed with the sequence of ones and zeros from the MTC direct output O (the third timing diagram of Figure 3.17(c)), and the resulting symbols (the fourth timing diagram of Figure 3.17(c)) are sent to the channel FC. In the Q channel, the differentially encoded symbols are directly fed into its FC. The FCs are identical and map the 1-bit symbols into 2-bit symbols of the same length T_b as shown in the fifth and sixth timing diagrams of Figure 3.17(c). In both channels, the 2-bit symbols enter the

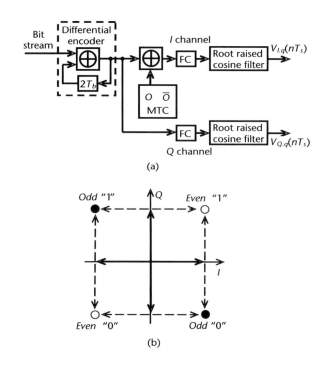

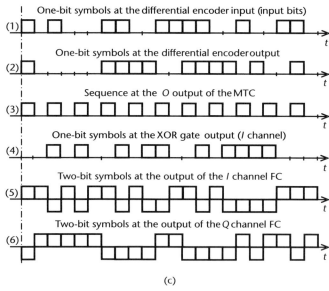

Figure 3.17 Second embodiment of AQ-DBPSK modulator: (a) modulator block diagram, (b) signal constellation, and (c) timing diagrams of signals at modulator blocks' outputs.

symbol-shaping filters. The filtered signals are I and Q components of the modulator's output digital complex-valued baseband signal.

Note that the digital symbols appear in turn at the outputs of the I and Q channels in the AQ-DBPSK modulator's first embodiment, but they appear together in its second embodiment (compare the fifth and eight timing diagrams of Figure 3.16(c) with the fifth and sixth timing diagrams of Figure 3.17(c), respectively). Therefore, the first embodiment always provides ±90° phase shifts between adjacent symbols, while the second one only when $|V_I(t)| = |V_Q(t)|$ (see the signal constellation in Figure 3.17(b)). Thus, although the second embodiment significantly simplifies the differential encoding, it is more sensitive to *IQ* imbalance than the first one. Consequently, the second embodiment is advantageous in the case of bandpass reconstruction in a Tx. Similar to AQ-DBPSK modulation, AQ-BPSK spreading can be used to reduce the crest factors of energy-efficient signals.

The above-demonstrated variety of effective techniques minimizing the energy-efficient signals' crest factors allows selecting the most appropriate one for virtually any type of Txs. Being realized in the digital domain, these techniques not only improve the Tx power utilization but also simplify the signal reconstruction. When the crest factors of energy-efficient signals are minimal, the requirements for Tx reconstruction circuits are determined by the Tx bandwidth and the required accuracy of symbol-shaping and interpolating filtering.

3.4.3 Power Utilization in Txs with Bandwidth-Efficient Modulation

Reduction of signals' crest factors for improving the Tx power utilization is used for both energy- and bandwidth-efficient signals. In general, however, the power utilization improvement methods and their impact on the reconstruction differ for these signals. For energy-efficient signals, digital methods of the crest factor reduction are diverse and improve both power utilization and reconstruction in Txs. For bandwidth-efficient signals, analog methods of improving Tx power utilization should also be taken into account due to the complexity of crest factor reduction for these signals and the possibility to improve Tx power utilization without reducing the crest factors. In this case, the power utilization improvement does not necessarily simplify the reconstruction.

Signal predistortion in a TDP, which is effective in improving the PA linearity, cannot substantially improve the Tx power utilization [43, 44]. A Doherty amplifier significantly improves the Tx power utilization by combining a Class-AB amplifier operating as a carrier amplifier and a Class-C amplifier operating as a peaking amplifier [45]. However, Doherty amplifiers, being employed in PAs of Txs, do not influence the signal reconstruction complexity. The situation with the envelope elimination and restoration technique [46] is similar.

The linear amplification with nonlinear components technique used for improving the energy efficiency of Txs (see, for instance, [47]) notably affects signal reconstruction in Txs. This technique is based on converting a signal $s(t)$ with a varying envelope into two constant-envelope signals $s_1(t)$ and $s_2(t)$ by properly modulating the phase of $s(t)$. After amplification in separate PAs, the amplified signals $s_1(t)$ and $s_2(t)$ can be summed, producing the amplified original signal $s(t)$. The phase modulations of $s(t)$ required for generating $s_1(t)$ and $s_2(t)$ can be almost ideally carried out in a TDP. Since the TDP output signals have constant envelopes, their crest factors are minimal, and their reconstruction is simplified. This simplification is still limited by the $s_1(t)$ and $s_2(t)$ bandwidths and the sensitivity of their constellations to distortion (both are usually increased by the phase modulations compared to those of $s(t)$). The separate PAs operating at saturation can provide maximum power efficiency, although making their parameters almost identical requires an effort. The major problem of this technique is combining the signals at the PAs' outputs. This combining may reduce the Tx efficiency or limit its linearity.

An interesting approach to the digital Tx design is the digital generation of two-level analog RF signals capable of representing not just binary but also multibit symbols at the TDP output [17, 48–52]. This eliminates the need for D/A converters and reduces the Tx analog and mixed-signal processing to filtering, amplification, and antenna coupling. As a result, the Tx PA operates in a switching mode, providing the highest power efficiency. Two major techniques are used to realize this approach. The first one utilizes bandpass sigma-delta modulation to generate binary signals at the Tx RF. The second technique employs pulse-width modulation (PWM). In both cases, the RF signals can be synthesized in the digital domain. In principle, this approach significantly simplifies signal reconstruction in Txs, but its overall realization is not that simple, and, despite significant efforts, its development is still at the research stage.

The reconstruction techniques based on the sampling theorem's direct interpretation (see Section 6.4.2) can improve the Tx power utilization, as well as its adaptivity and reconfigurability, without crest factor reduction or simplification of reconstruction circuits [53]. On the contrary, more complex but also more effective reconstruction circuits should be employed to enhance the overall Tx performance and power efficiency. The power efficiency enhancement is achieved not due to the signal crest factor reduction but due to varying the rail voltages of the reconstruction circuits and PAs proportionally to the signal level. This is possible because the reconstruction based on the sampling theorem's direct interpretation requires time-interleaved structures where the time intervals between neighboring samples in each channel are much longer than the sampling interval. Note that, in principle, many methods improving the Tx power utilization for bandwidth-efficient signals are also applicable to energy-efficient ones.

3.5 Summary

Modern IC and DSP technologies support a wide range of digital radios from multipurpose and/or multistandard SDRs and CRs to ubiquitous, inexpensive, low-power, single-purpose devices. These radios utilize most of the DSP advantages, but only SDRs and CRs can utilize all of them.

A digital Tx with substantial transmit power consists of two functionally dissimilar parts: a Tx drive (its "brains") and a PA (its "muscles"). From the D&R standpoint, four parts of the Tx can be identified: a digitizer of input analog signals, a TDP, an AMB, and a part that comprises the remaining analog and mixed-signal blocks.

The TDP input signals usually undergo source encoding to reduce unnecessary redundancy. For analog signals, a part of this encoding can be performed during their digitization. Further processing of digital and digitized analog input signals includes channel encoding and modulation. It may also include encryption (prior to channel encoding) as well as spreading (after or together with modulation) and multiple access (after modulation or spreadimg). Analog bandpass signals, carrying information over RF channels, are usually represented by their digital baseband complex-valued equivalents in the TDP and should be reconstructed at the TDP's output. The reconstruction can be bandpass or baseband. The operations performed in the TDP determine the reconstruction complexity.

In digital radios, accurate and stable frequencies are derived from MFSs by frequency synthesizers. Various DDSs are used there. Nonrecursive DDSs are discussed in this chapter due to their widespread use and the similarity of their algorithms to those of some WFGs employed for the D&R based on the direct and hybrid interpretations of the sampling theorem. DFC, the major block of both DDS and WFG, is optimized using Boolean algebra methods.

The digitization of input signals, considered in this chapter, is not combined with source encoding. It comprises antialiasing filtering, sampling, quantization, and digital operations (e.g., downsampling with decimating filtering). Both baseband and bandpass reconstructions of bandpass signals comprise digital operations (e.g., upsampling with digital interpolating filtering), D/A conversion, and analog interpolating filtering. Comparison of these reconstruction techniques shows that most drawbacks of baseband reconstruction are fundamental, whereas the weaknesses of bandpass reconstruction are provisional and will be eliminated by the technological progress.

Among conversion block architectures, the most promising ones are the offset upconversion architecture with bandpass reconstruction and the direct RF reconstruction architecture.

Digital methods of energy-efficient signals' crest factor reduction are diverse and effective. They allow improving the power utilization and simplifying the reconstruction in Txs. For bandwidth-efficient signals, improving Tx power utilization is more complex. It cannot always be performed in the digital domain, does not always simplify reconstruction, and does not always involve crest factor reduction. Still, methods developed for bandwidth-efficient signals are also applicable to energy-efficient ones.

References

[1] Eassom, R. J., "Practical Implementation of a HF Digital Receiver and Digital Transmitter Drive," *Proc. 6th Int. Conf. HF Radio Syst. & Techniques*, London, U.K., July 4–7, 1994, pp. 36–40.

[2] Sabin, W. E., and E. O. Schoenike (eds.), *Single-Sideband Systems and Circuits*, 2nd ed., New York: McGraw-Hill, 1995.

[3] Mitola, J. III, *Software Radio Architecture*, New York: John Wiley & Sons, 2000.

[4] Reed, J. H., *Software Radio: A Modern Approach to Radio Engineering*, Inglewood Cliffs, NJ: Prentice Hall, 2002.

[5] Johnson, W. A., *Telecommunications Breakdown: Concept of Communication Transmitted via Software Defined Radio*, New York: Pearson Education, 2004.

[6] Poberezhskiy, Y. S., and G. Y. Poberezhskiy, "Sampling and Signal Reconstruction Structures Performing Internal Antialiasing Filtering and Their Influence on the Design

of Digital Receivers and Transmitters," *IEEE Trans. Circuits Syst. I*, Vol. 51, No. 1, 2004, pp. 118–129.

[7] Poberezhskiy, Y. S., and G. Y. Poberezhskiy, "Flexible Analog Front-Ends of Reconfigurable Radios Based on Sampling and Reconstruction with Internal Filtering," *EURASIP J. Wireless Commun. and Netw.*, No. 3, 2005, pp. 364–381.

[8] Kenington, P., *RF and Baseband Techniques for Software Defined Radio*, Norwood, MA: Artech House, 2005.

[9] Vankka, J., *Digital Synthesizers and Transmitters for Software Radio*, New York: Springer, 2005.

[10] Mitola, J. III, *Cognitive Radio Architecture: The Engineering Foundations of Radio HML*, New York: John Wiley & Sons, 2006.

[11] Fette, B. A. (ed.), *Cognitive Radio Technology*, 2nd ed., New York: Elsevier, 2009.

[12] Grebennikov, A., *RF and Microwave Transmitter Design*, New York: John Wiley & Sons, 2011.

[13] Hueber, G., and R. B. Staszewski (eds.), *Multi-Mode/Multi-Band RF Transceivers for Wireless Communications: Advanced Techniques, Architectures, and Trends*, New York: John Wiley & Sons, 2011.

[14] Johnson, E. E., et al., *Third-Generation and Wideband HF Radio Communications*, Norwood, MA: Artech House, 2013.

[15] Grayver, E., *Implementing Software Defined Radio*, New York: Springer, 2013.

[16] Bullock, S. R., *Transceiver and System Design for Digital Communications*, 4th ed., Edison, NJ: SciTech Publishing, 2014.

[17] Nuyts, P. A. J., P. Reynaert, and W. Dehaene, *Continuous-Time Digital Front-Ends for Multi-standard Wireless Transmission*, New York: Springer, 2014.

[18] Lechowicz, L., and M. Kokar, *Cognitive Radio: Interoperability Through Waveform Reconfiguration*, Norwood, MA: Artech House, 2016.

[19] Goot, R., and M. Minevitch, "Some Indicators of the Efficiency of an Extreme Radio Link in Group Operation," *Telecommun. and Radio Engineering*, Vol. 32, No. 11, 1974, pp. 126–128.

[20] Goot, R., "Group Operation of Radiocommunication Systems with Channels Selection by Sounding Signals," *Telecommun. and Radio Engineering*, Vol. 35, No. 1, 1977, pp. 77–81.

[21] Tierney, J., C. Rader, and B. Gold, "A Digital Frequency Synthesizer," *IEEE Trans. Audio Electroacoust.*, Vol. 19, No. 1, 1971, pp. 48–57.

[22] Rabiner, L. R., and B. Gold, *Theory and Application of Digital Signal Processing*, Englewood Cliffs, NJ: Prentice-Hall, 1975.

[23] Rohde, U. L., J. Whitaker, and T. T. N. Bucher, *Communications Receivers*, 2nd ed., New York: McGraw Hill, 1997.

[24] Cordesses, L, "Direct Digital Synthesis: A Tool for Periodic Wave Generation," *IEEE Signal Process. Mag.*, Part 1: Vol. 21, No. 4, 2004, pp. 50–54; Part 2: Vol. 21, No. 5, 2004, pp. 110–112, 117.

[25] Poberezhskiy, Y. S., and M. N. Sokolovskiy, "The Logical Method of Phase-Sine Conversion for Digital Frequency Synthesizers," *Telecommun. and Radio Engineering*, Vol. 38/39, No. 2, 1984, pp. 96–100.

[26] Poberezhskiy, Y. S., "Method of Optimizing Digital Functional Converters," *Radioelectronics and Commun. Systems*, Vol. 35, No. 8, 1992, pp. 39–41.

[27] Crochiere, R. E., and L. R. Rabiner, *Multirate Digital Signal Processing*, Upper Saddle River, NJ: Prentice Hall, 1983.

[28] Poberezhskiy, Y. S., and M. V. Zarubinskiy, "Analysis of a Method of Fundamental Frequency Selection in Digital Receivers," *Telecommun. and Radio Engineering*, Vol. 43, No. 11, 1988, pp. 88–91.

[29] Poberezhskiy, Y. S., and S. A. Dolin, "Analysis of Multichannel Digital Filtering Methods in Broadband-Signal Radio Receivers," *Telecommun. and Radio Engineering*, Vol. 46, No. 6, 1991, pp. 89–92.

[30] Poberezhskiy, Y. S., S. A. Dolin, and M. V. Zarubinskiy, "Selection of Multichannel Digital Filtering Method for Suppression of Narrowband Interference" (in Russian), *Commun. Technol.*, TRC, No. 6, 1991, pp. 11–18.

[31] Vaidyanathan, P. P., *Multirate Systems and Filter Banks*, Englewood Cliffs, NJ: Prentice Hall, 1993.

[32] Harris, F. J, *Multirate Signal Processing for Communication Systems*, Englewood Cliffs, NJ: Prentice Hall, 2004.

[33] Vaidyanathan, P. P., S. -M. Phoong, and Y. -P. Lin, *Signal Processing and Optimization for Transceiver Systems*, Cambridge, U.K.: Cambridge University Press, 2010.

[34] Lin, Y. -P., S. -M. Phoong, and P. P. Vaidyanathan, *Filter Bank Transceivers for OFDM and DMT Systems*, Cambridge, U.K.: Cambridge University Press, 2011.

[35] Dolecek, G. J. (ed.), *Advances in Multirate Systems*, New York: Springer, 2018.

[36] Okunev, Y., *Phase and Phase-Difference Modulation in Digital Communications*, Norwood, MA: Artech House, 1997.

[37] Sklar, B., *Digital Communications, Fundamentals and Applications*, 2nd ed., Upper Saddle River, NJ: Prentice Hall, 2001.

[38] Middlestead, R. W., *Digital Communications with Emphasis on Data Modems*, New York: John Wiley & Sons, 2017.

[39] Poberezhskiy, Y. S., "Novel Modulation Techniques and Circuits for Transceivers in Body Sensor Networks," *IEEE J. Emerg. Sel. Topics Circuits Syst.*, Vol. 2, No. 1, 2012, pp. 96–108.

[40] Poberezhskiy, Y. S., "Alternating Quadratures Differential Binary Phase Shift Keying Modulation and Demodulation Method," U.S. Patent 7,627,058 B2, filed March 28, 2006.

[41] Poberezhskiy, Y. S., "Apparatus for Performing Alternating Quadratures Differential Binary Phase Shift Keying Modulation and Demodulation," U.S. Patent 8,014,462 B2, filed March 28, 2006.

[42] Poberezhskiy, Y. S, "Method and Apparatus for Synchronizing Alternating Quadratures Differential Binary Phase Shift Keying Modulation and Demodulation Arrangements," U.S. Patent 7,688,911 B2, filed March 28, 2006.

[43] Boumaiza, S., et al., "Adaptive Digital/RF Predistortion Using a Nonuniform LUT Indexing Function with Built-In Dependence on the Amplifier Nonlinearity," *IEEE Trans. Microw. Theory Tech.*, Vol. 52, No. 12, 2004, pp. 2670–2677.

[44] Woo, Y. Y., et al., "Adaptive Digital Feedback Predistortion Technique for Linearizing Power Amplifiers," *IEEE Trans. Microw. Theory Tech.*, Vol. 55, No. 5, 2007, pp. 932–940.

[45] Kim, B., et al., "The Doherty Power Amplifier," *IEEE Microw. Mag.*, Vol. 7, No. 5, 2006, pp. 42–50.

[46] Kahn, L. R., "Single-Sideband Transmission by Envelope Elimination and Restoration," *Proc. IRE*, Vol. 40, No. 7, 1952, pp. 803–806.

[47] Birafane, A., et al., "Analyzing LINC System," *IEEE Microw. Mag.*, Vol. 11, No. 5, 2010, pp. 59–71.

[48] Keyzer, K., et al., "Digital Generation of RF Signals for Wireless Communications with Bandpass Delta-Sigma Modulation," *Dig. IEEE MTT-S Int. Microw. Symp.*, Phoenix, AZ, May 20–24, 2001, pp. 2127–2130.

[49] Park, Y., and D. D. Wentzloff, "All-Digital Synthesizable UWB Transmitter Architectures," *Proc. IEEE Int. Conf. UWB*, Hannover, Germany, Vol. 2, September 10–12, 2008, pp. 29–32.

[50] Wurm, P., and A. A. Shirakawa, "Radio Transmitter Architecture with All-Digital

Modulator for Opportunistic Radio and Modern Wireless Terminals," *Proc. IEEE CogART*, Aalborg, Denmark, February 14, 2008, pp. 1–4.

[51] Hori, S., et al., "A Watt-Class Digital Transmitter with a Voltage-Mode Class-S Power Amplifier and an Envelope ΔΣ Modulator for 450 MHz Band," *Proc. IEEE CSICS*, La Jolla, CA, October 14–17, 2012, pp. 1–4.

[52] Cordeiro, R. F., A. S. R. Oliveira, and J. Vieira, "All-Digital Transmitter with RoF Remote Radio Head," *Dig. IEEE MTT-S Int. Microw. Symp.*, Tampa, FL, June 1–6, 2014, pp. 1–4.

[53] Poberezhskiy, Y. S., and G. Y. Poberezhskiy, "Impact of the Sampling Theorem Interpretations on Digitization and Reconstruction in SDRs and CRs," *Proc. IEEE Aerosp. Conf.*, Big Sky, MT, March 1–8, 2014, pp. 1–20.

CHAPTER 4
Digital Receivers

4.1 Overview

Initial information on digital communication Rxs is provided in Chapters 1 and 2. Signal flow in a typical multipurpose Rx is illustrated by the high-level block diagram in Figure 1.19(b) and is outlined in Section 1.4.1. The digital frequency translation of complex-valued signals and digital generation of baseband complex-valued equivalents in RDPs are described in Section 1.4.2 and illustrated by Figures 1.21 and 1.23, respectively. Several aspects of signal processing in digital Rxs are presented in Section 2.3.2. Thus, this chapter is based on the material of the first two chapters. It is also closely connected to the previous chapter because communication Rxs and Txs operate jointly, and, for instance, the advantages and limitations of the modulation techniques discussed in Chapter 3 cannot be assessed without analyzing the related demodulation techniques in this chapter. Similar technical solutions and common approaches to examining Rx and Tx D&R procedures also connect these two chapters.

Section 4.2 provides general information on digital Rxs. It describes the first steps of digital Rxs' development, its problems, their initial solutions, and the influence of these solutions on the subsequent progress of digital radios. It also explains why several Rx characteristics specify Rx performance, instead of the universal one that actually exists. The reception quality characteristics related to digitization of Rxs' input signals are analyzed. Structures and specifics of digital Rxs and transceivers are discussed.

Section 4.3 examines the dynamic range of digital Rxs, which reflects their capability to pick up a weak desired signal in the presence of strong unwanted ones. Various definitions of the dynamic range and factors limiting it are analyzed. The IMP parameters and their influence on the reception reliability are studied. Expressions for determining the minimum required dynamic range of a HF Rx are derived.

Digitization of Rxs' input signals is considered in Section 4.4. Both baseband and bandpass digitization techniques are discussed and illustrated by block and spectral diagrams. Since these techniques significantly influence AMF architectures, they and the architectures are compared in the same section.

Section 4.5 describes demodulation of several energy-efficient signals whose modulation techniques where discussed in Section 3.4. This allows the completion of their analysis and illustrates signal processing in RDPs.

4.2 Digital Rx Basics

4.2.1 First Steps of Digital Radio Development

The complexity of DSP and D&R in digital radios depends on the product of the bandwidth and logarithm of the dynamic range (D&R complexity also depends on the signals' carrier frequencies). Therefore, the first digital radios were designed for communications with submerged submarines where these products and carrier frequencies are minimal. The most significant next step in the digital radio development was the emergence of digital HF Rxs that require not only wider bandwidths but also much higher dynamic ranges than lower-frequency Rxs. Besides their own value, the first HF Rxs were the best proving ground for the digital technology implementation in radios. Indeed, reception of HF signals encounters all types of negative phenomena that can happen in other frequency bands: multipath propagation accompanied by Doppler shift and spread, mutual interference due to poor predictability of ionospheric conditions, and diurnal and seasonal changes in HF wave propagation. Flexibility and accuracy of DSP enable effective coping with these phenomena. Digital HF Rxs paved the way to DSP implementation in the radios of higher frequency bands.

Although the first experimental digital HF Rx was developed by TRW Inc. (United States) in the early 1970s [1], practically used HF digital radios were designed only in the mid-1980s due to the emergence of DSP chips and A/Ds with sufficient speed and resolution. Subsequent generations of DSPs, A/Ds, and D/As supported the fast progress of digital radios. FPGAs, which appeared on the market in the mid-1980s, proved to be the best platforms for the RDPs and TDPs of the most complex radios in the late 1990s. In the mid-1990s, application specific integrated circuit (ASIC) implementation of the RDPs and TDPs became practical for mass-produced digital radios. The quality of D&R in digital radios (and other applications) was significantly improved by the emergence of pipelined A/Ds in the late 1980s and bandpass sigma-delta A/Ds and D/As in the 1990s (although sigma-delta modulation principles have been known since 1954). The digital radios developed from the 1970s to the early 1990s created the foundation for the current progress in this field. They clarified the radios' fundamental design principles but also originated some misconceptions that hindered their development. Therefore, these radios are discussed below.

In the first experimental digital HF Rx [1], signals were digitized at the RF, and the preselector bandwidth was 1.5 MHz. Four different sampling rates $f_s \leq 14$ Msps facilitated the variable RF. Digitization was performed by a 9-bit A/D with an integrating sample-and-hold amplifier (SHA). The digital filter was bandpass and adjustable to the bandwidths and center frequencies of signals. Reference [1] showed clear understanding of the restrictions imposed on f_s and the existence of the optimal f_s for a given signal center frequency f_0. Yet it also demonstrated the necessity of more advanced hardware and additional knowledge on the subject. Subsequent technological progress and R&D efforts provided both. The practicality of signal digitization at the Rx IF and the ways of extending the frequency range from the HF band to 500 MHz were considered in [2]. Multichannel digital reception, mentioned in [1], was comprehensively discussed in [2].

Several equations needed for the design of digital Rxs, for example, (3.16), (3.23), and the equations for the required AMF gain and A/D resolution, were derived in [3]. The A/D resolution is dictated by the required Rx dynamic range that depends on the statistical characteristics of interference. The equations, connecting this range to the statistical characteristics of the HF band interference and to the Rx bandwidth, were derived in [4]. In [5], these equations were used for calculating the HF Rxs dynamic range in various RF environments. Several multichannel digital Rxs were also discussed there. The reception failure probabilities were estimated for various RF environments in [6]. Later, more precise estimates taking into account not only third-order but also fifth-order IMPs were derived in [7].

During the 1970s and 1980s, three mutually connected choices related to digital radios were studied: baseband D&R versus bandpass D&R, SHAs versus THAs, and representation of digital signals by their instantaneous values versus their I and Q components. The outcomes, which greatly impacted the subsequent developments, were as follows.

The comparison of baseband and bandpass D&R has resulted in the conclusion that while bandpass D&R provide better quality of reception and transmission, they are more complex, expensive, and less adaptive than baseband D&R. This conclusion, based on the contemporary technology, was, in principle, correct, but it exaggerated the complexity of bandpass D&R due to the insufficient insight into bandpass sampling and pulse shaping. Since the 1990s, most high-quality digital radios use bandpass D&R and most low-quality radios employ baseband D&R.

The choice between SHAs and THAs, made in favor of THAs, was incorrect. It was based on the erroneous assumption that bandpass sampling limits the SHA integration time T_i to $T_i \ll 1/f_0$ where f_0 is the bandpass signal center frequency. Actually, T_i should meet a much weaker condition: $T_i < 0.5/f_0$, as was shown in 1974. Although this result was later published in an internationally accessible journal [8, 9], it was overlooked. Moreover, the proof that weighted integration in SHAs removes the limitation related to f_0 and allows further increase in the integration time, which radically improves the Rx dynamic range [10–12], was also initially unnoticed.

The comparison of digital signals' representations by the samples of instantaneous values and by those of I and Q components has demonstrated the latter representation's advantage. The efficient digital methods of forming these components were suggested [10, 13–15]. These methods, as well as the corresponding methods of digital generation of bandpass signals from their I and Q components, have become common.

The first practically used digital HF Rx (HF-2050) was developed by Rockwell-Collins in 1984. It had superheterodyne architecture with dual frequency conversion and was tunable within the frequency range from 14 kHz to 30 MHz in 10-Hz increments. Its first and second IFs were, respectively, $f_{IF1} = 99$ MHz and $f_{IF2} = 3$ MHz. Bandpass digitization was performed at the second IF by a flash A/D with 7-bit resolution and sampling rate $f_{s1} = 12$ Msps. In the RDP, the samples of the instantaneous signal values were transformed into the samples of its I and Q components, and downsampling with decimating FIR filtering reduced the sampling rate to $f_{s2} = 48$ ksps for each component. The RDP also performed filtering with variable bandwidth from 0.3 kHz to 6 kHz, demodulation of several analog and digital signals, and forming automatic gain control (AGC) signals for the Rx AMF. The Rx in-band two-tone dynamic range was 40 dB. Its power consumption was 100W. The HF-2050 development was a significant technical achievement demonstrating the advantages of DSP in communication radios. The experience obtained during this development was partly reflected in [16–20].

The major drawback of HF-2050 was the insufficient (for the HF band) dynamic range limited by its digitization circuit and by the first IF crystal filter's nonlinearity. Although a high IF simplifies image rejection, reduces the number of spurious responses, and f_{IF1} = 99 MHz, specifically, eliminates the spurious responses to input signals at frequencies $(1/2)f_{IF1}$ and $(1/3)f_{IF1}$, crystal filters at frequencies ≥65 MHz exhibit significant nonlinearity. This required either lowering the IF or using a different type of filters. Both approaches were tried in digital radios later.

In the mid-1980s, a number of digital HF Rx were developed in Russia with the focus on providing sufficiently high dynamic range that determines the signal reception reliability in the HF band. Some of them were multichannel Rxs with channel bandwidths of 4 kHz and 8 kHz allowing spatial interference rejection and diversity combining. A Rx with a 40-kHz bandwidth was intended for the reception of spread spectrum (SS) signals and surveillance. Dual frequency conversion AMF with the first IF f_{IF1} = 62 MHz + f_{IF2} was used in all these Rxs. The second IFs were: f_{IF21} = 128 kHz in the narrowband Rxs and f_{IF22} = 540 kHz in the Rxs with the 40-kHz bandwidth. The in-band two-tone dynamic ranges of the AMFs slightly exceeded 70 dB. The digitization was bandpass with digital forming of the signal I and Q components. Mass-produced 12-bit A/Ds with custom-designed integrating SHAs were used for the digitization. The SHAs' design was based on the theory presented in [8–10], and their experimental investigation was later described in [21–23]. The SHAs' integration time $T_i = 1/(3f_0)$ was selected to reduce the third-order IMPs. Their in-band two-tone dynamic ranges without the A/Ds were >85 dB at f_{IF21} and >73 dB at f_{IF22}. With the A/Ds, they were >73 dB at f_{IF21} and >72 dB at f_{IF22}. The relations among f_{IF}, f_s, and the RDP bandwidth ensured the absence of second-order IMPs within this bandwidth. SHAs with weighted integration [10–12] were not used in these Rxs. They were experimentally investigated only in the 1990s.

Simultaneously with the development of digital communication Rxs, the research on digital broadcast Rxs was initiated. Therefore, the integrating SHAs intended for communication Rxs were also tested at the broadcast Rxs' IFs [23]. In connection with the R&D on the digital communication and broadcast Rxs mentioned above, several methods of digital filtering and demodulation were suggested and examined [24–29]. In the 1970s and 1980s, intensive R&D on digital radios for navigation, radar, and EW were also performed. During the 1980s, a number of digital HF Rxs with baseband digitization were developed, mostly for surveillance, in various countries [30]. Their in-band two-tone dynamic ranges did not exceed 40 dB, but they provided sufficiently wide bandwidths.

The experience in digital radio design obtained in the 1980s and the technological progress throughout that period allowed the development of several digital VLF-HF Rxs with high dynamic ranges (primarily for communications, surveillance, and direction finding) in the early 1990s. The most well-known ones were developed by Cubic Communications (United States), Marconi (United Kingdom), and Rohde & Schwarz (Germany). These Rxs had superheterodyne architectures with several frequency conversions and bandpass digitization. The difference in the approaches to providing high dynamic range can be illustrated by Rxs H2550 (Marconi) and EK895/EK896 (Rohde & Schwarz).

H2550 had dual frequency conversion with f_{IF1} = 62.5 MHz and f_{IF2} = 2.5 MHz [31]. To achieve better linearity, a two-cavity helical resonator instead of a crystal filter was used at f_{IF1}, and digitization was performed by a bandpass sigma-delta A/D with the sampling rate f_{s1} =10 Msps. The optimal relation between f_{s1} and f_{IF1} simplified forming of the I and Q components, while the downsampling with decimating FIR filtering reduced the sampling rate to $f_{s2} \approx$ 40 ksps, significantly increasing the A/D resolution. EK895/EK896 used triple frequency conversion with f_{IF1} = 41.44 MHz, f_{IF2} = 1.44 MHz, and f_{IF3} = 25 kHz [32]. Thus, the nonlinearity of the first IF crystal filters was avoided by reducing f_{IF1}, whereas very low f_{IF3} allowed selecting the A/D with sufficient resolution.

During the 1990s, the technology advancement and increased knowledge of DSP in general and digital radios specifically enabled significant improvement of the radios' capabilities, reduction of their size, weight, power consumption, and cost, as well as their expansion to higher frequency bands and more diverse applications. In that decade, the concepts of software defined radio (SDR), cognitive radio (CR), and sampling with internal antialiasing filtering were formulated. Even more significant progress in development and proliferation of digital radios was achieved in the twenty-first century.

4.2.2 Main Characteristics of Rxs

Properties of Rxs are reflected by a large number of characteristics. Below, only the reception quality characteristics are discussed. Note that the channel throughput reduction, caused by a Rx, fully describes the reception quality [33]. Indeed, an ideal Rx does not influence the channel throughput, while any real-world Rx reduces it by introducing noise, interference, and distortion. The smaller the throughput reduction, the better the reception quality. Despite the attractiveness of using only one characteristic instead of many, this characteristic is impractical because a huge variety of reception conditions make its statistically reliable measurement complex and long. In addition, the throughput reduction does not indicate its cause(s) and, consequently, does not allow determining the ways of its improvement. Therefore, several characteristics are used to describe the reception quality. Among them, sensitivity, selectivity, dynamic range, reciprocal mixing, and spurious outputs are or can be related to digitization in Rxs.

4.2.2.1 Sensitivity

The sensitivity of a Rx is determined by its internal noise and can be characterized by the minimum RF input level required for receiving a specific signal with given quality or by more general measures independent of a particular Rx mode. Two general measures are discussed below: (1) noise factor (F) that is called noise figure (NF) when expressed in decibels (NF = $10\log_{10}F$), and (2) minimum detectable signal (MDS). F is the ratio of the actual Rx output noise power to that of an ideal Rx that has the same gain and bandwidth but no internal noise, that is, it is the ratio of the total output noise power $P_{N.T}$ to the output noise power caused by an input source $P_{N.I}$:

$$F = \frac{P_{N.T}}{P_{N.I}} \quad \text{and} \quad \text{NF} = 10\log_{10} P_{N.T} - 10\log_{10} P_{N.I} \tag{4.1}$$

Thus, $F = 1$ and NF = 0 dB for an ideal Rx, whereas $F > 1$ and NF > 0 dB for a real-world Rx.

It is known that F of a linear Rx stage is the ratio of its input and output SNRs. Therefore, its NF is

$$\text{NF} = 10\log_{10}\left(\frac{\text{SNR}_{in}}{\text{SNR}_{out}}\right) \tag{4.2}$$

Consequently, NF = 0 dB for noiseless amplifiers and lossless passive circuits, and NF > 0 dB for nonideal circuits. The total Rx noise factor F can be expressed through the noise factors F_n and gains G_n of its stages with perfectly matched impedances [34]:

$$F = F_1 + \frac{F_2 - 1}{G_1} + \frac{F_3 - 1}{G_1 G_2} \cdots + \frac{F_n - 1}{G_1 G_2 \ldots G_{n-1}} + \ldots \tag{4.3}$$

where n is the stage number. The internal noise of the Rx input stages contains several components: shot noise and flicker noise of semiconductors and thermal (Johnson) noise of the impedance resistive part R. It may also contain Barkhausen noise produced by magnetic substances. Shot noise, which is the largest component of internal noise, can be considered white and Gaussian. Flicker noise is significant only at low frequencies, and its PSD is roughly inversely proportional to frequency. Thermal noise, produced by R, is white and Gaussian. Nyquist formula allows calculation of its mean square voltage within the Rx noise bandwidth B_N (in hertz):

$$\overline{V}_N^2 = 4k_B T R B_N \tag{4.4}$$

where k_B is Boltzmann's constant (1.38×10^{-23} J/K), and T is the resistor's absolute temperature (kelvin). Since the Nyquist model of a thermal noise source is a noise generator with open-circuit mean square voltage (4.4), the maximum noise power P_N that can be coupled from the generator into a Rx is

$$P_N = k_B T B_N \tag{4.5}$$

and its PSD $N(f)$ is

$$N(f) = N = k_B T \tag{4.6}$$

4.2 Digital Rx Basics

The MDS is the signal power equal to the noise power within the Rx noise bandwidth:

$$\text{MDS} = k_B T B_N F \quad (4.7)$$

The MDS defined according to (4.7) is also called Rx noise floor. At temperature $T = T_0 = 290\text{K}$ (adopted by the IEEE as standard), MDS expressed in dBm is

$$\text{MDS}_{\text{dBm}} = -174 + 10\log_{10} B_N + \text{NF} \quad (4.8)$$

where -174 dBm is the thermal noise power per hertz at 290K.

In a properly designed digital Rx, digitization should minimally degrade the sensitivity of input analog stages. The application of (4.3) to the sequence of the digital Rx circuits (see Figure 4.1) yields:

$$F = F_A + \frac{F_D - 1}{G_A} \quad (4.9)$$

where F_A and F_D are the noise factors of the input analog stages and the digitization circuits, respectively, whereas G_A and G_D are their gains. According to (4.9), the degradation can be minimized by reducing F_D and increasing G_A. Although both ways are used in practice, the second one is less desirable because it limits the Rx dynamic range (see Section 4.3). In any case, a minimally required G_A should be determined.

Imperfect antialiasing filtering, sampling, and quantization produce noise, interference, and distortion that can be characterized by the PSD of their sum. In correctly designed digitization circuits, the major component of this PSD is created by the A/D quantization noise, and its other components can be neglected. Since both quantization noise and internal noise of the input analog stages can be regarded as white with PSDs N_q and N_A, respectively, the minimally required G_A should make $N_q \ll N_A$. For determining the G_A lower bound, the fact that the quantization noise is non-Gaussian can be disregarded, but its dependence on the A/D input signal level must be kept in mind. For uniform quantization,

$$N_q \approx \frac{\Delta^2}{6 f_s} \quad \text{if} \quad \Delta < \sigma_{in} \quad (4.10)$$

where Δ is the A/D quantization step and σ_{in} is the A/D input signal rms value. If the Rx external noise certainly exceeds its internal noise, condition $N_A/N_q \geq 12$ is

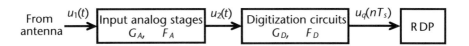

Figure 4.1 Sequence of circuits in a digital Rx.

sufficient to assume $N_q \ll N_A$. Taking into account that $N_A = N_0 G_A^2$ where N_0 is the PSD of the Rx internal noise referred to its input, we obtain [3, 5]:

$$G_A \geq \frac{\Delta}{(0.5 f_s N_0)^{0.5}} \quad (4.11)$$

As follows from (4.11), an increase in f_s allows reducing G_A. However, this option should be used with caution because this G_A reduction may violate condition $\Delta < \sigma_{in}$. To avoid this violation, (4.11) can be replaced with a more restrictive inequality:

$$G_A \geq \frac{\Delta}{(B_{a.f} N_0)^{0.5}} \quad (4.12)$$

where $B_{a.f}$ is the antialiasing filter bandwidth. Inequalities (4.11) and (4.12) are identical when $f_s = 2B_{a.f}$. Since in practice $f_s > 2B_{a.f}$, (4.12) requires a higher G_A.

When the Rx internal noise is the main component of the total noise, it is better to assume that $N_A/N_q \geq 24$. In this case, (4.11) and (4.12) should be replaced, respectively, with the inequalities:

$$G_A \geq \frac{2\Delta}{(f_s N_0)^{0.5}} \quad \text{and} \quad G_A \geq \frac{\Delta}{(0.5 B_{a.f} N_0)^{0.5}} \quad (4.13)$$

Signal attenuation by an AGC prior to the A/D may invalidate any of inequalities (4.11) through (4.13). The important task of digital Rx design is ensuring a proper N_A level at any time.

4.2.2.2 Selectivity

Frequency selectivity of a Rx characterizes its capability to avoid attenuation and distortion within the Rx passband and reject undesired signals within its stopband. Selectivity is expressed in decibels and is measured by comparing the strengths of output signals within the Rx passband and stopband for the same input signal level. Selectivity specification usually includes the widths of the passband and transition band, acceptable AFR distortion within the passband, and the minimum suppression within the stopband. In an ideal digital Rx, all these parameters are determined by the digital filters in its RDP. In a real-world digital Rx, they depend on the Rx architecture and are influenced by the characteristics of analog filters, digitization circuits, and digital filters.

All functional units of Rxs (e.g., analog and digital filters, amplifiers, mixers, samplers, quantizers, and LOs) produce undesired effects (such as linear and nonlinear distortions, noise, and spurious outputs) while performing their intended functions. Typically, each undesired effect influences several Rx characteristics. For example, nonlinear distortion can reduce sensitivity, selectivity, dynamic range, and produce spurious responses. Therefore, each Rx characteristic is influenced by several undesired effects. For instance, Rx selectivity can be degraded by linear and nonlinear distortions, spurious responses, and reciprocal mixing.

These effects degrade Rx performance and complicate measurements of Rx characteristics due to an increased number of parameters reflecting every specific characteristic, and the need to determine these parameters not only for an entire Rx but for its parts as well. For selectivity, these effects require measuring not only the adjacent channel rejection but also that of spurious channels, and most measurements must be performed not only for an entire Rx but also for its input analog stages, digitization circuits, and RDP. During the measurements of most Rx characteristics (including selectivity), some functions (e.g., AGC) necessary for Rx operation should be disabled. Digitization circuits contribute to Rx selectivity through analog antialiasing and digital decimating filtering they perform. To avoid the Rx performance degradation caused by linear and nonlinear distortions, spurious outputs, and noise of these circuits, these negative effects should be minimized or compensated in the RDP.

4.2.2.3 Dynamic Range

Among Rx characteristics, dynamic range, especially the in-band one, is particularly significant for two reasons. First, it determines the signal reception reliability in overcrowded frequency bands because it reflects the Rx capability to pick up a weak desired signal in the presence of strong unwanted ones within the Rx preselector bandwidth. Second, although all Rx characteristics are interdependent, dynamic range influences the largest number of other characteristics including sensitivity, selectivity, and the number and values of spurious responses. When the Rx dynamic range, which is limited by its AMF, is sufficient, virtually any required adaptivity and reconfigurability of the Rx can be achieved due to the DSP flexibility. Dynamic range is examined in detail in Section 4.3.

4.2.2.4 Reciprocal Mixing

Reciprocal mixing happens when a strong out-of-band interfering signal is converted to the IF as a result of its mixing with a part of the LO phase noise. Despite the low level of the phase noise, its mixing with strong interfering signals can reduce the actual sensitivity and dynamic range of a Rx. Reciprocal mixing takes place in the first mixer where RF filtering may insufficiently suppress out-of-band signals. Improving RF filtering and lowering LO phase noise reduce the negative consequences of reciprocal mixing. In a digital Rx with RF digitization, reciprocal mixing can be caused by sampling jitter if antialiasing filtering insufficiently suppresses out-of-band signals.

4.2.2.5 Spurious Outputs

There are two types of spurious outputs in Rxs: spurious signals and spurious responses. Spurious signals appear at a Rx output without any signal at its input. A spurious response happens when a signal at a frequency, to which a Rx is not tuned, produces an output signal. Both types of spurious outputs are concisely described below.

Spurious signals result from penetration of internal Rx signals into the main signal path. Internal signals of analog Rxs include output signals of MFSs and frequency synthesizers, as well as their harmonics, parasitic oscillations in amplifiers, harmonics of power supplies, and IF subharmonics (in Rxs with IFs above their frequency bands). In digital Rxs, the number of simultaneously generated internal signals is larger than in analog ones, and a part of them are generated for sampling and quantization. Even so, only AMFs of digital Rxs are susceptible to the penetration of internal signals. Proper layout, grounding, and shielding reduce the number and levels of spurious signals.

An analog superheterodyne Rx with an ideal mixer has two spurious responses: IF and image. Indeed, an ideal mixer is a multiplier that produces outputs only at two frequencies $f_{in} \pm f_{LO}$ where f_{in} and f_{LO} are, respectively, the frequencies of an input signal and LO. Since a real-world mixer is not a prefect multiplier, it produces outputs at frequencies $mf_{in} \pm nf_{LO}$ where m and n are integers that are either positive or have different signs, but the resulting frequencies must be positive in any case. When a resulting frequency is within the IF bandwidth or very close to it, a spurious response occurs. A large number of spurious responses can happen only at the first IF because the RF strip selectivity is usually low. Since the first IF strip selectivity is much higher, spurious responses at the second IF (if it exists) are not expected. In a digital Rx, digitization at its RF can cause even larger number of spurious responses if sufficient antialiasing filtering is not ensured.

4.2.3 Digital Rxs and Txs

Although RF signals entering digital Rxs are preprocessed and digitized in their AMFs first, most of the processing is performed in their RDPs. Technological progress increases the number of functions executed in RDPs and moves digitization closer to the antennas. Still, antenna coupling, preliminary amplification and/or attenuation, initial filtering (including antialiasing filtering), sampling, and quantization take place in the AMFs. Most output signals of RDPs are delivered to users in a digital form, but those that must be delivered in an analog form are converted to the analog domain, as illustrated by the block diagram in Figure 4.2, which is a more detailed version of the diagrams in Figures 1.19(b) and 2.2(b). The AMFs may have different architectures that are discussed in Section 4.4.3.

Independent of the AMF architecture, its passband is usually wider than bandwidths of desired signals (whose channel filtering is performed in the RDP) because it is currently more difficult and expensive to adaptively change the AMF passband than the RDP channel filter bandwidth. Signals digitized in the AMF are sums of desired signals, interfering signals (ISs), and noise. An excessive AMF bandwidth often makes the sums of ISs much larger than desired signals. Even when the AMF passband insignificantly exceeds the desired signal bandwidth, a strong IS can appear within the desired signal spectrum in the case of intentional jamming or poorly regulated frequency band. In multichannel Rxs, each channel usually regards signals of other channels as ISs. Thus, the requirements for digitization circuits in digital Rxs depend more on the IS statistics than on the desired signals' parameters. A sufficient dynamic range allows suppressing ISs using frequency and/or spatial filtering or other techniques.

AMF input circuits usually include antenna couplers, preselectors, and low-noise amplifiers (LNAs). They may contain an automatic attenuator reducing the LNA

4.2 Digital Rx Basics

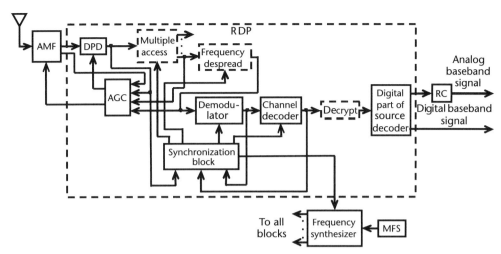

Figure 4.2 Block diagram of a multipurpose digital Rx.

input signals at the appearance of ISs capable of blocking Rxs. The input circuits may be preceded by a switch automatically disconnecting them from the antenna to prevent their damage by extremely strong ISs. The AMF frequency selectivity is distributed among its stages. Typically, its preselector has the widest passband and the lowest stopband suppression, whereas its antialiasing filter has the narrowest passband and the highest stopband suppression. There often are other filters between the preselector and the antialiasing filter. The presence of ISs within the AMF passband requires a high dynamic range and a high quality of AGC. Although the Rx dynamic range and factors affecting it are discussed in the next section, it should be noted here that digitization circuits are the bottleneck limiting it. The reasons are the highest power of analog signals at their input, much larger experience in developing preceding analog circuits than in developing digitization circuits, and the capability of RDPs to support virtually any required dynamic range. As follows from the first reason and mentioned in Section 4.2.2, it is undesirable to substantially exceed the minimum required gain G_A of the stages preceding the digitization circuits, determined by (4.11) through (4.13), and it is desirable to reduce the PSD of the sum of quantization noise and interference caused by nonideal antialiasing filtering and sampling. These conclusions are confirmed by the results of calculations in Table 4.1.

In a digital Rx, AGC automatically adjusts the gains of Rx stages to maximally utilize their dynamic ranges and maintain a proper signal level at the inputs of digitization and demodulation circuits despite the Rx input signal level variation. The Rx selectivity is distributed not only among the AMF stages but also among the RDP stages. In both AMF and RDP, the filters whose passbands are narrower than those of the previous stages reject a part of noise and can reject or weaken a part of the ISs. Therefore, the AGC must adjust signal levels at the outputs of these filters. Thus, the larger the number of such filters, the more complex the AGC functioning. When an automatic attenuator is installed at the LNA input, the AGC must interact with it. Thus, AGC structure and algorithms highly depend on the Rx architecture and purpose.

Digitization circuits largely determine the overall Rx performance. They include the analog and mixed-signal parts in the AMF and the DPD in the RDP. It was mentioned above that the requirements for these circuits depend more on the ISs than on the desired signals' parameters that may be distorted in communication channels. As noted in Section 2.3.2, considering digitization a special case of source encoding, which reduces the analog signals' redundancy prior to their digital processing, enriches both digitization and source coding theories and technologies. It is also shown there that digitization in Rxs can be baseband or bandpass, and both these types are analyzed in Section 4.4. To improve reception quality, the digitization circuits should have (besides high dynamic range) high flexibility and compatibility with the IC technology that are presently limited by the best bandpass antialiasing filters. As shown in Chapter 6, sampling based on the sampling theorem's direct interpretation enables simultaneous improvement of both dynamic range and flexibility.

Independent of the digitization type, received analog bandpass real-valued signals are converted into their digital baseband complex-valued equivalents in most RDPs. A high dynamic range is usually needed at the RDP input even if the signals undergo nonlinear transformations. They can be intended, for instance, to compensate nonlinear distortion in the AMF. Another example is robust AJ algorithms that necessitate a memoryless nonlinearity prior to a correlator or matched filter [35]. The adaptive robustness discussed in [35] requires accuracy and flexibility achievable only in RDPs. Processing in the RDPs depends on that performed in the corresponding TDPs: the signals that were multiplexed in the TDPs are demultiplexed, those that were spread are despread, and all received signals are demodulated and decoded. Despreading, demodulation, and channel decoding are the stages of generalized demodulation that can be performed separately or jointly. If the signals were encrypted, they are decrypted. Finally, a source decoder converts the retrieved information into a form convenient for recipients, and the signals expected to be received in analog form are reconstructed in an RC. The reconstruction of baseband signals at a Rx output is described in Section 5.3.3. Demodulation of energy-efficient signals is discussed in Section 4.5.

As explained in Section 2.3.2 and illustrated in Figure 4.2, demultiplexing, despreading, demodulation, channel decoding, and some other operations performed in RDPs require various types of synchronization (e.g., frame, code, word, symbol, frequency, and phase synchronization). Note that digitization in Rxs should not be synchronized with reconstruction in Txs. Although all synchronization procedures in Rxs are symbolically represented by a specialized block in Figure 4.2, they are usually distributed among various RDP blocks in practice. Similar to digital Txs (see Figure 3.1), digital Rxs require multiple internally generated frequencies for tuning and signal transformations. These frequencies are generated by frequency synthesizers that derive them from MFSs. The information on frequency synthesizers and MFSs in digital Txs, provided in Section 3.2, is also applicable to digital Rxs. The only difference is that RDPs may require a larger number of frequencies due to their higher complexity.

Depending on the purposes and operational conditions of Rxs, their RDP hardware platforms vary from GPP-based platforms in prototype Rxs to ASICs in mass-produced Rxs (see Section 3.2.1). In most complex SDRs and CRs, these platforms may combine FPGAs, GPPs, and SPUs. In the future, the share of RDPs with field-programmable ASICs as their platforms will significantly increase. Since the DSP complexity in Rxs is typically much higher than that in Txs, the RDP and TDP of the same communication system can use different hardware platforms. In transceivers, however, they usually have common platforms. Besides the publications on digital Rxs referenced above, information on the subject can be found in [36–53] and some other books and papers.

In duplex or half-duplex communication systems with low to moderate transmit power, Txs and Rxs are usually combined into transceivers where they share circuitry and housing. They typically have common power units, MFSs, frequency synthesizers, many control circuits, and built-in test equipment. In half-duplex transceivers, the Txs and Rxs can use the same IF strips for transmit and receive modes. Therefore, the cost, size, weight, and power consumption of transceivers are lower than those of separate Txs and Rxs with the same electrical characteristics. Transceivers are currently used in virtually all handheld and manpack radios, and in many onboard and amateur radios.

Digital transceivers may have properties and modes unrealizable in analog transceivers. A good example is implementation of full-duplex mode (transmission and reception on the same frequency, at the same time, and with the same antenna). Analog transceivers often use time- or frequency-division duplexing. In the first case, the Tx and Rx operate on the same frequency and with the same antenna, but at different time. In the second case, they operate at the same time and with the same antenna, but on different frequencies. The full-duplex mode increases the communication system throughput, but leakage of the Tx signal into the Rx prevents reception. In a transceiver, however, this leakage can, in principle, be compensated because its Rx knows the signal being transmitted by its Tx. While such compensation is impractical in analog transceivers due to inaccuracy and instability of analog processing, it became feasible in digital ones [54, 55].

Its possible realization is illustrated by the block diagram in Figure 4.3. Here, the Tx output signals entering the three-port circulator go to the antenna, while received signals from the antenna go to the Rx. Although ideal circulators would completely isolate Txs and Rxs, real-world circulators provide only ~20-dB attenuation between them. To cancel this self-interference, a properly scaled and delayed copy of the Tx output signal should be subtracted from the Rx input signal. The subtraction is performed in two stages. The first (analog) stage cannot be sufficiently accurate and is mostly intended to avoid overloading the AMF circuits. The residual self-interference is compensated at the second (digital) subtraction stage using the replica with more precise scaling, delay, and predistortion that matches the distortion of the self-interference signal in the analog circuits. Yet there are some obstacles to achieving the required level of self-interference cancellation (~130 dB in satellite communications, for example). Therefore, the full-duplex mode is still largely in the R&D phase. However, strong efforts are made to widely implement it due to the spectrum utilization benefits it brings.

Over the last decades, superconductor technology has become mature enough for implementation in advanced digital radios (see, for example, [56–58]). This technology is mostly based on Josephson junctions, superconducting quantum interference devices (SQUIDs), and rapid-single-flux-quantum logic (RSFQ), although there are

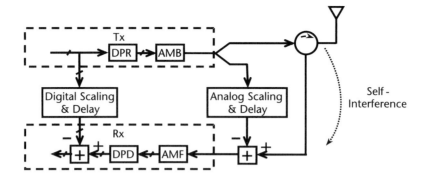

Figure 4.3 Block diagram of two-stage self-interference cancellation in a full-duplex transceiver.

promising devices based on different principles. Superconductor A/Ds, D/As, and DSP units operate at tens and even hundreds of gigahertz. They have low power consumption and high sensitivity. The best technical parameters are achieved at the temperature of ~4K. Significant progress has also been made in digital and mixed-signal superconductor devices operating at the temperature of ~60K (LNAs operating at the latter temperature have been known for a long time). Still, relatively high cost, technological complexity, and some vulnerabilities of superconductor radios currently limit their implementation to the areas where reception quality is more important than cost of equipment.

4.3 Dynamic Range of a Digital Rx

4.3.1 Factors Limiting Rx Dynamic Range

Most Rx characteristics called "dynamic range" inadequately represent it. On the one hand, they do not take into account the AGC capability to accommodate input signal level variations (caused, for instance, by slow fading), underestimating the actual dynamic range. On the other hand, they do not take into account reciprocal mixing and/or spurious outputs that reduce it. In addition, the measurements consistent with dynamic range definitions may actually reflect combined effects of several characteristics. For example, out-of-band dynamic range represents a combined effect of Rx dynamic range and selectivity. Therefore, Rx manufacturers readily provide this performance-inflating parameter to their customers.

The in-band dynamic range, as it is defined below, maximally reflects the effect of Rx nonlinearity on a sum of desired signals and ISs within the Rx preselector passband. This characteristic still depends on the selectivity distribution among the Rx stages and on the Rx sensitivity that determines its lower bound. It can still be reduced by reciprocal mixing or spurious outputs. Despite these limitations, the in-band dynamic range preserves the predominant influence of nonlinearity, and this allows its theoretical analysis that leads to some important practical conclusions. Due to the complexity of modern Rxs and variety of their operational modes, even the in-band dynamic range requires several parameters for its characterization. Some of them are more general, while others are architecture- and/or mode-specific. This section is focused on two most general parameters. Due to terminological differences in technical literature, their definitions are given first.

Rx functioning in the presence of one strong IS is characterized by single-tone dynamic range (also called blocking or 1-dB gain compression). Below, it is denoted as D_1 or $D_{1,dB}$ when expressed in decibels. It is usually defined as the ratio of the 1-dB compression point to the MDS. Recall that the 1-dB compression point is the signal level (referred to the input), at which the output signal is 1 dB lower than the expected linear response. At present, there are several techniques that enable reliable reception even when the desired signal spectrum is completely covered by a strong IS spectrum. Indeed, the IS can be rejected, for example, by spatial nulling. However, the spatial nulling and other IS-suppressing techniques can be realized only if D_1 is sufficient.

In most cases, however, several strong ISs can appear within a wideband AMF. Even if their spectra do not overlap with the desired signal spectrum, the ISs can reduce the Rx noise immunity due to the IMPs generated by the AMF nonlinearities. The Rx performance in such situations is adequately reflected by two-tone dynamic range that is the ratio of one of two sinusoids with identical levels, creating an IMP equal to the Rx MDS, to the MDS. Usually, a third-order IMP is used for its measurements, and the two-tone dynamic range determined this way is denoted as D_3, or $D_{3,dB}$ when expressed in decibels. If a second-order IMP is used, the two-tone dynamic range is denoted as D_2, or $D_{2,dB}$ when expressed in decibels. In both cases, the two-tone dynamic range can be equivalently characterized by the third- and second-order intercept points (IPs). In digital Rxs with baseband digitization, both second- and third-order IMPs must be taken into account. In Rxs with bandpass digitization, the role of second-order IMPs diminishes, especially if f_s is optimal, that is, satisfies (3.16). When an optimal f_s also meets condition $f_s \geq 6B$ where B is a signal bandwidth, the second-order IMPs can be neglected.

All definitions of a Rx dynamic range are ratios (typically expressed in decibels) of the upper and lower bounds of the input signal level. The lower bound is determined by the Rx sensitivity that is equal or proportional to the Rx MDS. The factors limiting the upper bound are clarified by several numeric examples below [33] using single-tone dynamic range D_1. Let us make the following assumptions: an IS is within the Rx preselector passband but out of the channel passband, and the Rx input resistance, its MDS voltage, and its MDS power are, respectively, $R_{in} = 50\Omega$, $V_{MDS} = 0.7\ \mu V = 0.7 \cdot 10^{-6}$ V, and $P_{MDS} = V^2_{MDS}/R_{in} \approx 1 \cdot 10^{-14}$ W. Then the maximum acceptable IS power at the Rx and its A/D inputs are, respectively,

$$P_{IS\max} = P_{MDS} \cdot D_1 \quad \text{and} \quad P_{IS\max A/D} = P_{IS\max} \cdot G_A = P_{MDS} \cdot D_1 \cdot G_A \quad (4.14)$$

where G_A is the gain of the Rx analog stages preceding the A/D. For $P_{MDS} = 1 \cdot 10^{-14}$ W, Table 4.1 presents $P_{IS\max}$ and $P_{IS\max A/D}$ as functions of $D_{1,dB}$ for several values of $G_{A,dB} = 10\log_{10} G_A$. Since the AMF power consumption must be at least larger than $P_{IS\max A/D}$ and the RDP computational load may also depend on $P_{IS\max A/D}$, Table 4.1 shows that the ultimate restriction on the input signal upper bound is imposed by the allowed Rx power consumption (or dissipation). For a given upper bound, the dynamic range can be increased by reducing the Rx MDS (i.e., improving its sensitivity) and/or G_A.

Table 4.1 Maximum IS Power as a Function of Dynamic Range for $P_{MDS} = 1 \cdot 10^{-14}$ W

$D_{1,dB}$	60	70	80	90	100	110	120	130	140
P_{ISmax}, W	10^{-8}	10^{-7}	10^{-6}	10^{-5}	10^{-4}	10^{-3}	10^{-2}	10^{-1}	10^{0}
$P_{ISmaxA/D}$, W ($G_{A,dB} = 20$ dB)	10^{-6}	10^{-5}	10^{-4}	10^{-3}	10^{-2}	10^{-1}	10^{0}	10^{1}	10^{2}
$P_{ISmaxA/D}$, W ($G_{A,dB} = 40$ dB)	10^{-4}	10^{-3}	10^{-2}	10^{-1}	10^{0}	10^{1}	10^{2}	10^{3}	10^{4}
$P_{ISmaxA/D}$, W ($G_{A,dB} = 60$ dB)	10^{-2}	10^{-1}	10^{0}	10^{1}	10^{2}	10^{3}	10^{4}	10^{5}	10^{6}

There is an opinion that improving Rx sensitivity is reasonable only if its internal noise is larger than the external one. That is incorrect. When the Rx internal noise is lower than the external one, the sum of the Rx input signals and external noise can be attenuated, thus expanding the dynamic range (at the cost of a slight SNR decrease). As noted in Section 4.2.2, G_A should be sufficient to make $N_q \ll N_A$ where N_q and N_A are, respectively, the PSDs of quantization noise and noise of the input analog stages. Inequalities (4.11) through (4.13) show that lowering G_A requires reducing the A/D quantization step Δ for given f_s and $B_{a.f.}$. As explained in the subsequent chapters, this in turn necessitates decreasing jitter and other negative phenomena caused by sampling. The G_A and Δ reduction must also be accompanied by increasing the number of A/D bits to improve the dynamic range.

The situation, considered above and intended to determine the major factors limiting the Rx dynamic range, is extreme due to the assumption that the ISs within the preselector passband cannot be rejected or weakened before the sampler. However, Rxs have antialiasing (and often other) filters with bandwidths narrower than that of the preselector. They can reject or weaken a part of the ISs, reducing the required dynamic range of subsequent circuits.

4.3.2 Intermodulation

In poorly regulated frequency bands (i.e., in the HF band), many narrowband ISs enter AMFs along with desired signals. The AMF nonlinearity, which becomes apparent at high IS levels, causes several effects reducing signal reception quality, and generation of IMPs is the major one. As proven theoretically in [6, 7] and seen in practice, even a moderate number of ISs within the AMF passband produce so many IMPs that they create an IMP floor when the AMF two-tone dynamic range is insufficient. Indeed, while the average number of ISs grows linearly with the increase of the preselector bandwidth, the number of their combinations that produce IMPs grows much faster, as shown below. In contrast with the noise floor whose PSD is uniform within the AMF passband, the IMP floor PSD is higher in the middle and lower at the edges. As shown in Figure 4.4(a), the ISs whose spectra do not overlap with that of the desired signal are rejected by the channel filter in the RDP when the AMF dynamic range is sufficient. When this range is insufficient (see Figure 4.4(b)), IMPs degrade the reception quality and can make the reception of weak signals impossible. Figure 4.4(b) also shows that a high IMP floor inhibits RF spectrum analysis and recognition of unutilized frequencies within an AMF passband, thus impeding the operation of CRs. In panoramic Rxs, a high IMP floor obstructs detection of low-power and SS signals. Since the IS levels can differ by many orders of magnitude, their logarithmic scale is assumed in the plots in Figure 4.4 and subsequent figures.

The necessity of sufficient dynamic range for the reception of SS signals in the presence of narrowband ISs is illustrated by Figure 4.5. The spectra of a desired DS SS signal and four narrowband ISs in Figure 4.5(a) show that the signal demodulation quality can be improved by rejection of ISs when the AMF dynamic range is sufficient. Figure 4.5(b) demonstrates that a high IMP floor, caused by insufficient dynamic range, reduces the demodulation quality because it may not allow identifying all ISs (IS_3 in this case), and even the rejected ISs contribute to the quality reduction through the IMPs. Figure 4.5 slightly simplifies the situation because only very strong ISs must be rejected for optimal demodulation. The ISs whose PSDs are comparable with that of a desired signal should be suppressed by a factor that depends on the signal type, reception mode, and accuracy of IS PSD estimation.

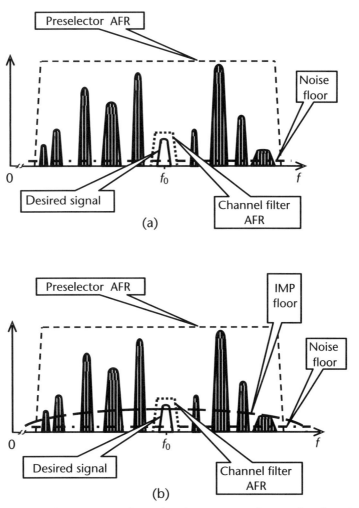

Figure 4.4 Reception of a narrowband signal in the presence of narrowband ISs: (a) sufficient dynamic range and (b) insufficient dynamic range.

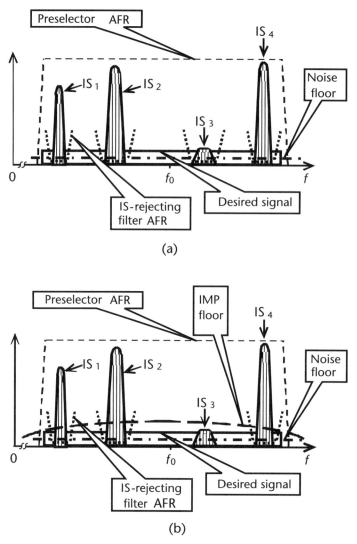

Figure 4.5 Reception of a wideband DS SS signal in the presence of narrowband ISs: (a) sufficient dynamic range and (b) insufficient dynamic range.

4.3.2.1 Suppressing ISs in Rxs with Sufficient Dynamic Range

Closed-form expressions for the transfer functions of IS-suppressing filters in different scenarios were obtained assuming that the sum of ISs and internal noise forms nonwhite Gaussian noise (see Appendix C). If the PSDs of nonwhite Gaussian noise and the desired signal across a 1-Ω resistance are denoted as $N(f)$ and $|S(f)|^2$, respectively, the transfer function $H_1(f)$ of a linear filter, placed before the correlators or matched filters of a demodulator to maximize the SNR at their outputs and minimize the demodulator bit error rate [59] is

$$H_1(f) = \frac{c}{N(f)} \exp(-j2\pi f t_0) \tag{4.15}$$

In (4.15) through (4.17), c is a scaling factor and t_0 is the filter delay; both are Rx- and filter-specific. Equation (4.15) is equivalent to (2.1). The transfer function $H_2(f)$ of a linear filter performing least-squares smoothing of a signal with PSD $|S(f)|^2$ in the presence of nonwhite Gaussian noise [60] is

$$H_2(f) = \frac{c|S(f)|^2}{|S(f)|^2 + N(f)} \exp(-j2\pi f t_0) \qquad (4.16)$$

A comparison of (4.15) and (4.16) shows that (4.16) requires weaker suppression of ISs than (4.15). The reason is that suppression of each IS also distorts the signal spectrum part covered by it, and the overall signal distortions are minimal when the distortions caused by ISs and by their suppression are balanced.

In DS SS demodulators, IS suppression optimal for code acquisition and tracking differs from that optimal for symbol demodulation. While $H_1(f)$ is optimal for symbol demodulation, its IS suppression excessively distorts the cross-correlation function, degrading code acquisition and tracking performance. Based on the results of [61, 62], the optimum transfer function $H_3(f)$ of a linear filter for code acquisition or tracking is

$$H_3(f) = \frac{c|S(f)|^2}{G_{ps}|S(f)|^2 + N(f)} \exp(-j2\pi f t_0) \qquad (4.17)$$

where G_{ps} is processing gain in the code acquisition or tracking mode. Thus, $H_3(f)$ suppresses ISs even less than $H_2(f)$ because G_{ps} reduces the signal distortion caused by ISs but cannot reduce the distortion caused by their suppression. This approach is also optimal for sounding and radar Rxs. Different filtering for synchronization and demodulation is possible only if these functions are either separated in time or performed in parallel by different channels. Otherwise, the filter transfer function is the result of a compromise.

Equations (4.15) through (4.17) correspond to the case when the PSDs of ISs and desired signals are accurately known. In practice, however, they vary (e.g., due to fading) and, therefore, require real-time channel estimation and adaptation. When fading is slow compared to symbol rate, channel estimation is more accurate than demodulation. The channel estimation gain G_e that can be easily implemented is equal to the ratio of channel coherence time to channel delay spread. In (4.15), $H_1(f)$ does not depend on a desired signal spectrum, but a filter with this transfer function must be placed before the correlators or matched filters whose transfer functions are determined by the signals' spectra influenced by communication channels. When the parameters are known, the IS suppression according to (4.15) is optimal. However, the necessity of estimating these parameters requires stronger suppression at the frequency intervals with low SNR to prevent their influence on the estimation [63]. When $G_e \geq 10$, the stronger suppression is unneeded [64].

IS-suppressing filters may have different realizations in Rxs' RDPs. A small number of strong narrowband ISs can be rejected by several adaptive notch IIR filters. Suppression of wideband ISs together with channel equalization can be performed by an adaptive FIR filter. Filter banks with adaptive gains represent the most versatile option. Any realization requires sufficient frequency resolution of the filters and sufficient dynamic range of the AMF.

4.3.2.2 Types of Nonlinear Products

The mathematical model of a nonlinear Rx signal path in Figure 4.6 is the simplest one (sometimes referred to as the Wiener-Hammerstein model), but the analytical results based on it have demonstrated good agreement with experimental data obtained in the HF band. In this model, the wideband linear filter can correspond to a preselector or an antialiasing filter depending on what part of the signal path has the dominating nonlinearity. The narrowband linear filter corresponds to a channel filter in the RDP. In fact, there can be more than one channel filter in a Rx, and their passbands can be located anywhere within the wideband filter passband.

To determine the impact of the memoryless nonlinear element on its output voltage $V_{s2}(t)$, let us expand $V_{s2}(t)$ into a power series with respect to the input voltage $V_{s1}(t)$:

$$V_{s2}(t) = \sum_{k=1}^{\infty} a_k V_{s1}^k(t) \tag{4.18}$$

where a_k are the power series coefficients. When the wideband linear filter passband is much wider than the average IS bandwidth, $V_{s1}(t)$ can be approximated by a sum of M sinusoids with different amplitudes V_m, frequencies f_m, and phases φ_m:

$$V_{s1}(t) = \sum_{m=1}^{M} V_m \sin(2\pi f_m t + \varphi_m) \tag{4.19}$$

This approximation can be considered a Fourier-series expansion of $V_{s1}(t)$ within a certain time interval. From (4.18) and (4.19),

$$V_{s2}(t) = \sum_{k=1}^{\infty} a_k \left[\sum_{m=1}^{M} V_m \sin(2\pi f_m t + \varphi_m) \right]^k \tag{4.20}$$

Thus, besides the term

$$a_1 \sum_{m=1}^{M} V_m \sin(2\pi f_m t + \varphi_m) \tag{4.21}$$

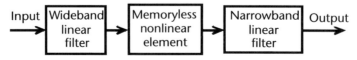

Figure 4.6 Simplified mathematical model of Rx signal path.

that is the desired linear output, (4.20) contains the terms

$$a_k \left[\sum_{m=1}^{M} V_m \sin(2\pi f_m t + \varphi_m) \right]^k \text{ with } k = 2, 3, \ldots \quad (4.22)$$

Terms (4.22) are unwanted nonlinear products. The higher the dynamic range, the smaller the coefficients a_k in (4.22). In an ideally linear Rx, $a_k = 0$ for $k \geq 2$. Trigonometric identities show that every term (4.22) produces IMPs with the frequencies that are combinations of those of the original sinusoids, and k determines the maximum order of IMPs. The higher k, the larger variety of IMPs it produces. Usually, coefficients a_k quickly diminish as k increases. Since only a small nonlinearity is tolerated in Rxs, low-order IMPs are used for determining two-signal dynamic range, and they are analyzed in detail below. However, higher-orders IMPs allow a more accurate evaluation of Rx performance in harsh RF environments, and fifth-order IMPs have been used for this purpose in [7].

There are two types of second-order IMPs with positive frequencies

$$f_{21} = f_i - f_j \text{ and } f_{22} = f_i + f_j \quad (4.23)$$

where f_i and f_j are frequencies of the original sinusoids, i and j are positive integers that meet conditions: $1 \leq i \leq M$, $1 \leq j \leq M$, and $i \neq j$. Besides IMPs, the second-power term in (4.22) produces second harmonics with frequencies $2f_i$ and dc components. Indeed, $\sin^2(2\pi f_i t) = 0.5\{1 - \cos[2(2\pi f_i t)]\}$. The harmonics and dc components can be considered special cases of second-order IMPs for $i = j$. The numbers and amplitudes of the second-order IMPs are presented in Table 4.2.

There are four types of third-order IMPs with positive frequencies

$$f_{31} = 2f_i - f_j, \; f_{32} = f_i + f_j - f_l, \; f_{33} = 2f_i + f_j, \text{ and } f_{34} = f_i + f_j + f_l \quad (4.24)$$

where f_i, f_j, and f_l are frequencies of the original sinusoids, i, j, and l are positive integers that meet conditions: $1 \leq i \leq M$, $1 \leq j \leq M$, $1 \leq l \leq M$, and $i \neq j \neq l$. The first two types are difference-frequency third-order IMPs, while the second two types are sum-frequency third-order IMPs. The third-order IMPs with frequency f_{31} can be considered a special case of the third-order IMPs with frequency f_{32} for $i = j$. Similarly, the third-order IMPs with frequency f_{33} can be considered a special case of the third-order IMPs with frequency f_{34} for $i = j$.

Besides the four types of third-order IMPs, the third-power term in (4.20) produces the third harmonics with frequencies $3f_i$ and nonlinear products with frequencies identical to those of the original sinusoids (i.e., original-frequency distortions

Table 4.2 Characteristics of Second-Order IMPs for $M \geq 2$

Type	$f_{21} = f_i - f_j$	$f_{22} = f_i + f_j$	dc Components	Second Harmonics
Number	$0.5M(M-1)$	$0.5M(M-1)$	M	M
Amplitude	$a_2 V_i V_j$	$a_2 V_i V_j$	$0.5 a_2 V_i^2$	$0.5 a_2 V_i^2$

(OFDs)). Indeed, $\sin^3(2\pi f_i t) = 0.25\{3\sin(2\pi f_i t) - \sin[3(2\pi f_i t)]\}$. The third harmonics and OFDs can be considered special cases of third-order IMPs for $i = j = l$. The numbers and amplitudes of the third-order IMPs of all type are presented in Table 4.3.

In general, even-order nonlinearities (k is even) generate difference-frequency and sum-frequency even-order IMPs whose orders do not exceed k, even harmonics (with frequencies $2f_i$, $4f_i$, ..., kf_i) of all original sinusoids, and dc components; whereas odd-order nonlinearities (k is odd) generate difference-frequency and sum-frequency odd-order IMPs whose orders do not exceed k, odd harmonics (with frequencies $3f_i$, $5f_i$, ..., kf_i) of all original sinusoids, and OFDs. As noted above, dc components, harmonics, and OFDs can be considered special cases of IMPs. Tables 4.2 and 4.3 show that the number of IMPs significantly exceeds the number of ISs even when the latter is moderate, confirming the notion of an IMP floor when the AMF dynamic range is insufficient.

4.3.2.3 IMPs Within Wideband Filter Passband

Since the passband of a narrowband filter shown in Figure 4.6 can be located anywhere within the wideband filter passband, it is important to determine what types of IMPs can fall within the wideband filter passband depending on the relation between its bandwidth B_w and center frequency f_{w0}. When $f_{w0}/B_w > 1.5$, sum-frequency IMPs and harmonics of any order as well as difference-frequency second-order IMPs cannot fall within the wideband filter passband. A further increase of f_{w0}/B_w prevents the appearance of difference-frequency IMPs of higher even orders within this passband. Specifically, difference-frequency fourth-order IMPs cannot fall within it when $f_{w0}/B_w > 2.5$, and difference-frequency sixth-order IMPs cannot fall within it when $f_{w0}/B_w > 3.5$. However, difference-frequency odd-order IMPs fall within this passband independently of f_{w0}/B_w. The condition $f_{w0} \gg B_w$ is sufficient for taking into account only difference-frequency odd-order IMPs. Differential structure of AMF stages significantly suppresses even-order IMPs. For such AMF stages, condition $f_{w0}/B_w > 3.5$ is usually sufficient for neglecting all even-order IMPs and taking into account only difference-frequency odd-order IMPs prior to sampling. When the differential stages are well balanced, even the condition $f_{w0}/B_w > 1.5$ could be sufficient for taking into account only difference-frequency odd-order IMPs prior to sampling.

Antialiasing filtering rejects the out-of-band IMPs generated before it. The out-of-band IMPs generated after it cannot be rejected. As a result of sampling, they can fall within the desired signal spectrum. Indeed, sampling maps the whole frequency axis $-\infty < f < \infty$ for an analog signal onto the interval $-0.5f_s \leq f < 0.5f_s$

Table 4.3 Characteristics of Third-Order IMPs for $M \geq 3$

Type	$f_{31} = 2f_i - f_j$	$f_{32} = f_i + f_j - f_l$	$f_{33} = 2f_i + f_j$	$f_{34} = f_i + f_j + f_l$	Third Harmonics	OFDs
Number	$M(M-1)$	$M(M-1)(M-2)/2$	$M(M-1)$	$M(M-1)(M-2)/6$	M	M
Amplitude	$0.75a_3 V_i^2 V_j$	$1.5a_3 V_i V_j V_l$	$0.75a_3 V_i^2 V_j$	$1.5a_3 V_i V_j V_l$	$0.25a_3 V_i^3$	$0.75a_3 V_i^3$

for the discrete-time one. This mapping translates an analog signal's spectral component with frequency f_i to frequency

$$f_{i1} = f_i - f_s \, \text{floor}\left(\frac{f_i}{f_s} + 0.5\right) \tag{4.25}$$

Mapping of the infinite frequency axis onto its relatively small region increases the IMPs density. Fortunately, only low-order IMPs should be taken into account. According to (4.25), the frequency of the kth harmonic of an analog sinusoid with frequency f_i after sampling becomes

$$f_{ki1} = kf_i - f_s \text{floor}\left(\frac{kf_i}{f_s} + 0.5\right) \tag{4.26}$$

In accordance with (4.26), the frequencies of second-order and third-order IMPs after sampling become

$$f_{211} = (f_i - f_j) - f_s \text{floor}\left(\frac{f_i - f_j}{f_s} + 0.5\right) \tag{4.27}$$

$$f_{221} = (f_i + f_j) - f_s \text{floor}\left(\frac{f_i + f_j}{f_s} + 0.5\right) \tag{4.28}$$

$$f_{311} = (2f_i - f_j) - f_s \text{floor}\left(\frac{2f_i - f_j}{f_s} + 0.5\right) \tag{4.29}$$

$$f_{321} = (f_i + f_j - f_l) - f_s \text{floor}\left(\frac{f_i + f_j - f_l}{f_s} + 0.5\right) \tag{4.30}$$

$$f_{331} = (2f_i + f_j) - f_s \text{floor}\left(\frac{2f_i + f_j}{f_s} + 0.5\right) \tag{4.31}$$

$$f_{341} = (f_i + f_j + f_l) - f_s \text{floor}\left(\frac{f_i + f_j + f_l}{f_s} + 0.5\right) \tag{4.32}$$

The equations for all other types of IMPs can be obtained similarly.

Thus, while the types and number of IMPs that can fall within the wideband filter passband in the stages preceding antialiasing filtering are determined by the ratio f_{w0}/B_w, they depend on the relations among f_{w0}, B_w, and f_s after antialiasing filtering. When f_s is optimal, that is, satisfies (3.16), and sampling is performed by THAs, both sum- and difference-frequency IMPs of odd orders fall within the RDP passband, as follows from (4.29) through (4.32). Sampling based on the sampling theorem's direct interpretation (see Chapter 5) allows avoiding or reducing sum-frequency odd-order IMPs within this passband. As to even-order IMPs, their number within it is reduced by the following additional conditions: $f_s/B_w > 6$ prevents second-order IMPs from emerging within the RDP passband, $f_s/B_w > 10$ prevents second-order and fourth-order IMPs from emerging within the RDP passband, and $f_s/B_w > 14$ prevents second-order, fourth-order, and sixth-order IMPs from emerging within it.

4.3.3 Required Dynamic Range of an HF Rx

Prior to [4], the recommendations for selecting Rx dynamic range in poorly regulated frequency bands could be reduced to the statement: the higher the dynamic range, the better. Although this statement is correct, a high cost of the "better" necessitates determining a sufficient dynamic range. It is intuitively clear that the higher the intensity of ISs and the wider the Rx AMF bandwidth, the higher dynamic range is needed. Since IMPs are the main factor degrading Rx performance in most practical situations, the goal was to derive closed-form equations for the minimum required two-tone dynamic range. This problem was solved for HF Rxs in [4], the levels of third-order and fifth-order IMPs for given preselector bandwidth and IS statistics in the case of insufficient dynamic range were determined in [6, 7, 10], and extending the approach used in [4] to other frequency bands was suggested in [33].

Below, the methodology used for determining the minimum required two-tone dynamic range is concisely discussed. To this end, the mathematical models of Rx signal paths and ISs as well as the dynamic range sufficiency criterion are selected first. The multistage structure of digital Rxs, large variety of their architectures, and diversity of ISs make determining the dynamic range a complex problem. Therefore, only simplification of the mathematical model and criterion allows deriving its closed-form solution. Due to the simplifications, the obtained equations cannot be precise. However, if properly used, they are sufficiently accurate for developing specifications of Rxs and estimating their performance in various RF environments. In addition, the availability of a closed-form approximation makes it easier to obtain more accurate results for specific Rx architectures and operational conditions using simulations.

A Rx signal path model in Figure 4.6 is the simplest one reflecting both its selectivity and nonlinearity. A variety of selectivity and nonlinearity distributions among Rx blocks make this model's accuracy dependent on correct identification of the stage with the dominant nonlinearity. In digital Rxs, the nonlinearity of digitization circuits usually dominates. In this case, the wideband filter represents the total selectivity of the preceding AMF stages (that are usually wideband in SDRs and CRs), and the narrowband filter represents the total selectivity of digital filtering in the RDP that is mostly determined by a channel filter (it may also be determined by the selectivity of the spectrum analyzer used for finding unutilized frequencies and/or frequencies of the strongest ISs). Since a desired signal is one of many signals within the AMF passband, it is reasonable to assume that the narrowband filter bandwidth is equal to the average IS bandwidth B_a. For simplicity, both wideband and narrowband filters are assumed to have rectangular AFRs and linear PFRs.

As mentioned in the previous section, the coefficients a_k of a power series approximating the nonlinear element diminish as k increases. Since the upper bound of the dynamic range corresponds to a relatively small nonlinearity, the nonlinear element can be approximated by a fairly short power series. When f_s is optimal, $f_s/B_w \geq 4$, $f_{w0}/B_w > 3.5$, and AMF stages have differential structures (these conditions are satisfied in most Rxs with bandpass digitization), only first-power and third-power terms are significant. If sum-frequency third-order IMPs are suppressed during sampling (for example, by a sampler based on the sampling theorem's direct interpretation), only IMPs with frequencies f_{311} and f_{321} (see (4.29) and (4.30)) can fall within the narrowband filter's passband. These IMPs were used for calculating the minimum required dynamic range of an HF Rx in [4].

When the wideband filter passband is much wider than the average IS bandwidth, that is, $B_w \gg B_a$, the PSD of ISs within this passband is a two-dimensional stochastic function $N(f, t)$ of frequency f and time t. During short time intervals (~200 ms in the HF band), $N(f, t)$ can be considered a time-invariant stationary stochastic function $N(f)$ with frequency correlation interval B_a. According to the central limit theorem (see Appendix C), the probability distribution of $N(f)$ should be close to log-Gaussian, because the value of $N(f)$ at any frequency is determined by many independent or weakly dependent multiplicative factors with comparable variances, such as differences among Txs' powers, distances from the Rx, ISs' bandwidths, propagation conditions, and types of Txs' antennas and their orientations. Results of experimental measurements of $N(f)$ in the HF band do not contradict this hypothesis.

The following assumptions further simplify the IS statistical model. Let us divide the wideband filter passband B_w into n elementary frequency intervals, each with the bandwidth equal to the average IS bandwidth B_a. The intervals can be considered potential positions of the narrowband filter's passband. Let us additionally assume that: ISs can be considered sinusoids because $B_a \ll B_w$, only one IS can appear in each interval with probability p_0, ISs in different intervals are mutually independent, and, as explained above, have log-Gaussian distribution of their powers and, consequently, voltages:

$$W(V_{\text{IS,dB}}) = \frac{\exp\left[-(V_{\text{IS,dB}} - \mu)^2 / 2\sigma^2\right]}{\sqrt{2\pi}\sigma} \quad (4.33)$$

where $W(V_{\text{IS,dB}})$ is the PDF of IS voltages expressed in decibels relative to 1 μV, while μ and σ are the PDF parameters also expressed in decibels. When ISs have log-Gaussian distribution, their IMPs also have log-Gaussian distribution with parameters that can be calculated based on the formulas in the third rows of Tables 4.2 and 4.3. The simplified model of ISs is illustrated by Figure 4.7.

It is clear that when the dynamic range is sufficient, the probability of interference appearing in an elementary interval is determined only by the IS statistics. Simultaneously, as noted above and shown in Figures 4.4 and 4.5, even a moderate number of ISs create an IMP floor within the AMF passband if the dynamic range is noticeably exceeded by the sum of a desired signal, ISs, and noise (usually, the ISs are the major terms in this sum). In this case, IMPs appear in all elementary intervals. At the upper bound, IMPs appear only in a few elementary intervals. Thus, a situation, in which the number of difference-frequency third-order IMPs with levels

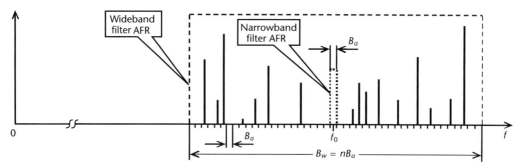

Figure 4.7 Simplified model of ISs.

exceeding the Rx MDS is a small fraction θ of the average number of ISs within the AMF passband, indicates crossing the dynamic range upper bound by ISs. The smaller θ, the closer a real Rx to the ideally linear one. Although the choice of θ seems subjective, its selection within the interval $0.05 \leq \theta \leq 0.25$ does not make a significant difference. This allows deriving a closed-form solution for the minimum required two-tone dynamic range D_3 or $D_{3,\text{dB}}$ (see Section 4.3.1) from the inequality:

$$p_{31}n_{31} + p_{32}n_{32} \leq \theta n_{\text{IS}} \tag{4.34}$$

where n_{IS}, n_{31}, and n_{32} are the average numbers of ISs and difference-frequency third-order IMPs with frequencies f_{31} and f_{32}, respectively, within the AMF bandwidth, while p_{31} and p_{32} are the probabilities that the IMPs with frequencies f_{31} and f_{32}, respectively, exceed the Rx MDS set at $V_{MDS} = 1\ \mu V$. As shown in [4],

$$n_{\text{IS}} = np_0,\ n_{31} = 0.5n(n-1)p_0^2,\ \text{and}\ n_{32} = \frac{n(n-1)(n-2)p_0^3}{3} \tag{4.35}$$

$$p_{31} = 0.5\left\{1 - \Phi\left[\frac{1.34(D_{3,\text{dB}} - \mu)}{\sigma}\right]\right\}\ \text{and}$$

$$p_{32} = 0.5\left\{1 - \Phi\left[\frac{1.73(D_{3,\text{dB}} - \mu - 2)}{\sigma}\right]\right\} \tag{4.36}$$

where

$$\Phi(x) = \frac{2}{\sqrt{2\pi}}\int_0^x \exp\left(\frac{-t^2}{2}\right)dt \tag{4.37}$$

The transformation of (4.34) with regard to (4.35) through (4.37) yields:

$$\frac{\left[4(p_{31} - 2p_0 p_{32})^2 - 21.3(0.667 p_0^2 p_{32} - 0.5 p_0 p_{31} - \theta)p_{32}\right]^{0.5} + 4p_0 p_{32} - 2p_{31}}{2.67 p_{32} p_0} \geq n \tag{4.38}$$

4.3 Dynamic Range of a Digital Rx

The minimum required $D_{3,\text{dB}}$ corresponds to (4.38) being an equality. The results of $D_{3,\text{dB}}$ calculations according to (4.35) through (4.38) for various n, p_0, μ, and σ are plotted in Figure 4.8 [5].

Even with all simplifications described above, the closed-form expression for $D_{3,\text{dB}}$ as a function of n, p_0, μ, and σ is too complex to show it here. Only for $n \geq 40$ and $p_0 \geq 0.1$, a sufficiently accurate closed-form approximation of the minimum required $D_{3,\text{dB}}$ was obtained in [5]. When $\theta = 0.2$, it is

$$D_{3,\text{dB}} = \{[0.72 - 0.185\log_{10}(10p_0)]\log_{10} n + 0.14 + \log_{10}(10p_0)\}\sigma \\ + \mu + 2 \quad (4.39)$$

Although (4.36) through (4.39) were derived for $V_{MDS} = 1\ \mu\text{V}$, they can be recalibrated for another V_{MDS}. Note that the required Rx dynamic range is determined mostly by IMPs only when $n \geq 5$. When $n < 5$, it is determined by strong ISs appearing within the AMF passband.

When second-order IMPs cannot fall within the AMF passband, $D_{3,\text{dB}}$ is sufficient. Otherwise, both $D_{3,\text{dB}}$ and $D_{2,\text{dB}}$ should be calculated. The approach described above also allows determining the minimum required $D_{2,\text{dB}}$, and expressions for $D_{2,\text{dB}}$ are simpler than those for $D_{3,\text{dB}}$. Identity [32]:

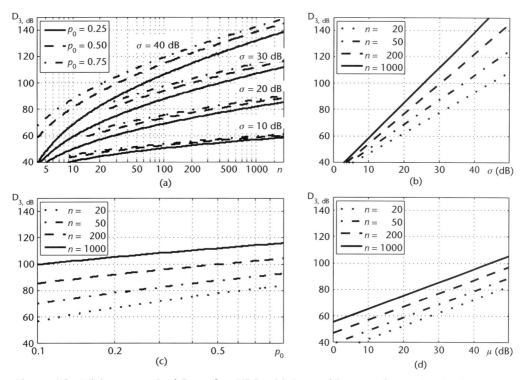

Figure 4.8 Minimum required $D_{3,\text{dB}}$ of an HF Rx: (a) $D_{3,\text{dB}} = f_1(n, p_0, \sigma)$ for $\mu = 30$ dB, (b) $D_{3,\text{dB}} = f_2(\sigma, n)$ for $\mu = 30$ dB and $p_0 = 0.5$, (c) $D_{3,\text{dB}} = f_3(p_0, n)$ for $\mu = 30$ dB and $\sigma = 30$ dB, and (d) $D_{3,\text{dB}} = f_4(\mu, n)$ for $\sigma = 20$ dB and $p_0 = 0.5$.

$$\text{IP}_{n,\text{dBm}} = \frac{nD_{n,\text{dB}}}{n-1} + \text{MDS}_{\text{dBm}} \qquad (4.40)$$

allows translating $D_{3,\text{dB}}$ and $D_{2,\text{dB}}$ into the corresponding IPs:

$$\text{IP}_{2,\text{dBm}} = 2D_{2,\text{dB}} + \text{MDS}_{\text{dBm}} \qquad (4.41)$$

$$\text{IP}_{3,\text{dBm}} = 1.5D_{3,\text{dB}} + \text{MDS}_{\text{dBm}} \qquad (4.42)$$

While approximation of the Rx nonlinearity by a third-order power series is sufficient for determining the minimum required $D_{3,\text{dB}}$ and $D_{2,\text{dB}}$, power series of higher orders are needed for estimating the performance of Rxs with given dynamic ranges. For example, the fifth-order power series is used for this purpose in [7].

The methodology described above can be employed for determining the minimum required dynamic ranges of Rxs operating in other overcrowded frequency bands with known IS statistics. However, even without accurate calculations, this section's material confirms the intuitive expectation that the minimum required Rx dynamic range is at least approximately proportional to the AMF bandwidth and the intensity of ISs.

4.4 Digitization in a Digital Rx

4.4.1 Baseband Digitization

Baseband digitization of Rx input signals is described below, and their bandpass digitization is considered in the next section. The baseband digitization, reflected by the block diagram in Figure 4.9, comprises: (1) translation of an input bandpass real-valued signal $u_{in}(t)$ to zero frequency simultaneously with forming the I and Q components $I_{in}(t)$ and $Q_{in}(t)$ of its analog baseband complex-valued equivalent $Z_{in}(t)$; (2) separate sampling of $I_{in}(t)$ and $Q_{in}(t)$ that transforms $Z_{in}(t)$ into discrete-time $Z_1(nT_{s1})$ represented by $I_1(nT_{s1})$ and $Q_1(nT_{s1})$; (3) separate quantization of $I_1(nT_{s1})$ and $Q_1(nT_{s1})$ that produces the digital baseband complex-valued equivalent $Z_{q1}(nT_{s1})$ of $u_{in}(t)$ represented by $I_{q1}(nT_{s1})$ and $Q_{q1}(nT_{s1})$; and (4) downsampling of $I_{q1}(nT_{s1})$ and $Q_{q1}(nT_{s1})$ with their decimating filtering (see Appendix B) that produces $Z_{q2}(mT_{s2})$ represented by $I_{q2}(mT_{s2})$ and $Q_{q2}(mT_{s2})$. While the first three procedures are performed in the AMF, the fourth one is executed in the RDP. The downsampling is needed because the A/Ds' sampling rate $f_{s1} = 1/T_{s1}$ is usually selected excessively high due to wide transition bands of the analog antialiasing LPFs, but the RDP signal processing should be performed at the minimum acceptable sampling rate to efficiently utilize the RDP hardware.

When an even AFR and a linear PFR of the digital decimating filter in the RDP are desirable, FIR filters with real-valued coefficients are usually used for downsampling, reducing the complex-valued decimating filter to two identical real-valued FIR LPFs. If the downsampling factor $L = f_1/f_2$ is equal to 2 or to a power of 2, the HBFs or cascade structures of HBFs, respectively, are typically used for decimating filtering (see Section B.4). When the digital decimating filter is also employed

4.4 Digitization in a Digital Rx

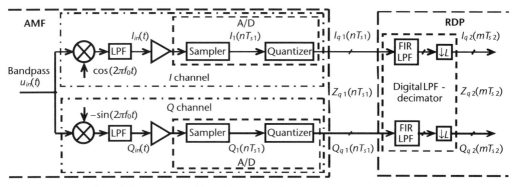

Figure 4.9 Baseband digitization of bandpass signals.

for compensating linear distortion of signals in the AMF, its coefficients can be complex-valued, and it consists of four real-valued LPFs.

Figure 4.10 shows the signal spectrum transformations during the baseband digitization and the required AFRs of the antialiasing and decimating filters. The amplitude spectrum $|S_{in}(f)|$ of $u_{in}(t)$ is depicted in Figure 4.10(a). It contains the amplitude spectra of three ISs, besides the spectrum $|S(f)|$ of a desired signal $u(t)$ centered at f_0. The amplitude spectrum $|S_{Zin}(f)|$ of $Z_{in}(t)$ in Figure 4.10(b) comprises the amplitude spectra of the analog baseband complex-valued equivalent $Z(t)$ of

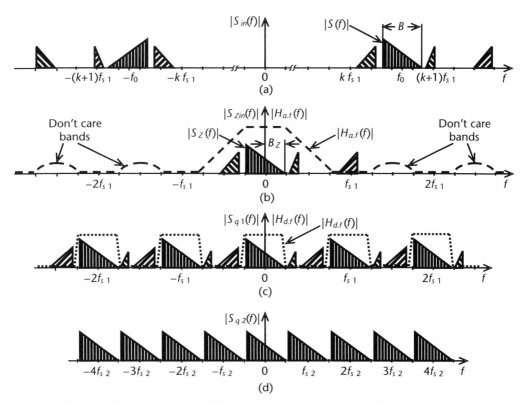

Figure 4.10 Amplitude spectra and AFRs for baseband digitization of bandpass signals: (a) $|S_{in}(f)|$, (b) $|S_{Zin}(f)|$ and $|H_{a.f}(f)|$ (dashed line), (c) $|S_{q1}(f)|$ and $|H_{d.f}(f)|$ (dotted line), and (d) $|S_{q2}(f)|$.

$u(t)$ and those of the ISs. While the amplitude spectrum $|S_Z(f)|$ of $Z(t)$ is located within the passband of the antialiasing filter with AFR $|H_{a.f}(f)|$, the spectra of two ISs are located within the filter transition bands and the spectrum of the third IS within the filter stopband.

As noted in Section 3.3.1, the sampling of baseband signals (like $Z_{in}(t)$) causes proliferation of their spectra ($S_{Zin}(f)$ in the considered case), and the spectrum $S_{d1}(f)$ of the discrete-time signal $Z_1(nT_{s1})$ at the sampler output is a periodic function of frequency with period f_{s1} according to (3.13). Hence, $|H_{a.f}(f)|$ must suppress noise and ISs within the frequency intervals in which the $S_{Zin}(f)$ replicas appear after sampling:

$$\left[kf_{s1} - B_Z, kf_{s1} + B_Z \right] \quad (4.43)$$

where $B_Z = 0.5B$ (see Figure 4.10) and k is any nonzero integer. The gaps between intervals (4.43) are "don't care" bands (see Section 3.3) where the suppression is, in principle, unnecessary since noise and ISs there can be rejected in the RDP. However, some weakening of ISs within these bands by antialiasing filtering may lower the required resolution of the quantizer and subsequent DSP. Traditional analog filters do not utilize the don't care bands, but these bands allow increasing the efficiency of antialiasing and interpolating filtering based on the sampling theorem's direct and hybrid interpretations (see Chapters 5 and 6).

Figure 4.10(c) shows the amplitude spectrum $|S_{q1}(f)|$ of $Z_{q1}(nT_{s1})$ and the AFR $|H_{d.f}(f)|$ of the digital decimating FIR filter. In this spectral diagram, the IS located within the antialiasing filter stopband is already rejected and those located within its transition bands are weakened. When quantization in the A/Ds is accurate, $|S_{q1}(f)|$ is virtually identical to the amplitude spectrum $|S_{d1}(f)|$ of the discrete-time complex-valued equivalent $Z_1(nT_{s1})$. The digital decimating filter suppresses noise and ISs within the frequency intervals where the $S_{Zin}(f)$ replicas appear after the downsampling. The amplitude spectrum $|S_{q2}(f)|$ of $Z_{q2}(nT_{s2})$ is depicted in Figure 4.10(d). It is easy to notice that $L = 2$, and $|H_{d.f}(f)|$ corresponds to the AFR of an HBF in Figure 4.10. As shown in Section B.4, HBFs significantly reduce the computational load of decimating filtering.

4.4.2 Bandpass Digitization

In digital Rxs, bandpass digitization of bandpass signals can be performed at the RF or IF. As shown in Figure 4.11, it comprises: (1) antialiasing filtering of an input bandpass real-valued signal $u_{in}(t)$ by a BPF, (2) sampling that produces discrete-time $u(nT_{s1})$, (3) quantization of $u(nT_{s1})$ that generates digital bandpass real-valued $u_q(nT_{s1})$, (4) translation of $u_q(nT_{s1})$ to zero frequency simultaneously with forming the I and Q components $I_{q1}(nT_{s1})$ and $Q_{q1}(nT_{s1})$ of the digital baseband complex-valued equivalent $Z_{q1}(nT_{s1})$ of $u_q(nT_{s1})$, and (5) downsampling of $I_{q1}(nT_{s1})$ and $Q_{q1}(nT_{s1})$ with their decimating filtering (see Appendix B) that produces $Z_{q2}(mT_{s2})$ represented by $I_{q2}(mT_{s2})$ and $Q_{q2}(mT_{s2})$. Stages (4) and (5) are actually combined because the decimating LPF is involved in both forming $Z_{q1}(nT_{s1})$ and its downsampling. Bandpass digitization requires more digital processing than baseband one. The major challenge of the AMF processing is sampling, which is

4.4 Digitization in a Digital Rx 143

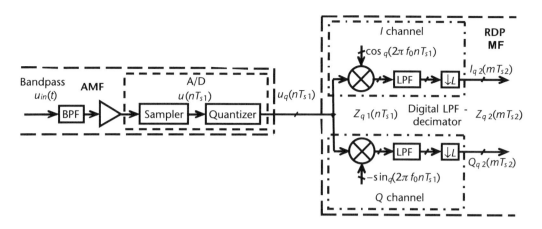

Figure 4.11 Bandpass digitization of bandpass signals.

more complex for bandpass signals than for baseband signals. In Figure 4.11, the complex-valued decimating LPF is reduced to two identical real-valued FIR LPFs due to its even AFR and linear PFR.

Figure 4.12 shows the signal spectrum transformations during the bandpass digitization and the required AFRs of the antialiasing and decimating filters. The amplitude spectrum $|S_{in}(f)|$ of $u_{in}(t)$ and the AFR $|H_{a.f}(f)|$ of the antialiasing BPF are depicted in Figure 4.12(a). Besides the spectrum $|S(f)|$ of a desired signal $u(t)$ centered at f_0, $|S_{in}(f)|$ includes the amplitude spectra of two ISs located within the

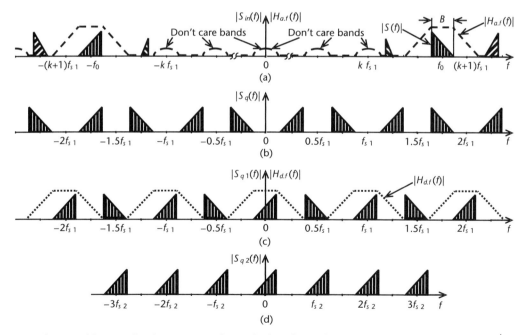

Figure 4.12 Amplitude spectra and AFRs for bandpass digitization of bandpass signals: (a) $|S_{in}(f)|$ and $|H_{a.f}(f)|$ (dashed line), (b) $|S_q(f)|$, (c) $|S_{q1}(f)|$ and $|H_{d.f}(f)|$ (dotted line), and (d) $|S_{q2}(f)|$.

antialiasing BPF stopbands and, therefore, rejected by this filter. The antialiasing BPF purpose is suppressing noise and ISs within the frequency intervals where the $|S(f)|$ replicas appear after sampling. The amplitude spectrum $|S_q(f)|$ of $u_q(nT_{s1})$ is shown in Figure 4.12(b). It is virtually identical to the amplitude spectrum $|S_d(f)|$ of discrete-time $u(nT_{s1})$ if quantization in the A/D is accurate. Figure 4.12(c) shows the amplitude spectrum $|S_{q1}(f)|$ of $Z_{q1}(nT_{s1})$ and the AFR $|H_{d.f}(f)|$ of the digital decimating LPF. The amplitude spectrum $|S_{q2}(f)|$ of $Z_{q2}(mT_{s2})$ is depicted in Figure 4.12(d). As follows from the last spectral diagram, $Z_{q2}(mT_{s2})$ can be further decimated.

The spectral diagrams in Figure 4.12 correspond to an optimal f_s that meets (3.16). In this case, the values of $\cos_q(2\pi f_0 nT_{s1})$ and $\sin_q(2\pi f_0 nT_{s1})$ can be equal only to +1, 0, −1 if they are sampled in-phase, as shown in Figure 3.12. For the block diagram in Figure 4.11, this means that the multiplication of $u_q(nT_{s1})$ by $\cos_q(2\pi f_0 nT_{s1})$ and by $-\sin_q(2\pi f_0 nT_{s1})$ can be reduced to zeroing the odd samples of $u_q(nT_{s1})$ and alternating the signs of its even samples in the I channel, and zeroing the even samples of $u_q(nT_{s1})$ and alternating the signs of its odd samples in the Q channel. These operations are performed by the sign alternator (SA) and demultiplexer (Dmx) in Figure 4.13 that illustrates the transformation of the block diagram shown in Figure 4.11 for an optimal f_s and $L = 2$. The SA periodically (with period $4T_{s1} = 2T_{s2}$) changes the signs of sequential pairs of digital samples (comprising an odd sample and an even sample), whereas the Dmx sends even samples to the I channel and odd samples to the Q channel. Thus, the samples enter each channel at the rate $f_{s2} = 0.5 f_{s1}$.

Still, the samples in the Q channel are shifted by $T_{s1} = 0.5T_{s2}$ relative to those in the I channel. Representing the I and Q components by the samples corresponding to the same time instants simplifies RDP processing. This is achieved at the LPFs' outputs in the I and Q channels. When these LPFs are HBFs (see Section B.4), the HBF in the I channel turns into the digital delay line corresponding to its center tap because all the samples at its other taps with nonzero coefficients are sent to the Q channel, and the HBF in the Q channel turns into the interpolating FIR LPF that contains all the nonzero coefficients of the original HBF except the center one.

In the spectral diagrams of this section, the passbands of antialiasing filters were equal to the bandwidths of the desired signals, and there were no ISs within the desired signals' bandwidths. Practical scenarios often differ from these situations. In Rxs with frequency multiplexing, for instance, the passbands of antialiasing filters can be significantly wider than the bandwidth of any channel signal. Many

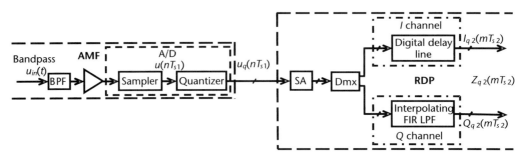

Figure 4.13 Bandpass digitization of bandpass signals with optimal f_s.

ISs can be located within the desired signal spectrum, especially in the case of SS signals (see Figure 4.5). This means that a high dynamic range is needed not only before antialiasing filters but also after them.

4.4.3 Comparison of Digitization Techniques and Architectures of AMFs

Since digitization of bandpass signals at the RDP input can be performed using baseband or bandpass techniques, they must be compared. The block diagrams in Figures 4.9, 4.11, and 4.13 show that baseband digitization requires separate I and Q channels in the analog and discrete-time domains, whereas bandpass digitization requires a single channel in these domains. Specifically, baseband digitization requires two A/Ds, whereas bandpass digitization requires only one. However, the sampling rate of the A/D used for bandpass digitization is two times higher than that of the A/Ds used for baseband digitization if the transition bands of the antialiasing filters are identical. Indeed, one-sided bandwidths of bandpass signals are two times wider than those of their complex-valued equivalents (see Figures 4.10 and 4.12). In addition, it is difficult to achieve accurate sampling at a high f_0 using conventional THAs, and the best antialiasing BPFs (e.g., crystal, electromechanical, ceramic, surface acoustic wave (SAW), and bulk acoustic wave (BAW) filters) are inflexible and incompatible with the IC technology. Sampling based on the hybrid or direct interpretation of the sampling theorem (see Chapter 5) allows reducing or eliminating these drawbacks of bandpass digitization.

The drawbacks of baseband digitization have more fundamental causes. First, the baseband position of the analog signal spectrum (compare the position of $|S_Z(f)|$ in Figure 4.10(b) to that of $|S(f)|$ in Figure 4.12(a)) makes dc offset, flicker noise, and much larger power and number of IMPs inevitable. Second, separate I and Q channels in the analog and discrete-time domains (see Figure 4.9) make IQ imbalance unavoidable. Adaptive compensation, which reduces the dc offset and IQ imbalance, increases the complexity and cost of Rxs.

Thus, bandpass D&R (see Section 3.3.3) are more promising for digital radios. They are also consistent with the major trend in the digital radio development: increasing the number of functions performed in the digital domain and reducing this number for the analog domain (compare Figures 3.6 and 4.9 with Figures 3.9 and 4.11, respectively).

As mentioned in Section 4.2.3, AMFs in digital Rxs necessarily perform: antenna coupling, preliminary amplification and/or attenuation, initial filtering (including antialiasing filtering), sampling, and quantization. They usually execute AGC-related functions and can perform other operations (e.g., frequency conversion) depending on their architectures. These architectures are contingent on the Rx digitization techniques but are not completely determined by them. The AMF's main purpose is to create optimum conditions for the Rx input signal digitization. Simplified block diagrams of the most widely used AMF architectures are shown in Figure 4.14.

In the direct conversion (homodyne) AMF architecture (see Figure 4.14(a)), the RF strip performs preliminary filtering and amplification of the Rx input signals that are then converted to the baseband simultaneously with forming their I and Q components. The LO, which generates $\cos(2\pi f_r t)$ and $-\sin(2\pi f_r t)$ signals at the RF f_r, is tunable within the Rx frequency range. The LPFs provide antialiasing filtering

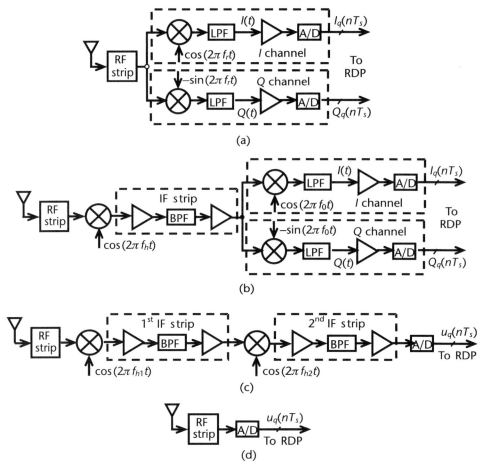

Figure 4.14 AMF architectures: (a) homodyne, (b) superheterodyne with baseband digitization, (c) superheterodyne with bandpass digitization, and (d) direct RF digitization.

of the signals' I and Q components, while the A/Ds perform their sampling and quantization. Channel filtering is carried out in the RDP.

The major advantages of this architecture (compared to others in Figure 4.14) are: simplicity of IC implementation, low requirements for A/Ds, and adjustability of the LPFs' bandwidths. Therefore, Rxs with such AMF architectures have higher flexibility and smaller size, weight, and cost. Simultaneously, homodyne architecture not only preserves all the drawbacks of baseband digitization (e.g., IQ imbalance, dc offset, flicker noise, and the largest number and power of IMPs) but aggravates many of them. For example, minimizing the IQ imbalance within the Rx frequency range is harder than at a single frequency. An additional problem is the LO leakage through the antenna that creates interference to other Rxs and contributes to the AMF dc offset.

In the superheterodyne AMF architecture with baseband digitization in Figure 4.14(b), input signals are converted to the IF f_0 after image rejection and preliminary amplification in the RF strip. Antialiasing filtering performed at f_0 enables the use of high-selectivity BPFs. Then the signal is converted to baseband and its I and Q

components are formed. This architecture solves the LO leakage problem because the LO frequency differs from the RF. It provides better antialiasing filtering due to the use of high-selectivity BPFs, such as crystal, electromechanical, ceramic, SAW, and BAW filters. However, the improved antialiasing filtering is achieved at the expense of lower flexibility and reduced scale of integration because these BPFs cannot change their parameters and are incompatible with the IC technology. This architecture also slightly reduces the impact of flicker noise and dc offset due to a lower gain at zero frequency. It also decreases *IQ* imbalance because the *I* and *Q* components are formed from signals at a constant f_0. The power and number of IMPs are still high.

An example of the superheterodyne AMF architecture with bandpass digitization is shown in Figure 4.14(c). In this example, the AMF has two frequency conversions. The first IF is selected high enough to simplify image rejection and reduce the number of spurious responses, whereas the second (lower) IF is chosen to simplify digitization. Double frequency conversion also divides the AMF gain between the first and second IF strips. Translation to baseband and forming complex-valued equivalents of received signals are executed in the RDP. As mentioned in Section 4.2.1, this architecture was used in most digital HF Rxs starting from Rockwell-Collins HF-2050, and the first IF was selected above the HF band. While crystal filters and helical resonators (in Marconi H2550) were initially used at the first IF, SAW resonator filters later emerged as a better option in some designs. The superheterodyne AMF architecture with bandpass digitization excludes *IQ* imbalance, dc offset, and flicker noise. It minimizes the power and number of IMPs within the signal spectrum.

Initially, the main problem of this architecture was the digitization at a relatively high last IF, and the low-IF architecture was proposed to reduce requirements for the A/Ds. However, image rejection at a low IF is difficult. Although several solutions of the latter problem (e.g., the use of dual-quadrature converters) were suggested, the progress of the A/D technology made this architecture less attractive. Implementation of sampling based on the sampling theorem's hybrid or direct interpretation (see Chapter 5) eliminates the need for the low-IF architecture. Currently, the major drawbacks of the superheterodyne AMF architecture with bandpass digitization are its low flexibility and incompatibility with the IC technology. Again, sampling based on the sampling theorem's hybrid or direct interpretation allows eliminating or reducing these drawbacks.

The apparent advantage of the direct RF digitization architecture in Figure 4.14(d) is that almost all signal processing is executed in the RDP, providing the highest accuracy and flexibility. However, this architecture greatly increases the requirements for the analog antialiasing BPF, A/D, and DPD, compared to the architecture in Figure 4.14(c). Indeed, it is currently difficult to design compact tunable antialiasing BPFs with adequate stopband suppression and sufficiently narrow transition bands. A wider BPF passband requires higher Rx dynamic range (and, consequently, its A/D dynamic range), as shown in the previous section. The requirements for the DPD are increased due to the frequency tuning in the digital domain. Although sampling based on the sampling theorem's hybrid or direct interpretation helps solving many problems of this architecture, it will not completely replace the superheterodyne architecture with bandpass digitization. The latter architecture will probably be the best one for the Rxs operating in the highest frequency bands or in multiple frequency bands.

4.5 Demodulation of Energy-Efficient Signals

4.5.1 Demodulation of Differential Binary Phase-Shift Keying Signals with DS Spreading

As explained in Section 3.4, reduction of the signals' crest factors in TDPs simplifies reconstruction of these signals and improves power utilization in digital Txs. Several such modulation techniques were described and analyzed there. Demodulation of the signals described in Section 3.4.1 is presented and examined below, and demodulation of AQ-DBPSK signals described in Section 3.4.2 is discussed in the next section. This material also illustrates signal processing in RDPs, outlined in Section 4.2.3.

As mentioned in Section 3.4.1, BPSK requires minimum E_b for a given P_b among binary modulations in AWGN channels. Its major problem, the initial phase ambiguity, was solved by the invention of DBPSK and several alternative techniques. DBPSK is widely used due to its simplicity and insignificance of the energy loss caused by the replacement of coherently demodulated BPSK with noncoherently demodulated DBPSK. Among quaternary modulation techniques, QPSK requires minimum E_b for a given P_b. DQPSK is less common because of its relative complexity and higher energy loss caused by the replacement of coherently demodulated QPSK with noncoherently demodulated DQPSK. Therefore, QPSK or dual independent BPSK in I and Q channels are used instead of DQPSK when coherent demodulation is acceptable.

Signals of both QPSK and dual BPSK poorly utilize Tx power because they have high crest factors after suppressing their spectral sidelobes. A half-symbol offset between the signals' I and Q components reduces the crest factors, simplifying the signals' reconstruction and improving the Tx power utilization. When modulation is followed by DS spreading, the Tx power utilization is determined not by the modulation but by the spreading, and a half-chip offset between the signals in I and Q channels should be used instead of a half-symbol offset. Output signals of the dual independent DBPSK modulator with DS spreading shown in Figure 3.15 allow both coherent and noncoherent demodulation when PN sequences in the Tx I and Q channels are orthogonal or quasi-orthogonal.

At the demodulator input, the digital baseband complex-valued equivalent $Z_{rq}(nT_s)$ of a received signal with AWGN is represented by its I and Q components:

$$Z_{rq}(nT_s) = V_{Irq}(nT_s) + jV_{Qrq}(nT_s) \tag{4.44}$$

where $V_{Irq}(nT_s)$ and $V_{Qrq}(nT_s)$ are multibit numbers, whose subscripts r and q stand, respectively, for received and quantized. DBPSK transmits data by the phase difference between the kth and $(k-1)$th symbols, which can be equal only to 0° or 180°, and its "zero" and "one" are, respectively:

4.5 Demodulation of Energy-Efficient Signals

$$s_1(t) = \begin{cases} s_0 \sin(2\pi f_0 t + \psi_0) & 0 < t \leq T_b \\ s_0 \sin(2\pi f_0 t + \psi_0) & T_b < t \leq 2T_b \end{cases}$$

and (4.45)

$$s_2(t) = \begin{cases} s_0 \sin(2\pi f_0 t + \psi_0) & 0 < t \leq T_b \\ s_0 \sin(2\pi f_0 t + \psi_0 \pm \pi) & T_b < t \leq 2T_b \end{cases}$$

where ψ_0 is the initial phase.

The block diagram of a coherent demodulator of dual independent DBPSK with DS spreading is shown in Figure 4.15. It comprises two identical channel demodulators, one of which processes $V_{Irq}(nT_s)$ and another $V_{Qrq}(nT_s)$. The only blocks shared by these demodulators are the generator of PN sequences and synchronization block (the latter is not depicted in Figure 4.15). The root raised cosine filters in I and Q channels mitigate ISI jointly with those of the modulator in Figure 3.15. The sample selectors pick samples in the middle of every chip. The multipliers and subsequent integrators despread $V_{Irq}(nT_s)$ and $V_{Qrq}(nT_s)$ using the corresponding PN sequences. Each differential decoder, comprising a multibit digital memory with the T_b delay and a digital multiplier, operates according to the rule:

$$\alpha_1 = \beta_1, \text{ and } \alpha_k = \beta_k \times \beta_{k-1} \text{ for } k \geq 2 \tag{4.46}$$

One should be written to the memory of each decoder upon initialization. Note that n is the sample number and k is the symbol number in the equations and figures of this section. The decision stage of each channel demodulator uses the sign and magnitude of α_k for soft decision and only its sign for hard decision. If a half-chip offset between the signals in I and Q channels is introduced in the modulator, it should be compensated in the demodulator (this compensation is not reflected in Figure 4.15).

As noted in Section 3.4.1, the use of mutually orthogonal PN sequences in the modulator's I and Q channels allows noncoherent demodulation of dual independent DBPSK. Moreover, a half-chip offset between the signals in these channels insignificantly reduces their orthogonality when the spreading factor $G_{ps} \geq 16$. Indeed, even after the offset, the PN sequences' orthogonality is preserved during

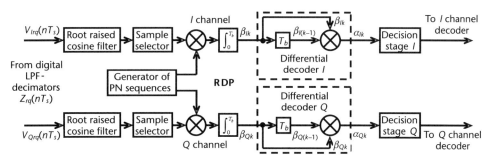

Figure 4.15 Coherent demodulator of dual independent DBPSK with DS spreading.

every half-chip, while the cross-correlation between these sequences is low during another half-chip. Therefore, the signals are quasi-orthogonal when the number of chips per symbol is sufficiently large.

The block diagram of a noncoherent demodulator of dual independent DBPSK with DS spreading is shown in Figure 4.16. Similar to the coherent demodulator in Figure 4.15, it comprises two identical demodulators separately processing signals of the Tx I and Q channels. The input signals $V_{Irq}(nT_s)$ and $V_{Qrq}(nT_s)$ enter both demodulators. Each demodulator selects its Tx channel signals during despreading because the mutually orthogonal or quasi-orthogonal local PN sequences (copies of those used for the Tx I and Q channels) are sent to the different demodulators. The sum of the differential decoders' output signals in each demodulator is processed the same way as the differential decoders' output signals of the demodulator in Figure 4.15.

In general, a half-symbol offset for modulation without spreading and a half-chip offset for modulation with DS spreading between the signals' I and Q components is an effective crest factor reduction method for many energy-efficient signals. The dual independent DBPSK with DS spreading is selected to explain it and illustrate the signal processing operations in TDPs and RDPs due to its relative simplicity.

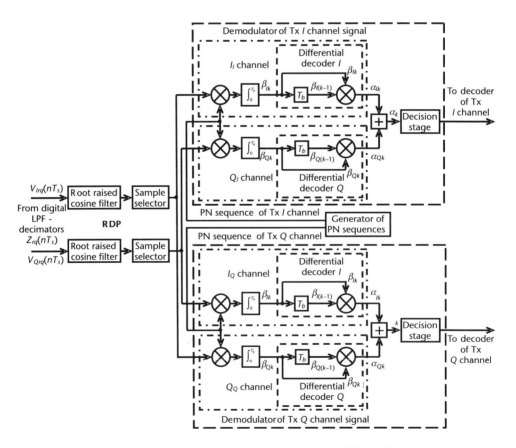

Figure 4.16 Noncoherent demodulator of dual independent DBPSK with DS spreading.

4.5.2 Demodulation of AQ-DBPSK Signals

As explained in Section 3.4.2, AQ-DBPSK signals have lower crest factors than BPSK, DBPSK, and QPSK signals because the phase transitions between their adjacent symbols are only ±90°. The reduced crest factors improve power utilization and simplify signal reconstruction in Txs [65–68]. AQ-DBPSK and DBPSK require the same E_b for a given P_b because they both transmit data by phase differences equal to 0° or 180°. Better utilization of the Tx power makes AQ-DBPSK more energy-efficient than DBPSK. It is shown in this section that AQ-DBPSK allows noncoherent demodulation with minimum energy loss compared to coherent one (in contrast with OQPSK, MSK, GMSK, and other energy-efficient signals with small crest factors), frequency-invariant demodulation (important for channels with large frequency offsets between Rxs and corresponding Txs), and much faster signal acquisition than alternative techniques. Below, AQ-DBPSK demodulation is first discussed for AWGN channels and then for channels with frequency offsets between the Rxs and Txs.

The AQ-DBPSK binary 0 and 1 are, respectively,

$$s_1(t) = \begin{cases} s_0 \sin(2\pi f_0 t + \psi_0) & 0 < t \le T_b \\ s_0 \sin(2\pi f_0 t + \psi_0) & 2T_b < t \le 3T_b \end{cases}$$

and (4.47)

$$s_2(t) = \begin{cases} s_0 \sin(2\pi f_0 t + \psi_0) & 0 < t \le T_b \\ s_0 \sin(2\pi f_0 t + \psi_0 \pm \pi) & 2T_b < t \le 3T_b \end{cases}$$

where ψ_0 is the initial phase. Since the intervals between data-carrying symbols in DBPSK and AQ-DBPSK are, respectively, T_b and $2T_b$, the noncoherent AQ-DBPSK demodulator optimum in AWGN channels in Figure 4.17 differs from that of DBPSK only by a longer delay in the differential decoders. Here, the digital baseband complex-valued equivalent $Z_{rq}(nT_s)$, represented by its I and Q components $V_{Irq}(nT_s)$ and $V_{Qrq}(nT_s)$, is the sum of a received AQ-DBPSK signal and noise. It is assumed that the number of samples per symbol is a constant integer.

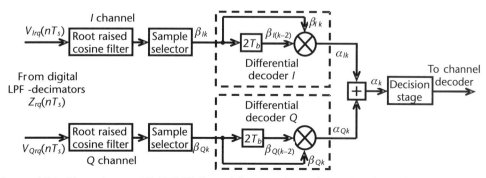

Figure 4.17 Noncoherent AQ-DBPSK demodulator optimum in AWGN channels.

The root raised cosine filters mitigate ISI jointly with the root raised cosine filters of the AQ-DBPSK modulator (see Figures 3.16(a) and 3.17(a)) and perform matched filtering of received signals. The sample selectors pick samples in the middle of every symbol and send them to the differential decoders of same-parity symbols. Each decoder comprises a multibit digital memory with the $2T_b$ delay and a digital multiplier. The decoders transform sequences $\{\beta_{Ik}\}$ and $\{\beta_{Qk}\}$ into sequences $\{\alpha_{Ik}\}$ and $\{\alpha_{Ik}\}$, respectively, according to the rule:

$$\alpha_1 = \beta_1, \; \alpha_2 = \beta_2, \text{ and } \alpha_k = \beta_k \times \beta_{k-2} \text{ for } k \geq 3 \tag{4.48}$$

Two ones should be written to the memory of each decoder upon initialization. The multipliers' outputs are summed:

$$\alpha_k = \alpha_{Ik} + \alpha_{Qk} \tag{4.49}$$

The demodulator decision stage uses the sign and magnitude of α_k for soft decision and only its sign for hard decision. Since both AQ-DBPSK and DBPSK convey information by the phase differences equal to 0° or 180°, the Euclidean distances between their signals are the same for a given energy per symbol. Their demodulators are essentially identical: the different delays in the differential decoders only compensate the different delays in the differential encoders (compare the differential encoders in Figure 3.15 to those in Figures 3.16 and 3.17). Yet AQ-DBPSK provides higher noise immunity than DBPSK due to better utilization of the Tx power.

The optimum noncoherent AQ-DBPSK demodulators have lower tolerance to frequency offsets between Rxs and corresponding Txs than the optimum noncoherent DBPSK demodulators due to the two times longer delay between data-carrying symbols (although this tolerance is still higher than that of DQPSK, for example). However, this is a drawback of the described demodulation technique, whereas AQ-DBPSK allows frequency-invariant demodulation with the fastest signal acquisition and reliable reception in channels with large frequency offsets between Rxs and corresponding Txs. It employs the approaches similar to those in the second-order DBPSK proposed by Y. Okunev in the early 1960s [69] and later investigated in many publications (see references in [69, 70]). The frequency-invariant demodulators of both second-order DBPSK and AQ-DBPSK compensate frequency offsets between Rxs and Txs equally effectively, but AQ-DBPSK signals better utilize Tx power. This compensation involves every three consecutive symbols. While the information is carried only by the phase differences between the first and third symbols, the phase differences between the first and second as well as between the second and third symbols are used for the frequency offset compensation and elimination of the second symbol's influence on the demodulation result.

The phase shifts $\zeta = \pm 90°$ between all adjacent AQ-DBPSK symbols complicate their frequency-invariant demodulation compared to that of the second-order DBPSK. The following four assumptions simplify the explanation of AQ-DBPSK frequency-invariant demodulation without loss of generality: (1) the initial phase of the demodulator input signal at the center frequency f_0 is zero, (2) all operations with phases are modulo 360°, (3) $f_0 T_b = m$ where m is an integer, and

(4) the frequency offset Δf between the modulator and demodulator, caused by the instability of LOs and/or Doppler shift, is constant during at least three consecutive symbols. The first assumption is logical in demodulators insensitive to the signal initial phase, the second one is fair because $F(\theta \pm 2\pi) = F(\theta)$ for any trigonometric function $F(\theta)$, and the last two assumptions correspond to most practical situations. The first assumption allows writing the phases of three consecutive AQ-DBPSK symbols as follows:

$$\begin{cases} \theta_k = 2\pi f_0 t + \varphi_k, \\ \theta_{k+1} = 2\pi f_0 (t + T_b) + \varphi_{k+1} + \zeta, \\ \theta_{k+2} = 2\pi f_0 (t + 2T_b) + \varphi_{k+2} \end{cases} \quad (4.50)$$

where k is the symbol number. Values of φ_k, φ_{k+1}, and φ_{k+2}, determined by the transmitted data, can only be equal to 0° or 180°. The sign of ζ is unimportant because the impact of θ_{k+1} on the demodulation result is eliminated:

$$\theta_{k+2} - \theta_k = 4\pi f_0 T_b + \varphi_{k+2} - \varphi_k \quad (4.51)$$

According to the second and third assumptions, (4.51) can be rewritten as:

$$\theta_{k+2} - \theta_k = \varphi_{k+2} - \varphi_k \quad (4.52)$$

In the presence of frequency offset Δf, the phases of the same three consecutive bandpass symbols are:

$$\begin{cases} \theta_k = 2\pi (f_0 + \Delta f) t + \varphi_k, \\ \theta_{k+1} = 2\pi (f_0 + \Delta f)(t + T_b) + \varphi_{k+1} + \zeta, \\ \theta_{k+2} = 2\pi (f_0 + \Delta f)(t + 2T_b) + \varphi_{k+2}. \end{cases} \quad (4.53)$$

The phase difference between the $(k + 2)$th and kth symbols in (4.53) is

$$\theta_{k+2} - \theta_k = 4\pi \Delta f T_b + \varphi_{k+2} - \varphi_k \quad (4.54)$$

A comparison of (4.54) and (4.52) demonstrates that Δf can distort the demodulation results when odd and even symbols are processed separately like in Figure 4.17. Several demodulation algorithms that use even symbols for compensating the Δf impact on the odd ones and vice versa are discussed below.

The first algorithm comprises three steps. At the first step, the phase of the $(k + 1)$th symbol is shifted by $\xi = \pm 90°$. The sign of ξ is unimportant if it is used consistently. Thus, θ_{k+1} in (4.53) is converted into

$$\theta_{k+1} = 2\pi (f_0 + \Delta f)(t + T_b) + \varphi_{k+1} + \zeta + \xi \quad (4.55)$$

At the second step, the phase differences $\Delta\theta_{k+1}$ and $\Delta\theta_{k+2}$ are calculated using (4.53) and (4.55):

$$\Delta\theta_{k+1} = \theta_{k+1} - \theta_k = 2\pi\Delta f T_b + \left(\varphi_{k+1} + \zeta + \xi\right) - \varphi_k \qquad (4.56)$$

$$\Delta\theta_{k+2} = \theta_{k+2} - \theta_{k+1} = 2\pi\Delta f T_b + \varphi_{k+2} - \left(\varphi_{k+1} + \zeta + \xi\right) \qquad (4.57)$$

Although (4.56) and (4.57) still contain the terms caused by Δf, they are eliminated in the second-order difference

$$\Delta\theta^{(2)} = \Delta\theta_{k+2} - \Delta\theta_{k+1} = \varphi_{k+2} + \varphi_k - 2\left(\varphi_{k+1} + \zeta + \xi\right) \qquad (4.58)$$

at the third step. Since φ_k, φ_{k+1}, and φ_{k+2}, and $\zeta + \xi$ can only be equal to $0°$ or $\pm 180°$, and operations with phases are modulo $360°$ (which means the equivalency of $+180°$ and $-180°$),

$$\Delta\theta^{(2)} = \Delta\theta_{k+2} - \Delta\theta_{k+1} = \varphi_{k+2} + \varphi_k = \varphi_{k+2} - \varphi_k \qquad (4.59)$$

This result is not affected by Δf and φ_{k+1}.

While the first algorithm shifts the phase of each $(k + 1)$th symbol by $\xi = \pm 90°$ at its first step, the second algorithm shifts the phase of every delayed symbol. Thus, the phase differences calculated at the second step are

$$\Delta\theta_{k+1} = \theta_{k+1} - \theta_k = 2\pi\Delta f T_b + \varphi_{k+1} - \varphi_k + \zeta - \xi \qquad (4.60)$$

$$\Delta\theta_{k+2} = \theta_{k+2} - \theta_{k+1} = 2\pi\Delta f T_b + \varphi_{k+2} - \varphi_{k+1} - \zeta - \xi \qquad (4.61)$$

The second-order difference obtained at the third step is

$$\Delta\theta^{(2)} = \Delta\theta_{k+2} - \Delta\theta_{k+1} = \varphi_{k+2} + \varphi_k - 2\varphi_{k+1} - 2\zeta = \varphi_{k+2} - \varphi_k \pm \pi \qquad (4.62)$$

A comparison of (4.59) and (4.62) shows that the sign of the second algorithm's output signal is opposite to that of the first one. The second algorithm is illustrated by the block diagram in Figure 4.18(a). The digital bandpass signal $u_{rq}(nT_s)$ at the demodulator input is an additive mixture of a desired signal and noise. The block diagram shows that adding and subtracting phases of bandpass real-valued signals require multiplying these signals and integrating the product. Here, the output values $V_{Irq}(nT_s - 2T_b)$ and $V_{Qrq}(nT_s - 2T_b)$ of the I and Q channels' integrators are, respectively, the differential decoders' input samples $\beta_{I(k-2)}$ and $\beta_{Q(k-2)}$.

Moving the $90°$ phase shifter from the input of the first digital memory with the delay T_b to its output transforms the second algorithm into the third one whose block diagram is shown in Figure 4.18(b). In that algorithm, the first step is omitted, and, at the second step, the first-order phase differences between adjacent symbols are calculated jointly with forming the I and Q components from $u_{rq}(nT_s)$ using autocorrelation.

4.5 Demodulation of Energy-Efficient Signals

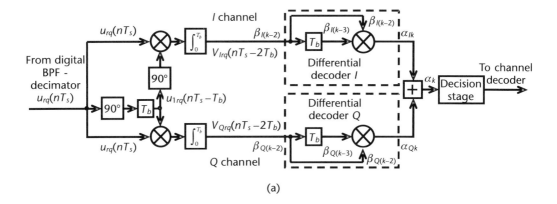

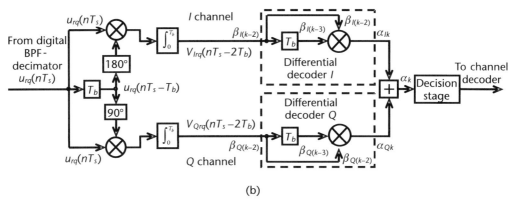

Figure 4.18 Bandpass frequency-invariant AQ-DBPSK demodulators corresponding to: (a) second algorithm and (b) third algorithm.

In most RDPs, the sum of signal and noise at the demodulator input is already represented by its complex-valued baseband equivalent $Z_{rq}(nT_s)$ (see (4.44)). To derive the structure of a baseband frequency-invariant AQ-DBPSK demodulator realizing the second algorithm for that case, let us determine the results of several operations with $Z_{rq}(nT_s)$. The one-symbol delay of $Z_{rq}(nT_s)$ produces

$$Z_{rq}\left(nT_s - T_b\right) = V_{Irq}\left(nT_s - T_b\right) + jV_{Qrq}\left(nT_s - T_b\right) \tag{4.63}$$

The ±90° phase shift of each delayed symbol is equivalent to multiplying $Z_{rq}(nT_s - T_b)$ by ±j:

$$\pm jZ_{rq}\left(nT_s - T_b\right) = \pm\left[-V_{Qrq}\left(nT_s - T_b\right) + jV_{Irq}\left(nT_s - T_b\right)\right] \tag{4.64}$$

Calculation of the first-order phase differences at the algorithm's second step requires integrating the product $[\pm jZ_{rq}(nT_s - T_b)]^*Z_{rq}(nT_s)$ over the interval $[t, t + T_b]$, that is, summing all resulting samples with numbers from n_1 to $n_2 = n_1 + n_b - 1$ where n_1 is the number of the first sample within $[t, t + T_b]$, and $n_b = T_b/T_s$. Therefore,

$$\beta_{k-2} = \sum_{n=n_1}^{n_2} \left\{ \left\{ \pm j Z_{rq}\left[(n-n_b)T_s\right] \right\}^* Z_{rq}(nT_s) \right\}$$

$$= \mp j \sum_{n=n_1}^{n_2} \left\{ Z_{rq}^*\left[(n-n_b)T_s\right] Z_{rq}(nT_s) \right\} \qquad (4.65)$$

$$= \beta_{I(k-2)} + j\beta_{Q(k-2)}$$

where

$$\beta_{I(k-2)} = \mp \sum_{n=n_1}^{n_2} \left\{ V_{Qrq}\left[(n-n_b)T_s\right] \cdot V_{Irq}(nT_s) - V_{Irq}\left[(n-n_b)T_s\right] \cdot V_{Qrq}(nT_s) \right\} \quad (4.66)$$

$$\beta_{Q(k-2)} = \mp \sum_{n=n_1}^{n_2} \left\{ V_{Irq}\left[(n-n_b)T_s\right] \cdot V_{Irq}(nT_s) + V_{Qrq}\left[(n-n_b)T_s\right] \cdot V_{Qrq}(nT_s) \right\} \quad (4.67)$$

Since the phase of β_{k-2} is the first-order phase difference between the $(k-1)$th and $(k-2)$th input symbols, the second-order difference between the phases of β_{k-2} and β_{k-3} is calculated as

$$\beta_{k-3}^* \beta_{k-2} = \left[\beta_{I(k-3)}\beta_{I(k-2)} + \beta_{Q(k-3)}\beta_{Q(k-2)} \right] + j\left[\beta_{I(k-3)}\beta_{Q(k-2)} - \beta_{Q(k-3)}\beta_{I(k-2)} \right] \quad (4.68)$$

The phase of product (4.68) is the second-order phase difference that can be only 0° or 180° in the absence of noise and distortions. Therefore, it is sufficient to calculate just the product's real component:

$$\alpha_k = \beta_{I(k-3)}\beta_{I(k-2)} + \beta_{Q(k-3)}\beta_{Q(k-2)} \qquad (4.69)$$

The block diagram of the baseband frequency-invariant AQ-DBPSK demodulator is depicted in Figure 4.19. Here, the phase shift of every delayed symbol by $\pm 90°$ and transition to the complex conjugate of $\pm j Z_{rq}(nT_s - T_b)$ are reduced to swapping $V_{Irq}(nT_s - T_b)$ and $V_{Qrq}(nT_s - T_b)$ because the signs before sums (4.66) and (4.67) can be neglected, as follows from (4.62). The demodulator section that includes the integrators and all the blocks located to the right of them is identical to those of the demodulators in Figure 4.18.

The frequency invariance of the AQ-DBPSK demodulators in Figures 4.18 and 4.19 is achieved at the cost of the reduced noise immunity in AWGN channels compared to the demodulator in Figure 4.17, caused by the algorithms' nonoptimality for these channels and increased bandwidth of the input filters. In channels with both frequency offsets and additive white Gaussian noise, the frequency-invariant demodulators provide higher noise immunity than the optimum noncoherent demodulator without frequency synchronization when $\Delta f \geq 1/(6T_b)$.

In AWGN channels, it is preferable to distribute symbol-shaping filtering between modulators and demodulators as illustrated by root raised cosine filters in Figures 3.16(a), 3.17(a), and 4.17. In channels with significant frequency offsets between

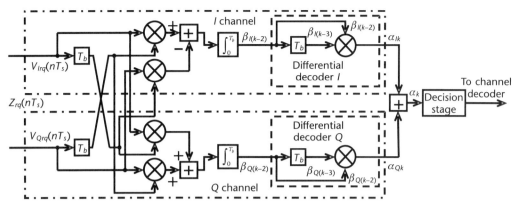

Figure 4.19 Baseband frequency-invariant AQ-DBPSK demodulator.

Rxs and Txs, such filtering is distorted if frequency synchronization is not used. In this case, symbol-shaping filtering should be performed exclusively in modulators. In particular, the root raised cosine filters in the AQ-DBPSK modulators in Figures 3.16(a) and 3.17(a) should be replaced with raised cosine filters. The flexibility of DSP allows adjusting symbol-shaping filtering to various scenarios. The crest factor reduction provided by AQ-DBPSK compared to DBPSK depends on the types and parameters of symbol-shaping filters in AQ-DBPSK modulators. When root raised cosine or raised cosine filters with roll-off factors between 0.1 and 0.25 are used, the reduction is slightly larger than 3 dB, doubling the Tx power utilization. This is also true for the transition from BPSK to AQ-BPSK spreading.

The synchronization of noncoherent demodulators is faster and simpler than that of coherent ones due to the absence of phase synchronization. Still, noncoherent demodulators require frequency synchronization in the case of insufficient stability of the radios' LOs and/or significant Doppler shifts. Frequency-invariant demodulators do not need frequency synchronization even in this case. At the same time, AQ-DBPSK allows the fastest symbol acquisition due to the regularity and the highest rate of transitions between symbols. Similarly, AQ-BPSK spreading allows the fastest code acquisition. The negative side of the regularity and the highest rate of transitions between symbols or chips is that they simplify determining the symbol and chip rates by unauthorized Rxs.

4.6 Summary

The most significant initial step in digital radio development was the emergence of digital HF Rxs with relatively wide bandwidths and high dynamic ranges. The main principles of RF signals' D&R, their representation in RDPs and TDPs, as well as numerous DSP procedures, initially developed for these radios, were later employed in the radios of other frequency bands used for communications, broadcasting, navigation, radar, and EW.

During the 1970s and 1980s, three mutually connected choices related to digital radios were considered: (1) baseband D&R versus bandpass D&R, (2) SHAs versus THAs, and (3) representation of digital signals by their instantaneous values versus their *I* and *Q* components. The progress of digital radios was accelerated by correctly made choices (1) and (3), but hindered by the erroneous choice (2) in favor of THAs.

Technological progress increases the number of functions executed in RDPs and moves signal digitization closer to the antenna. Still, antenna coupling, preliminary amplification and/or attenuation, initial filtering (including antialiasing filtering), sampling, and quantization take place in AMFs.

In duplex or half-duplex communication systems with low to moderate transmit power, combining Txs and Rxs into transceivers provides many advantages due to sharing the circuitry and housing. Digital transceivers may have some capabilities and modes impossible in analog transceivers. For instance, they allow realization of full-duplex mode, that is, transmission and reception on the same frequency, at the same time, and with the same antenna.

Rx performance is reflected by several characteristics despite the existence of the universal one that is the channel throughput reduction caused by the Rx. Although this reduction fully reflects the reception quality, it is not used because its statistically reliable measurement is virtually impossible, and it does not indicate the cause(s) of the throughput reduction and, consequently, does not allow determining the ways of its improvement.

The sensitivity, selectivity, dynamic range, reciprocal mixing, and spurious outputs of a Rx are directly or indirectly affected by the digitization quality in it.

Reduction of the Rx input stages' noise factor and the summative PSD of quantization noise and interference, caused by nonideal antialiasing filtering and sampling, as well as proper selection and distribution of the AMF gain among its stages are necessary for maximizing the Rx sensitivity.

The presence of ISs within the AMF passband requires high dynamic range and high quality of AGC. Dynamic range reflects the Rx capability to pick up a weak desired signal in the presence of strong unwanted ones. The in-band dynamic range maximally reflects the effect of Rx nonlinearity on a sum of desired signals and ISs.

When the AMF bandwidth is close to the average IS bandwidth, single-tone dynamic range is the best characteristic of the reception reliability in the presence of strong ISs. When the AMF bandwidth significantly exceeds the average IS bandwidth, two-tone dynamic range more adequately characterizes this reliability.

The ultimate restriction on the input signal's upper bound is imposed by the allowed Rx power consumption (or dissipation). For a given upper bound, the dynamic range can be increased by improving the Rx sensitivity.

Knowledge of ISs' statistical characteristics allows deriving expressions for calculating the minimum required dynamic range. Such expressions, obtained for HF Rxs, confirm the intuitive expectation that this range is at least approximately proportional to the AMF bandwidth and the intensity of ISs.

Digitization circuits are the bottleneck limiting the Rx dynamic range due to the highest power of analog signals at their input, much larger experience in developing preceding analog circuits than in developing digitization circuits, and capability of RDPs to support virtually any required dynamic range.

Besides high dynamic range, the Rxs' digitization circuits should have high flexibility and compatibility with the IC technology that are presently limited by the best bandpass antialiasing filters. In many RF channels, the requirements for Rxs' digitization circuits depend less on desired signals than on ISs' statistical parameters.

To support high Rx dynamic range, it is undesirable to substantially exceed the minimum required gain of the stages preceding the digitization circuits, determined by (4.11) through (4.13), and it is desirable to reduce the summative PSD of quantization noise and interference caused by nonideal antialiasing filtering and sampling.

Bandpass digitization of Rx input signals has significant advantages over baseband one, and sampling based on the hybrid or direct interpretation of the sampling theorem amplifies these advantages. The superheterodyne (with bandpass digitization) and direct RF digitization AMF architectures are most advantageous and most consistent with current technological trends.

Demodulation of energy-efficient signals described in this chapter completes the explanation of the modulation techniques (presented in Chapter 3) that simplify the signal reconstruction and improve power utilization in digital Txs. This material also illustrates signal processing in RDPs.

References

[1] Stephenson, A. G., "Digitizing Multiple RF Signals Requires an Optimum Sampling Rate," *Electronics*, Vol. 45, No. 7, 1972, pp. 106–110.

[2] Chiffy, F. P., and B. E. Bjerede, "Communication Receivers of the Future," *Signal*, Vol. 30, No. 3, 1975, pp. 16–21.

[3] Poberezhskiy, Y. S., "Equations for HF Receiver with Digital Heterodyning" (in Russian), *Commun. Technol.*, TRC, No. 2, 1975, pp. 3–15.

[4] Poberezhskiy, Y. S., "Determining the Relationship Between Preselector Bandwidth and Required Dynamic Range of an HF Receiver" (in Russian), *Commun. Technol.*, TRC, No. 6, 1976, pp. 56–65.

[5] Poberezhskiy, Y. S., "Digital Short-Wave Radio Receivers," *Telecommun. and Radio Engineering*, Vol. 32/33, No. 5, 1978, pp. 72–78.

[6] Poberezhskiy, Y. S., and M. N. Sokolovskiy, "Influence of Intermodulation Noise on Reception Noise Immunity in the SW Band," *Telecommun. and Radio Engineering*, Vol. 33/34, No. 12, 1979, pp. 91–93.

[7] Poberezhskiy, Y. S., and M. N. Sokolovskiy, "The Effect of Fifth-Order Intermodulation Interference on Reception Noise Immunity in the Decameter Band," *Telecommun. and Radio Engineering*, Vol. 47, No. 11, 1992, pp. 119–123.

[8] Poberezhskiy, Y. S., "Gating Time for Analog-to-Digital Conversion in Digital Reception Circuits," *Telecommun. and Radio Engineering*, Vol. 37/38, No. 10, 1983, pp. 52–54.

[9] Poberezhskiy, Y. S., "Digital Radio Receivers and the Problem of Analog-to-Digital Conversion of Narrow-Band Signals," *Telecommun. and Radio Engineering*, Vol. 38/39, No. 4, 1984, pp. 109–116.

[10] Poberezhskiy, Y. S., *Digital Radio Receivers* (in Russian), Moscow, Russia: Radio & Communications, 1987.

[11] Poberezhskiy, Y. S., and M. V. Zarubinskiy, "Sample-and-Hold Devices Employing Weighted Integration in Digital Receivers," *Telecommun. and Radio Engineering*, Vol. 44, No. 8, 1989, pp. 75–79.

[12] Poberezhskiy, Y. S., and G. Y. Poberezhskiy, "Optimizing the Three-Level Weighting Function in Integrating Sample-and-Hold Amplifiers for Digital Radio Receivers," *Radio and Commun. Technol.*, Vol. 2, No. 3, 1997, pp. 56–59.

[13] Rader, M., "A Simple Method for Sampling In-Phase and Quadrature Components," *IEEE Trans. Aerosp. Electron. Syst.*, Vol. AES-20, No. 6, 1984, pp. 821–824.

[14] Zarubinskiy, M. V., and Y. S. Poberezhskiy, "Formation of Readouts of Quadrature Components in Digital Receivers," *Telecommun. and Radio Engineering*, Vol. 40/41, No. 2, 1986, pp. 115–118.

[15] Mitchell, R. L., "Creating Complex Signal Samples from a Band-Limited Real Signal," *IEEE Trans. Aerosp. Electron. Syst.*, Vol. 25, No. 3, 1989, pp. 425–427.

[16] Anderson, T., and J. W. Whikohart, "A Digital Signal Processing HF Receiver," *Proc. Third Int. Conf. Commun. Syst. & Techn.*, London, U.K., February 26–28, 1985, pp. 89–93.

[17] Groshong, R., and S. Ruscak, "Undersampling Techniques Simplify Digital Radio," *Electronic Design*, No. 10, 1991, pp. 67–78.

[18] Groshong, R., and S. Ruscak, "Exploit Digital Advantages in an SSB Receiver," *Electronic Design*, No. 11, 1991, pp. 89–96.

[19] Frerking, M. E., *Digital Signal Processing in Communication Systems*, New York: Van Nostrand Reinhold, 1994.

[20] Sabin, W. E., and E. O. Schoenike (eds.), *Single-Sideband Systems and Circuits*, 2nd ed., New York: McGraw-Hill, 1995.

[21] Poberezhskiy, Y. S., M. V. Zarubinskiy, and B. D. Zhenatov, "Large Dynamic Range Integrating Sampling and Storage Device," *Telecommun. and Radio Engineering*, Vol. 41/42, No. 4, 1987, pp. 63–66.

[22] Poberezhskiy, Y. S., et al., "Design of Multichannel Sampler-Quantizers for Digital Radio Receivers," *Telecommun. and Radio Engineering*, Vol. 46, No. 9, 1991, pp. 133–136.

[23] Poberezhskiy, Y. S., et al., "Experimental Investigation of Integrating Sampling and Storage Devices for Digital Radio Receivers," *Telecommun. and Radio Engineering*, Vol. 49, No. 5, 1995, pp. 112–116.

[24] Poberezhskiy, Y. S., and M. V. Zarubinskiy, "Analysis of a Method of Fundamental Frequency Selection in Digital Receivers," *Telecommun. and Radio Engineering*, Vol. 43, No. 11, 1988, pp. 88–91.

[25] Poberezhskiy, Y. S., and S. A. Dolin, "Analysis of Multichannel Digital Filtering Methods in Broadband-Signal Radio Receivers," *Telecommun. and Radio Engineering*, Vol. 46, No. 6, 1991, pp. 89–92.

[26] Poberezhskiy, Y. S., S. A. Dolin, and M. V. Zarubinskiy, "Selection of Multichannel Digital Filtering Method for Suppression of Narrowband Interference" (in Russian), *Commun. Technol., TRC*, No. 6, 1991, pp. 11–18.

[27] Khvetskovich, E. B., and Y. S. Poberezhskiy, "Analysis of a Method of Demodulating Frequency-Modulated Signals in a Digital Radio Receiver," *Telecommun. and Radio Engineering*, Vol. 48, No. 6, 1993, pp. 96–102.

[28] Poberezhskiy, Y. S., and S. A. Dolin, "The Design of an Optimal Incoherent Demodulator of Frequency-Shift Keyed Signals in a Digital Receiver," *Telecommun. and Radio Engineering*, Vol. 49, No. 6, 1995, pp. 14–19.

[29] Poberezhskiy, Y. S., and G. Y. Poberezhskiy, "The Effect of Binary Quantization of the Reference Oscillations on the Noise Immunity of Digital Demodulation of Frequency Shift Keying Signals," *Telecommun. and Radio Engineering*, Vol. 49, No. 11, 1995, pp. 28–31.

[30] Tsui, J. B., *Digital Microwave Receivers: Theory and Concepts*, Norwood, MA: Artech House, 1989.

[31] Eassom, R. J., "Practical Implementation of a HF Digital Receiver and Digital Transmitter Drive," *Proc. 6th Int. Conf. HF Radio Syst. & Techn.*, London, U.K., July 4–7, 1994, pp. 36–40.

[32] Rohde, U. L., J. C. Whitaker, and T. T. N. Bucher, *Communications Receiver: Principles and Design*, 2nd ed., New York: McGraw-Hill, 1996.

[33] Poberezhskiy, Y. S., "On Dynamic Range of Digital Receivers," *Proc. IEEE Aerosp. Conf.*, Big Sky, MT, March 3–9, 2007, pp. 1–17.

[34] Friis, H. T., "Noise Figures of Radio Receivers," *Proc. IRE*, Vol. 32, No. 7, 1944, pp. 419–422.

[35] Poberezhskiy, Y. S., and G. Y. Poberezhskiy, "On Adaptive Robustness Approach to Anti-Jam Signal Processing," *Proc. IEEE Aerosp. Conf.*, Big Sky, MT, March 2–9, 2013, pp. 1–20.

[36] Mitola, J. III, *Software Radio Architecture*, New York: Wiley-Interscience, 2000.

[37] Reed, J. H., *Software Radio: A Modern Approach to Radio Engineering*, Inglewood Cliffs, NJ: Prentice Hall, 2002.

[38] Poberezhskiy, Y. S., and G. Y. Poberezhskiy, "Sampling and Signal Reconstruction Structures Performing Internal Antialiasing Filtering and Their Influence on the Design of Digital Receivers and Transmitters," *IEEE Trans. Circuits Syst. I*, Vol. 51, No. 1, 2004, pp. 118–129.

[39] Poberezhskiy, Y. S., and G. Y. Poberezhskiy, "Flexible Analog Front-Ends of Reconfigurable Radios Based on Sampling and Reconstruction with Internal Filtering," *EURASIP J. Wireless Commun. Netw.*, No. 3, 2005, pp. 364–381.

[40] Mitola, J. III, *Cognitive Radio Architecture: The Engineering Foundations of Radio HML*, New York: John Wiley & Sons, 2006.

[41] Bard, J., *Software Defined Radio: The Software Communications Architecture*, New York: John Wiley & Sons, 2007.

[42] Abidi, A. A., "The Path to the Software-Defined Radio Receiver," *IEEE J. Solid-State Circuits*, Vol. 42, No. 5, 2007, pp. 954–966.

[43] Fette, B. A. (ed.), *Cognitive Radio Technology*, 2nd ed., New York: Elsevier, 2009.

[44] Venosa, E., F. J. Harris, and F. A. N. Palmieri, *Software Radio: Sampling Rate Selection, Design and Synchronization*, New York: Springer, 2012.

[45] Bullock, S. R., *Transceiver and System Design for Digital Communications*, 5th ed., London, U.K.: IET, 2014.

[46] Betz, J. W., *Engineering Satellite-Based Navigation and Timing: Global Navigation Satellite Systems, Signals, and Receivers*, New York: John Wiley & Sons, 2016.

[47] Das, S. K., *Mobile Terminal Receiver Design: LTE and LTE-Advanced*, New York: John Wiley & Sons, 2017.

[48] Rouphael, T. J., *Wireless Receiver Architectures and Design: Antennas, RF, Synthesizers, Mixed Signal, and Digital Signal Processing*, Waltham, MA: Academic Press, 2018.

[49] Grayver, E., *Implementing Software Defined Radio*, New York: Springer, 2013.

[50] Jamin, O., *Broadband Direct RF Digitization Receivers*, New York: Springer, 2014.

[51] Poisel, R. A., *Electronic Warfare Receivers and Receiving Systems*, Norwood, MA: Artech House, 2014.

[52] Lechowicz, L., and M. Kokar, *Cognitive Radio: Interoperability Through Waveform Reconfiguration*, Norwood, MA: Artech House, 2016.

[53] Tsui, J. B., and C. H. Cheng, *Digital Techniques for Wideband Receivers*, 3rd ed., Raleigh, NC: SciTech Publishing, 2016.

[54] Choi, J. I., et al., "Achieving Single Channel, Full Duplex Wireless Communication," *Proc. MobiCom*, Chicago, IL, September 20–24, 2010, pp. 1–12.

[55] Grayver, E., "Full-Duplex Communications for Noise-Limited Systems," *Proc. IEEE Aerosp. Conf.*, Big Sky, MT, March 3–10, 2018, pp. 1–10.

[56] Fujimaki, A., et al., "Broadband Software-Defined Radio Receivers Based on Superconductor Devices," *IEEE Trans. Appl. Supercond.*, Vol. 11, No. 1, 2001, pp. 318–321.

[57] Gupta, D., et al., "Digital Channelizing Radio Frequency Receiver," *IEEE Trans. Appl. Supercond.*, Vol. 17, No. 2, 2007, pp. 430–437.

[58] Mukhanov, O. A., et al., "Hybrid Semiconductor-Superconductor Fast-Readout Memory for Digital RF Receivers," *IEEE Trans. Appl. Supercond.*, Vol. 21, No. 3, 2011, pp. 797–800.

[59] Kotelnikov, V. A., *The Theory of Optimum Noise Immunity*, New York: McGraw-Hill, 1959.

[60] Bode, H., and C. Shannon, "A Simplified Derivation of Linear Least Square Smoothing and Prediction Theory," *Proc. IRE*, Vol. 38, No. 4, 1950, pp. 417–425.

[61] Poberezhskiy, Y. S., "Derivation of the Optimum Transfer Function of a Narrowband Interference Suppressor in an Oblique Sounding System" (in Russian), *Problems of Radio-Electronics*, TRC, No. 9, 1969, pp. 3–11.

[62] Poberezhskiy, Y. S., "Optimum Filtering of Sounding Signals in Non-White Noise," *Telecommun. and Radio Engineering*, Vol. 31/32, No. 5, 1977, pp. 123–125.

[63] Poberezhskiy, Y. S., "Optimum Transfer Function of a Narrowband Interference Suppressor for Communication Receivers of Spread Spectrum Signals in Channels with Slow Fading" (in Russian), *Problems of Radio-Electronics*, TRC, No. 8, 1970, pp. 104–110.

[64] Poberezhskiy, Y. S., "Comparative Analysis of Methods of Narrowband Interference Suppression in Wideband Receivers" (in Russian), *Problems of Radio-Electronics*, TRC, No. 1, 1974, pp. 44–50.

[65] Poberezhskiy, Y. S., "Alternating Quadratures Differential Binary Phase Shift Keying Modulation and Demodulation Method," U.S. Patent 7,627,058 B2, filed March 28, 2006.

[66] Poberezhskiy, Y. S., "Apparatus for Performing Alternating Quadratures Differential Binary Phase Shift Keying Modulation and Demodulation," U.S. Patent 8,014,462 B2, filed March 28, 2006.

[67] Poberezhskiy, Y. S, "Method and Apparatus for Synchronizing Alternating Quadratures Differential Binary Phase Shift Keying Modulation and Demodulation Arrangements," U.S. Patent 7,688,911 B2, filed March 28, 2006.

[68] Poberezhskiy, Y. S., "Novel Modulation Techniques and Circuits for Transceivers in Body Sensor Networks," *IEEE J. Emerg. Sel. Topics Circuits Syst.*, Vol. 2, No. 1, 2012, pp. 96–108.

[69] Okunev, Y., *Phase and Phase-Difference Modulation in Digital Communications*, Norwood, MA: Artech House, 1997.

[70] Simon, M. K., and D. Divsalar, "On the Implementation and Performance of Single and Double Differential Detection Schemes," *IEEE Trans. Commun.*, Vol. 40, No. 2, 1992, pp. 278–291.

CHAPTER 5
Sampling Theory Fundamentals

5.1 Overview

This chapter's material clarifies many aspects of D&R in digital radios described in Chapters 3 and 4. As noted there, D&R remain the persistent bottleneck of the radio design despite their significant improvements over the last decades. These improvements have been achieved mainly due to new IC technologies, as well as novel methods of signal quantization and analog decoding of code words (the latter is the main part of D/A conversion). Advancement in S&I techniques has been less substantial due to incomplete understanding of their theoretical basis.

The S&I techniques used in digital radios are based on the classical sampling theorem that can be interpreted in different ways. However, only one of its interpretations, which is called indirect in this book, is clearly explained in technical literature, and that interpretation has become an obstacle to the development of S&I techniques. This chapter explicates possible interpretations of the theorem, its constructive nature, and the necessity to widen the theoretical basis of S&I techniques and circuits. It also demonstrates that the required mathematics is quite simple, and the theoretical results are in good agreement with engineering intuition.

Section 5.2 concisely presents the history of the sampling theorem and explains the technical needs as well as the technological and mathematical premises for the theorem origination and subsequent development. It also outlines the theoretical results most important for digital radios.

Section 5.3 introduces the uniform sampling theorem for baseband signals in a way that demonstrates its constructive nature. Different forms of the theorem's equations corresponding to its direct, indirect, and hybrid interpretations are derived. This section also presents some information on S&I of baseband signals, in addition to that provided in Chapters 3 and 4.

Section 5.4 supplements the information, provided in Chapters 3 and 4, on the S&I of bandpass signals based on the indirect interpretation of the sampling theorem.

To simplify the comprehension of the chapter material and avoid the distraction from the physical substance of S&I, the formal proofs of several versions of the uniform sampling theorem have been moved to Appendix D.

5.2 S&I from a Historical Perspective

5.2.1 Need for S&I at the Dawn of Electrical Communications

In communications, the problem of sampling and interpolation emerged in connection with TDM in telegraphy and telephony. The efforts to implement TDM of telegraph signals were made very soon after the inventions of electromagnetic telegraphs by P. Schilling (1832) and by S. Morse (1837). Successful attempts to transmit several telegraph signals simultaneously over a single wire using this type of multiplexing were made by F. Bakewell (1848), A. Newton (1851), M. Farmer (1853), B. Meyer (1870), and J. Baudot (1874), as well as P. Lacour and P. Delany (1878) [1, 2]. For TDM of telegraph signals, it was relatively easy to empirically determine the number of samples per symbol. The empirical approach worked well for the reconstruction of symbols received with high signal-to-noise ratio (SNR) because the simplest interpolation technique, stepwise interpolation, was acceptable.

The most urgent theoretical problem for the transmission of TDM and other signals in telegraphy was determining the maximum rate of distortionless transmission. This problem was solved first by H. Nyquist [3, 4] and, slightly later, by K. Küpfmüller [5, 6]. They proved that when the symbol rate R_t in a communication channel with one-sided bandwidth B meets condition

$$R_t \leq 2B \tag{5.1}$$

the symbols can be transmitted without distortion. Thus, the maximum rate of distortionless transmission is

$$R_{t\max} = 2B \tag{5.2}$$

Nowadays, the physical substance of (5.1) and (5.2) is easily understandable even for undergraduate students. Indeed, the symbol length should be determined by the settling time of the channel, and this time in a linear time invariant (LTI) system is inversely proportional to its bandwidth B. As follows from [4, 7], H. Nyquist and another great researcher, R. Hartley, clearly understood in the late 1920s that $R_{t\max}$ does not limit the data transmission rate because the latter can be raised by increasing the number of symbol levels.

In contrast with telegraphy where the implementation of TDM was successful without theoretical substantiation of S&I procedures, the practical realization of TDM in telephony was critically delayed by the absence of such substantiation. The first patent on TDM in telephony was received by W. Miner (1903) [8]. He used electromechanical devices, and sampling was performed by a rotating commutator. Miner experimented with various sampling rates and achieved the best intelligibility at the rate of 4,300 samples per second (sps). It is retrospectively clear that this rate corresponds to the cutoff frequency 2.15 kHz. At the time of the invention, however, Miner thought that the sampling rate should be approximately equal to the highest frequency of the speech components.

Like Miner in the early 1900s, most engineers and researchers had a very vague understanding of the requirements for sampling rate until the late 1940s [1, 2].

Meanwhile, as shown below, the information on the sampling theorem has been available since 1915. This means that urgently needed theoretical results were unnoticed and, consequently, unused by engineers for decades. The unawareness about the existing theory critically delayed the implementation of TDM in telephony and provided technologically and economically unjustifiable advantages to the systems with FDM that were more complex and expensive.

5.2.2 Discovery of Classical Sampling Theorem

In some publications, the roots of the sampling theorem have been traced back to as early as the eighteenth century. From this book's perspective, it was originated and proven in 1915 by E. Whittaker, who derived the interpolation equation for reconstructing continuous bandlimited functions from their samples [9]. In mathematics, his work was continued by W. Ferrar and, to a larger extent, by J. Whittaker [10–12]. With regard to communication theory, this theorem for bandlimited signals was first formulated and independently proven again by V. Kotelnikov in 1933 [13]. Because [13] was not published in an internationally accessible form until much later, this result was unknown, and the theorem was deduced once again by H. Raabe in 1939 [14]. W. Bennett cited [14] in his publication on TDM [15]. Since the sampling theorem emerged within the scope of interpolation theory and was later investigated from the standpoint of approximation theory, the substance of these theories is outlined below.

In practical activities and scientific research, results of measurements of continuous functions are typically represented by their discrete values (samples) that are used to reconstruct the original functions or determine their unknown samples. The reconstruction within the interval of the known samples' location is called interpolation, and it is called extrapolation outside this interval. The interpolation problem of determining the maximum increment T_s of an independent variable t, which still allows reconstructing a function $s(t)$ with given accuracy from its samples $s(nT_s)$, exists in many fields. This problem is formulated differently depending on the application. For instance, how often should the ambient temperature and/or humidity be measured to allow reconstructing these processes without significant loss of valuable information? Also, how often should the position and/or velocity of a moving object be determined to restore its trajectory with required accuracy? In communications, as mentioned above, it requires determining the minimum sampling rate needed for precise reconstruction of voice and other signals. The interpolation theory examines such problems for deterministic and stochastic functions.

A different mathematical discipline, namely approximation, is often combined with interpolation and extrapolation. It is usually understood as replacing a function described by a relatively complex equation with a function described by a simpler one. This inevitably reduces the function representation precision, but it is justified when the benefits of the achieved simplicity exceed the drawbacks caused by the loss of accuracy. There can be various reasons for combining the approximation and interpolation approaches. In the sampling theory, they were combined for two reasons: the necessity of replacing accurate but physically unrealizable interpolation (sampling) functions with less accurate but physically realizable ones and limited accuracy of the known samples' assessment.

To achieve the required accuracy of interpolation, some constraints should be imposed on the interpolated and interpolating functions. However, these constraints should not prevent interpolation of all or most functions used in the likely applications. The success of [9] should mostly be attributed to the constraints properly selected by E. Whittaker. Many types of functions (e.g., piecewise, linear, polynomial, spline, trigonometric, wavelet) of one or several variables are currently used for interpolation. In this book, the attention is focused on the interpolating functions used in the uniform sampling theorem.

Despite the publications [9–14], the sampling theorem remained virtually unknown to the engineering and scientific community until its introduction by C. Shannon in his revolutionary articles [16, 17] in 1948 to 1949. Being familiar with [9, 15] and unfamiliar with [13, 14], he mentioned that mathematicians had proven this theorem in a different form and presented it as "common knowledge in communication art," which "in spite of its evident importance seems not to have appeared explicitly in the literature of communication theory." Shannon also mentioned that the possibility of representing signals with bandwidth B and duration T by $2BT$ samples had been noted by H. Nyquist [4] and D. Gabor [18]. He coined the term "Nyquist sampling interval" based on the equality of the minimum acceptable sampling rate to the maximum distortionless transmission rate (5.2) determined by H. Nyquist. In 1949, the sampling theorem was also discussed by I. Someya [19] and J. Weston [20]. Now it is recognized that Shannon's statement about the sampling theorem as "common knowledge in communication art" in the late 1940s was overly optimistic. In reality, only a few people mentioned above had more or less clear understanding of this theorem. However, the situation changed dramatically very soon after his publications.

Thus, the uniform sampling theorem for bandlimited signals was independently or almost independently derived by several outstanding scientists, and the greatest contribution to its origination was made by both Whittakers, V. Kotelnikov, and C. Shannon. For this reason, naming this theorem the WKS sampling theorem after them, as in [21], seems fair. Sometimes, however, it is called the Nyquist-Shannon sampling theorem. The question of Nyquist authorship of this theorem goes far beyond attributing the theorem to one or another scientist. It is related to the correct understanding of its essence.

H. Nyquist was an outstanding scientist and engineer. He authored classical works on thermal noise, stability of feedback amplifiers, and initial investigation of channel throughput. As an engineer, he significantly contributed to the development of telegraphy, facsimile, television, and other branches of communications. Nowadays, the Nyquist stability criterion is included in any textbook on feedback control theory. However, Nyquist did not participate in the origination or subsequent development of the sampling theorem. Although he clearly understood that a signal with duration T and bandwidth B can be represented by $2BT$ discrete values even in the 1920s, his understanding was based on the fact that such a signal can be represented by $2BT$ coefficients of the trigonometric Fourier series. This understanding can be traced back to the late nineteenth century, and it is not directly related to the sampling theorem.

The originators of the sampling theorem clearly understood that it has constructive nature because it provides the ways of representing $u(t)$ by its samples

$u(nT_s)$ or by the samples $y(nT_s)$ of its function $y(t) = y[u(t)]$, and the optimal, in the least-squares sense, interpolation algorithms. There is no evidence that, prior to the Shannon's publications, Nyquist knew how to represent an analog signal by its time-domain samples or reconstruct it from these samples. At the same time, the terms "Nyquist sampling interval" or "Nyquist sampling rate" are fair because he deduced the maximum rate of distortionless transmission (5.2) that is equal to the minimum sampling rate, and these rates have a common physical cause: the limited signal or channel bandwidth. Thus, the term "Nyquist-Shannon sampling theorem" is, in the authors' opinion, inappropriate, whereas the terms "Nyquist sampling interval" and "Nyquist sampling rate" are logical.

At present, all communication engineers know that the sampling theorem determines the ways of representing bandlimited functions by their samples. Many also recognize that this theorem provides the optimal interpolation procedures. However, a very small number of engineers know that it reveals the optimal sampling procedures as well, and even a smaller number understand that the optimal S&I procedures depend on the interpretation of the sampling theorem. This situation has significantly hindered the progress of D&R technology.

5.2.3 Sampling Theory After Shannon

Shannon's publications attracted attention of many researchers and engineers to the sampling theorem, and its theoretical and practical importance was quickly recognized. The intensive research in the sampling theory as well as the invention and subsequent evolution of S&I techniques that started in the late 1940s were initially stimulated by the development of communication, control, and measurement equipment with TDM. The implementation of PCM and DSP, started in the late 1950s, greatly widened the use of S&I circuits. Digital audio, ultrasound, radio, and video systems accelerated their progress. Although the first experimental digital radios were developed in the early 1970s, their implementation in communications, broadcasting, navigation, radar, direction finding, and surveillance started only in the 1980s. The speed of this implementation was astonishing: virtually all radios developed in the 1990s were already digital. During all these years, the number of publications on the sampling theory and circuits grew very fast, reflecting substantial progress in the field. The analysis of these publications can be found in [1, 21–26]. Below, only the aspects of the sampling theory directly related to the subjects discussed in this book are outlined.

From the time of the sampling theorem origination, researchers made a lot of efforts to extend, generalize, and further substantiate it. Interestingly, its first extensions and generalizations were made by its founding fathers: Kotelnikov proved this theorem not only for baseband but also for bandpass signals [13], and Shannon suggested two additional ways of the discrete-time signal representation [17]. One of them is representing signals by their nonuniformly spaced samples, and another way is representing them by the samples of their values and those of their derivatives. Shannon even extended this approach to higher derivatives. He predicted that the total number of samples required for representing a signal with bandwidth B and duration T should be equal to $2BT$ in all these cases. Since the Kotelnikov's results were initially unknown, and Shannon did not prove his findings, the sampling methods proposed by them were rediscovered and/or proven later by other researchers.

In addition to these extensions, the sampling theorem was proven for random processes. This was important because most functions that undergo S&I in applications are random (e.g., information signals, noise, and interference in digital radios). The sampling theorem was also extended to functions of several variables (multidimensional case). It was proven for time-varying systems, including systems with time-varying bandwidths. Its several extensions were related to the prediction of bandlimited processes from the past samples. The sampling theorem was also proven for integral transforms more general than the Fourier transform.

The interest to nonuniform sampling (NUS), which was suggested by Shannon as early as 1949, was revitalized several times since then for different reasons. Sometimes, the interest was caused by new theoretical findings. In other cases, it was stimulated by technological needs. For uniform and nonuniform sampling, it was proven in [27] that the minimally acceptable average sampling rate in stationary conditions must be twice the bandwidth occupied by the desired signal spectrum, independent of the spectrum location if this location is known. As shown in [28], the absence of knowledge of the signal spectrum location at least doubles the minimal sampling rate. The latter result was derived within the scope of compressive sampling (or sensing). As to the technological needs, NUS was motivated by the desire to reduce the requirements for analog antialiasing filtering and for voltage quantization by combining voltage and time quantization techniques [29–33]. The existence of radio systems whose signals are sparse in the time domain, such as ultrawideband radio and some types of sensor networks, as well as systems with adaptive transmission rates, stimulated the development of various types of NUS.

The sparsity of signals in the frequency domain can be utilized by compressive sampling that is a relatively new direction of the sampling theory development (see, for instance, [28, 34–41]). It allows reducing the sampling rate below $2B$ for signals with sparse spectrum. No knowledge of the signal spectrum distribution within the overall bandwidth B is required. Due to the penalty for the absence of this knowledge, compressive sampling is effective only if the signal spectrum occupies less than one-third of B. In general, compressive sampling allows reducing the average number of samples required for representing signals and images by exploiting their sparsity in some domains.

An approach suggested in [42] introduces a new optimality criterion of the combined S&I procedures, according to which the reconstructed signal, being reinjected into the sampler, should produce the same samples as the original input signal. This approach shifts the burden of constraints from the input signals to the sampling or weight functions used for S&I. It is convenient if the weight functions generation is flexible. The use of adjustable weight functions for S&I in digital radios was first proposed in [43, 44] and later developed in [45–55]. As follows from [46–55], the sampling theorem provides the optimal algorithms not only for interpolation but also for sampling.

During the first two decades after Shannon's publications, theorists were intensely involved in the implementation of S&I techniques. Some of the sampling theorem extensions mentioned above were responses to the industry needs. For example, the extensions related to S&I of random processes and functions of several variables were required almost immediately after the beginning of S&I implementation.

Other urgent theoretical problems were related to the fact that the theorem had been proven for the signals with strictly limited bandwidth and infinite duration, using sampling functions that are physically unrealizable, and presuming that samples are determined precisely and generated exactly at the expected sampling instants. Two approaches were used to resolve these problems. The first one was proving the sampling theorem extensions for more relaxed assumptions, while the second approach was investigating the errors caused by nonideal realizations of S&I. Specifically, the latter approach analyzed aliasing errors caused by the impossibility of perfect bandlimiting, time-domain truncation errors caused by representing finite-duration signals by infinite-duration functions, jitter-induced errors caused by deviation of actual sampling instants from the expected ones, and round-off errors caused by inaccurate estimation of samples' values.

The research related to practical implementation of S&I techniques has significantly deepened the understanding of the essence, capabilities, and limitations of the sampling theorem. It also allowed considering S&I from different viewpoints. Some of them proved to be very convenient and productive. For example, interpretation of sampling as double-sideband suppressed carrier (DSB-SC) of a train of short pulses greatly simplified the spectral analysis of sampled signals.

The major approaches to the S&I became clear by the late 1950s, and the basic circuits performing these operations reached a high degree of maturity by the late 1960s. Around that time, most theorists started to consider the research potential of this field exhausted and shifted their scientific interests from it. The subsequent development of S&I algorithms and circuits, carried out by practical engineers, was initially successful. The absence of the theoretical guidance made itself felt only in the mid-1970s, after the emergence of the IC technology and digital radios. At that point, some conceptually incorrect decisions, related to S&I of bandpass signals, were made. Along with other topics, these decisions are analyzed in this and subsequent chapters.

The historical sketch above shows, in particular, that the gap between theorists and practical engineers delayed the development of TDM and S&I techniques in the first half of the twentieth century. This happened when communications and broadcasting were relatively small fields, whereas television, radar, sonar, navigation, and direction-finding were in their cradles. At present, a lot of theorists work in various areas of electrical engineering. However, the sampling theory and applications are used in many technical fields that have become huge and require deep specialization of researchers and engineers. The gaps among them, created by the specialization, impede the progress of the sampling theory and S&I techniques. This book is intended to bridge the gaps.

5.3 Uniform Sampling Theorem for Baseband Signals

5.3.1 Sampling Theorem and Its Constructive Nature

Although the WKS sampling theorem is discussed for both baseband and bandpass signals in this chapter, the baseband version is considered first due to its relative simplicity and the possibility to reduce S&I of bandpass signals to two-channel baseband operations.

5.3.1.1 Theorem Statement and Explanation

According to this theorem, an analog baseband real-valued signal $u(t)$ with one-sided bandwidth B can be represented by its instantaneous values $u(nT_s)$ sampled uniformly with period $T_s = 1/(2B)$ and reconstructed from $u(nT_s)$ using the equation

$$u(t) = \sum_{n=-\infty}^{\infty} u(nT_s)\varphi_{nBB}(t) = \sum_{n=-\infty}^{\infty} u(nT_s)\varphi_{0BB}(t - nT_s) \qquad (5.3)$$

where n is an integer, and $\varphi_{nBB}(t)$ are the baseband sampling functions

$$\varphi_{nBB}(t) = \operatorname{sinc}\left[2\pi B(t - nT_s)\right] = \frac{\sin\left[2\pi B(t - nT_s)\right]}{2\pi B(t - nT_s)} \qquad (5.4)$$

(see Section D.1). When $n = 0$, it follows from (5.4) that

$$\varphi_{0BB}(t) = \operatorname{sinc}(2\pi Bt) = \frac{\sin(2\pi Bt)}{2\pi Bt} \qquad (5.5)$$

Functions $\varphi_{nBB}(t)$ are mutually orthogonal, and their squared norm, that is, the $\varphi_{nBB}(t)$ energy dissipated in the 1Ω resistor, is

$$\left\|\varphi_{nBB}(t)\right\|^2 = \int_{-\infty}^{\infty} \frac{\sin^2\left[2\pi B(t - nT_s)\right]}{\left[2\pi B(t - nT_s)\right]^2} dt = T_s \qquad (5.6)$$

The spectral density of a rectangular signal is the sinc function (1.50). Therefore, according to the time-frequency duality of the Fourier transform (see Section 1.3.3 and Section A.1), the spectral density of $\varphi_{nBB}(t)$ is

$$S_{\varphi nBB}(f) = \begin{cases} T_s \exp(-j2\pi fnT_s) & \text{for } f \in [-B, B] \\ 0 & \text{for } f \notin [-B, B] \end{cases} \qquad (5.7)$$

Thus, $\varphi_{nBB}(t)$ has a uniform amplitude spectrum $|S_{\varphi nBB}(f)| = T_s$ and a linear phase spectrum $-j2\pi fnT_s$ within the frequency interval $[-B, B]$. Its spectral density $S_{\varphi nBB}(f) = 0$ outside this interval. For $n = 0$,

$$S_{\varphi 0BB}(f) = |S_{\varphi nBB}(f)| = \begin{cases} T_s & \text{for } f \in [-B, B] \\ 0 & \text{for } f \notin [-B, B] \end{cases} \qquad (5.8)$$

The plots of $\varphi_{0BB}(t)$ and $S_{\varphi 0BB}(f)$ are shown in Figure 5.1.

5.3 Uniform Sampling Theorem for Baseband Signals

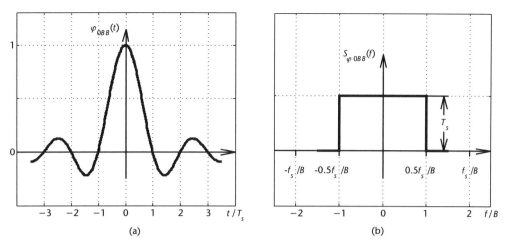

Figure 5.1 (a) Baseband sampling function $\varphi_{0BB}(t)$ and (b) its spectrum $S_{\varphi 0BB}(f)$.

In Figure 5.2, the analog baseband signal $u(t)$ (dashed line) is represented by its samples $u(nT_s)$ (solid lines). The sampling functions $\varphi_{nBB}(t)$ (dotted lines) have values $\varphi_{nBB}(t) = 1$ for $t = nT_s$, and $\varphi_{nBB}(t) = 0$ for $t = (n \pm m)T_s$ where m is a nonzero integer. Consequently, (5.3) is true at any instant $t = nT_s$. Equation (5.3) is also true for any t if $u(t)$ has no spectral components outside $[-B, B]$ (see Section D.1). This theoretical result is intuitively acceptable for smooth signals, similar to the one in Figure 5.2, but questionable for abruptly changing signals. To clarify the situation, recall that the limited bandwidth of $u(t)$ makes its abrupt changes impossible. Moreover, this limitation restricts the rate of change in $u(t)$ the same way as in $\varphi_{nBB}(t)$. Indeed, the wider bandwidth B is, the steeper the variation of $u(t)$ can be, and the shorter the $\varphi_{nBB}(t)$ lobes are.

The discrete-time signal $u_d(t)$ obtained after sampling of $u(t)$ is a train of samples $u(nT_s)$ that ideally should be a product of $u(t)$ and a uniform train of delta functions $\delta_{T_s}(t)$ (see Section A.2):

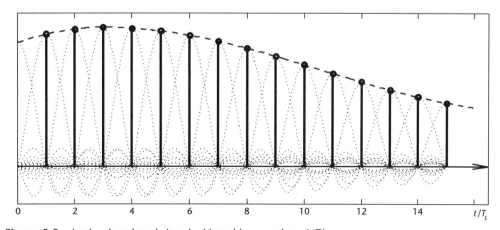

Figure 5.2 Analog baseband signal $u(t)$ and its samples $u(nT_s)$.

$$u_d(t) = u(t)\delta_{T_s}(t) = u(t)\sum_{n=-\infty}^{\infty}\delta(t-nT_s) = \sum_{n=-\infty}^{\infty}u(t)\delta(t-nT_s)$$
$$= \sum_{n=-\infty}^{\infty}u(nT_s)\delta(t-nT_s) \quad (5.9)$$

According to the frequency convolution property of the Fourier transform, the time-domain product $u(t)\delta_{T_s}(t)$ in (5.9) corresponds to the convolution $S(f) * S_{\delta T_s}(f)$ in the frequency domain (see Section 1.3.3 and Section A.2):

$$S_d(f) = S(f) * S_{\delta T_s}(f) = \frac{1}{T_s}\sum_{k=-\infty}^{\infty}S(f-kf_s) \quad (5.10)$$

where $S_d(f)$, $S(f)$, and $S_{\delta T_s}(f)$ are, respectively, the spectral densities of $u_d(t)$, $u(t)$, and $\delta_{T_s}(t)$, whereas $f_s = 2B = 1/T_s$ is the sampling rate. Thus, $S_d(f)$ is a periodic function with period f_s, which means that, sampling causes replication of the original $u(t)$ spectrum $S(f)$ as shown in Figure 5.3. This explains the origin of (3.13).

5.3.1.2 Constructive Nature of Sampling Theorem

As noted in Section 5.2.2, the sampling theorem has a constructive nature because it provides the ways of representing a bandlimited signal $u(t)$ by its samples $u(nT_s)$ or by the samples $y(nT_s)$ of its function $y(t) = y[u(t)]$, and the optimal, in the least-squares sense, interpolation algorithms. These two components follow from the theorem's statement. Indeed, $u(t)$ can be represented by its samples $u(nT_s)$ and reconstructed from them using interpolation equation (5.3). As shown below, the theorem's constructive nature includes the third component: the optimal sampling algorithms.

To deduce the optimal sampling algorithm from the sampling theorem, recall that sampling is an expansion of $u(t)$ by the generalized Fourier series with respect to the set of sampling functions $\{\varphi_{nBB}(t) = \varphi_{0BB}(t-nT_s)\}$ where samples $u(nT_s)$ are

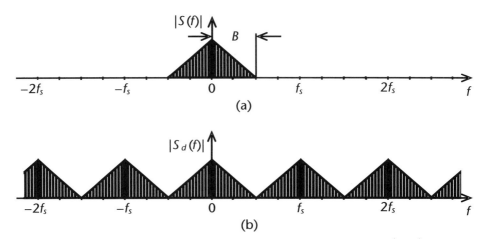

Figure 5.3 Ideal baseband sampling, amplitude spectra of $u(t)$ and $u_d(t)$: (a) $|S(f)|$ and (b) $|S_d(f)|$.

the coefficients of the series. The minimum rms error of sampling is achieved (see Section 1.3.1) when these coefficients are calculated as

$$u(nT_s) = \frac{1}{\|\varphi_{nBB}(t)\|^2} \int_{-\infty}^{\infty} u(t)\varphi^*_{nBB}(t)dt \qquad (5.11)$$

where $\varphi^*_{nBB}(t)$ is the complex conjugate of $\varphi_{nBB}(t)$. Since $\varphi_{nBB}(t)$ is real-valued, $\varphi^*_{nBB}(t) = \varphi_{nBB}(t)$. As follows from (5.6), all $\varphi_{nBB}(t)$ have the same squared norm $\|\varphi_{nBB}(t)\|^2 = T_s$. Therefore, (5.11) can be rewritten as:

$$u(nT_s) = \frac{1}{T_s} \int_{-\infty}^{\infty} u(t)\varphi_{nBB}(t)dt = c \int_{-\infty}^{\infty} u(t)\varphi_{0BB}(t - nT_s)dt \qquad (5.12)$$

where $c = 1/T_s$ is a constant. According to (5.12), the optimal sampling accumulates $u(t)$ with weight $\varphi_{nBB}(t)$ for the nth sample. It also performs internal antialiasing filtering. Indeed, in (5.12), replacing $u(t)$, whose spectrum $S(f)$ is located within the interval $[-B, B]$, with an input signal $u_{in}(t)$, whose spectrum $S_{in}(f)$ includes $S(f)$ but is wider than $[-B, B]$, does not change the result of $u(nT_s)$ calculation because (5.12) multiplies $S_{in}(f)$ by the spectrum $S_{\varphi nBB}(f)$ of $\varphi_{nBB}(t)$ (see Figure 5.1(b)), as follows from Section 1.3.5. The multiplication rejects all the spectral components of $S_{in}(f)$ outside $[-B, B]$, performing antialiasing filtering. This allows rewriting (5.12) in a more general form:

$$u(nT_s) = c \int_{-\infty}^{\infty} u_{in}(t)\varphi_{nBB}(t)dt = c \int_{-\infty}^{\infty} u_{in}(t)\varphi_{0BB}(t - nT_s)dt \qquad (5.13)$$

Thus, the sampling theorem not only shows the way of discrete-time representation of analog signals, but also provides the least-squares algorithms for their S&I reflected, respectively, by (5.13) and (5.3). The constructive nature demonstrated by the sampling theorem's baseband version is also demonstrated by the bandpass one. Despite the advantages of sampling algorithm (5.13), that is, efficient accumulation of the signal energy and internal antialiasing filtering, it still does not attract sufficient attention of researchers and engineers.

In many publications, sampling is still described as an operation performed by a multiplier, multiplying an original analog signal $u(t)$ by a uniform train $u_{s,p}(t)$ of short pulses, or by an electronic switch controlled by $u_{s,p}(t)$, as illustrated, respectively, in Figures 5.4(a, b). Both methods produce identical results if $u_{s,p}(t)$ is the same. Therefore, the timing diagram in Figure 5.4(c) corresponds to both methods. Note that this sampling performs PAM and wastes most of $u(t)$ energy.

5.3.1.3 Antialiasing Filtering

Due to the importance of antialiasing filtering, it is concisely discussed below. In communications and signal processing, aliasing refers to a phenomenon that makes different signals identical after their sampling or distortions that cause the divergence of interpolated signals from the original ones whose samples were used for

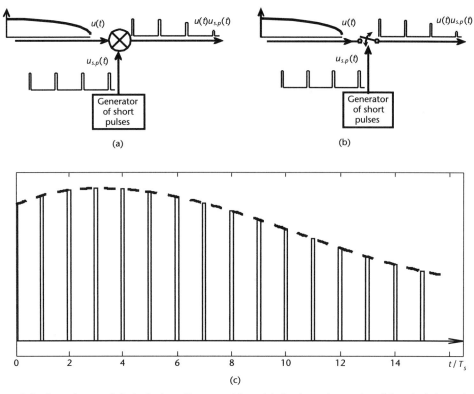

Figure 5.4 Samplers and their timing diagram: (a) multiplier-based sampler, (b) switch-based sampler, and (c) timing diagram of $u(t)$ (dashed line) and its samples $u(t)u_{s,p}(t)$ (solid line).

the interpolation. In digital radios, aliasing is typically temporal (i.e., related to time-domain sampling).

Figure 5.5 demonstrates aliasing of three cosine signals of different frequencies represented by the same samples. The frequencies of the first (dotted line), second (dashed line), and third (solid line) signals are, respectively, $f_1 = (1/3)f_s$, $f_2 = f_1 + f_s$, and $f_3 = f_1 + 2f_s$ where f_s is the sampling rate. It is clear that all cosine signals with the same amplitudes and initial phases, which have frequencies $f_k = f_1 + (k-1)f_s$, produce the same samples if k is a natural number. The fact that a train of cosine samples can represent an infinite number of cosine signals with properly related frequencies has the same nature as the replication of the analog signal spectrum caused by sampling (see Figure 5.3). Returning to Figure 5.5, it is easy to notice that an unlimited number of periodic signals different from cosine can also be represented by the same samples. However, if only the signals whose spectra are located within the frequency interval $[-0.5f_s, 0.5f_s[$ are selected by appropriate filtering, the samples represent only the cosine signal with frequency $f_1 = (1/3)f_s$, depicted with a dotted line.

Aliasing that refers to the distortions causing the divergence of the interpolated signals from the original ones is shown in Figure 5.6. Spectrum $S_{in}(f)$ of $u_{in}(t)$ in Figure 5.6(a) contains the desired signal $u(t)$, whose spectrum $S(f)$ is located within the interval $[-B, B]$, and out-of-band spectral components within the intervals $[-f_1, -f_2]$ and $[f_1, f_2]$. After sampling at the rate $f_s = 2B$, the out-of-band spectral components fall within $S(f)$ as a result of the $S_{in}(f)$ replication, as shown in Figure

5.3 Uniform Sampling Theorem for Baseband Signals

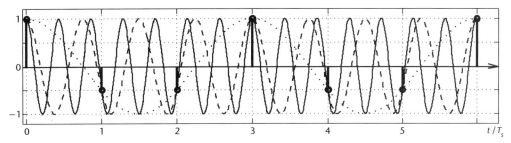

Figure 5.5 Aliasing of cosine signals.

5.6(b). Figure 5.6 demonstrates that damaging effects of aliasing are easier to prevent than to compensate.

Antialiasing filtering rejects all undesired out-of-band spectral components of $u_{in}(t)$ prior to or during its sampling, as illustrated in Figure 5.7. The AFR $|H_{a.f}(f)|$ of the ideal antialiasing filter is depicted with a dashed line in Figure 5.7(a). As a result, the amplitude spectrum $|S_d(f)|$ of $u_d(t)$, shown in Figure 5.7(b), allows accurate reconstruction of the desired signal $u(t)$. The transfer function of the ideal antialiasing filter should have a rectangular AFR with the cutoff frequency $0.5f_s = B$ and a linear PFR:

$$H_{a.f}(f) = \begin{cases} H_0 \exp(-j2\pi f t_0) & \text{for } f \in [-B, B] \\ 0 & \text{for } f \notin [-B, B] \end{cases} \quad (5.14)$$

where t_0 is a group delay. Comparison of (5.14) with (5.7) shows that (5.13) provides ideal antialiasing filtering.

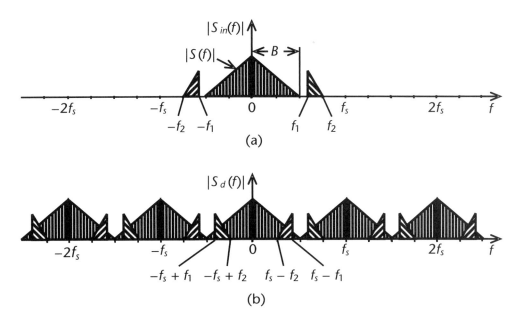

Figure 5.6 Effects of temporal aliasing in the frequency domain: (a) amplitude spectrum $|S_{in}(f)|$ of $u_{in}(t)$ and (b) amplitude spectrum $|S_d(f)|$ of $u_d(t)$.

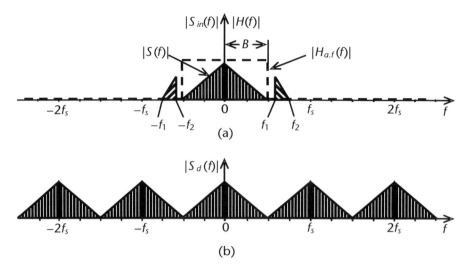

Figure 5.7 Ideal antialiasing filtering: (a) amplitude spectrum $|S_{in}(f)|$ of $u_{in}(t)$ and AFR $|H_{a.f}(f)|$ of the antialiasing filter, and (b) amplitude spectrum $|S_d(f)|$ of $u_d(t)$.

To illustrate some specifics of antialiasing filtering, let us consider sampling of a musical signal at a Tx input. If $f_s = 16$ kHz, the cutoff frequency is $0.5f_s = 8$ kHz, and the signal spectral components above 8 kHz will be folded over the signal components with lower frequencies, creating aliasing. Without antialiasing filtering (see the spectral diagrams in Figure 5.8), the signal is corrupted by two types of distortion: the first type is caused by the loss of the spectral components above 8 kHz and the second type is caused by aliasing.

Figure 5.9 illustrates the situation when $u_{in}(t)$ undergoes antialiasing filtering. Figure 5.9(a) presents its amplitude spectrum $|S_{in}(f)|$ and the AFR $|H_{a.f}(f)|$ of an antialiasing filter. The amplitude spectrum $|S(f)|$ of $u(t)$ obtained after this filtering is shown in Figure 5.9(b). The amplitude spectrum $|S_d(f)|$ of the discrete-time signal $u_d(t)$ is depicted in Figure 5.9(c). Figure 5.9 shows that antialiasing filtering excludes the distortion of the second type, but does not change the distortion of

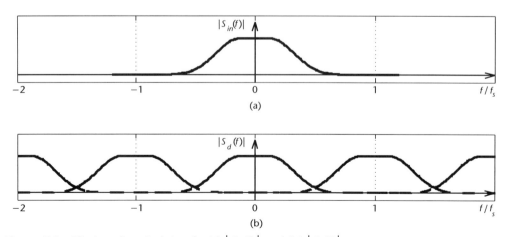

Figure 5.8 Aliasing of musical signals: (a) $|S_{in}(f)|$ and (b) $|S_d(f)|$.

the first type, which can be reduced by increasing f_s. When f_s exceeds the highest audible frequency of the musical signal by at least factor of 2, the distortion of the first type disappears.

This condition is met if $f_s \geq 32$ kHz. Antialiasing filtering is needed even with this f_s because music may have spectral components above the audible range. Although $f_s \geq 32$ kHz (combined with antialiasing filtering) eliminates both types of distortions, any increase of f_s expands the transmitted signal bandwidth and requires higher transmit power to provide the same noise immunity of the transmission. Both bandwidth and power are valuable resources. Therefore, selection of f_s and parameters of antialiasing filters requires a thorough analysis that takes into account the necessary quality of transmission, type of source coding, and available bandwidth and power. In mass-produced equipment, these factors are taken into account in advance.

5.3.2 Interpretations of Sampling Theorem

As shown above, sampling of baseband signals according to (5.13) and their interpolation according to (5.3) constitute a way to perform S&I that is optimal in the least-squares sense. Simultaneously, it is easy to notice a different way of S&I that is currently well known. Indeed, Section A.2 makes it clear that applying the sifting property of a train of delta functions to a signal $u(t)$ allows its sampling at the output of an antialiasing filter, and Figure 5.4 demonstrates the possibility to replace the delta functions by short pulses. Figures 5.3 and 5.7 make it evident that rejection of the spectral replicas outside the frequency interval $[-B, B]$ in $S_d(f)$ reconstructs $u(t)$.

It is shown below that the latter way of S&I, which seems heuristic, also follows from the sampling theorem and, consequently, minimizes the rms error. These two

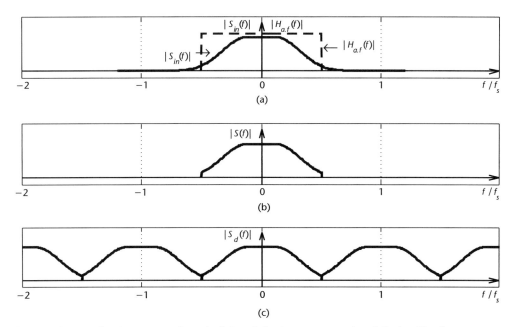

Figure 5.9 Amplitude spectra of musical signals in the presence of antialiasing filtering: (a) $|S_{in}(f)|$ and $|H_{a.f}(f)|$ (dashed line), (b) $|S(f)|$, and (c) $|S_d(f)|$.

ways correspond to different interpretations of the sampling theorem. It is logical to call the first interpretation, which is based on the original theorem's equations (5.3) and (5.13), the direct interpretation. Respectively, all interpretations that are based on the transformed equations are indirect. In this book, however, the term "indirect interpretation" refers only to the interpretation based exclusively on the use of traditional analog antialiasing and interpolating filters.

5.3.2.1 Derivation of the Sampling Theorem's Indirect Interpretation

Let us prove that sending $u_{in}(t)$ through a filter with the impulse response $h(t) = c\varphi_{0BB}(-t)$ and applying the sifting property of the delta function to the filter output produces the same result as (5.13). Since $\varphi_{0BB}(t)$ is even,

$$h(t) = c\varphi_{0BB}(-t) = c\varphi_{0BB}(t) \tag{5.15}$$

From (1.79), (5.15), and (A.14), we get

$$u(nT_s) = \int_{-\infty}^{-\infty}\left[\int_{-\infty}^{\infty} u_{in}(\tau)h(t-\tau)d\tau\right]\delta(t-nT_s)dt = \int_{-\infty}^{\infty} u_{in}(\tau)h(nT_s-\tau)d\tau$$

$$= c\int_{-\infty}^{\infty} u_{in}(t)\varphi_{0BB}(t-nT_s)dt \tag{5.16}$$

Since (5.13) and (5.16) produce the same result, sampling according to (5.16) also minimizes the rms error. This sampling is illustrated in Figure 5.10(a) where the ideal antialiasing LPF with bandwidth B rejects all out-of-band spectral components of $u_{in}(t)$ without distorting $u(t)$, which is multiplied by the train of $\delta(t - nT_s)$. Subsequent circuits implicitly perform the outer integration in the second part of (5.16).

To show that (5.3) is equivalent to sending $u_d(t)$ through the ideal interpolating LPF with impulse response $h(t) = \varphi_{0BB}(t)$, recall (1.79). Modifying it with regard to (5.9) yields

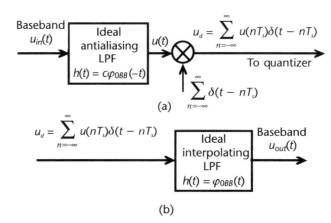

Figure 5.10 Ideal S&I of baseband signals (indirect interpretation): (a) sampling structure and (b) interpolating structure.

$$u_{out}(t) = \int_{-\infty}^{\infty} \left\{ \sum_{n=-\infty}^{\infty} \left[u(nT_s)\delta(\tau - nT_s) \right] \right\} h(t-\tau)\, d\tau$$

$$= \sum_{n=-\infty}^{\infty} u(nT_s) \int_{-\infty}^{\infty} h(t-\tau)\delta(\tau - nT_s)\, d\tau \qquad (5.17)$$

$$= \sum_{n=-\infty}^{\infty} u(nT_s) h(t - nT_s)$$

Since $h(t) = \varphi_{0BB}(t)$, (5.17) can be rewritten:

$$u_{out}(t) = \sum_{n=-\infty}^{\infty} u(nT_s)\varphi_{0BB}(t - nT_s) = \sum_{n=-\infty}^{\infty} u(nT_s)\varphi_{nBB}(t) = u(t) \qquad (5.18)$$

as was to be proved. Thus, summing the sampling functions $\varphi_{0BB}(t - nT_s)$ with the weights equal to samples $u(nT_s)$ is equivalent to the ideal interpolating filtering. Interpolation that corresponds to the indirect interpretation of the sampling theorem is illustrated in Figure 5.10(b) where the ideal interpolating LPF with impulse response $h(t) = \varphi_{0BB}(t)$ rejects all spectral replicas in $S_d(f)$ except the one corresponding to $S(f)$ of $u(t)$.

The intermediate position that is between the direct and indirect interpretations of the sampling theorem is occupied by its hybrid interpretation derived in Section 5.3.2.4.

5.3.2.2 Ideal and Nonideal S&I

The ideal S&I based on either direct or indirect interpretation are physically unrealizable because the WKS sampling theorem uses many assumptions that cannot be met in actuality. First, strictly bandlimited signals have infinite duration, whereas all practical signals are time-limited. Second, sampling functions $\varphi_{nBB}(t)$ defined by (5.4) and a filter with the impulse response $h(t) = c\varphi_{0BB}(-t)$ contradict the causality principle and therefore are physically unrealizable. Third, the delta functions $\delta(t - nT_s)$ and signal samples of infinitesimal duration also cannot be generated. This makes only nonideal S&I possible.

Nonideal realization of S&I based on the indirect interpretation means replacing the delta functions with short gating pulses and ideal filters with nonideal physically realizable ones. The AFRs of nonideal filters are nonuniform within the passbands and provide finite suppression within the stopbands, their PFRs may be nonlinear, they introduce losses and have nonzero transition bands that require $f_s > 2B$. Sufficient suppression of out-of-band interference and tolerable distortions within the passbands make them acceptable for practical purposes. Nonideal realization of S&I based on the direct interpretation necessitates replacing physically unrealizable $\varphi_{nBB}(t) = \varphi_{0BB}(t - nT_s)$ in (5.3) and (5.13) with realizable weight functions $w_{nBB}(t) = w_{0BB}(t - nT_s)$ that also require $f_s > 2B$.

Besides physical realizability, implementation of S&I circuits requires taking into account linear and nonlinear distortions, jitter, and interference there, as well as similar phenomena in the preceding and subsequent stages. It also requires considering the adaptivity and reconfigurability of S&I circuits and their compatibility with the IC technology. Therefore, while all interpretations derived from the sampling theorem's original equations are equally optimal in the ideal case, they provide different performance and encounter different implementational challenges in real-world situations (see their initial comparison in Section 5.3.2.5). Since the optimality of nonideal realizations of S&I circuits cannot be determined within the scope of the sampling theorem, their theoretical basis should include, besides the sampling theory, the theories of linear and nonlinear circuits, optimal filtering, and so forth.

5.3.2.3 On Implementation of Innovations

S&I techniques illustrate that three major factors determine how and when innovations are implemented: practical demand, technological level, and theoretical basis. The absence of at least one of these factors makes the implementation impossible. While the implementation of TDM was substantially delayed by the absence of its theoretical basis (since the publications on the sampling theorem were initially overlooked), the selection of this theorem's indirect interpretation for the early-day TDM systems was justified by both contemporary technology and application. Indeed, generation of proper $w_{nBB}(t)$ required for the direct interpretation was virtually impossible then, whereas filters with acceptable $h(t)$ were available. Simultaneously, the loss of most signal energy caused by the use of short pulses for sampling was unimportant, as shown below.

In the earliest TDM communication systems (see Figure 5.11), the analog signals from K information sources S_k (where $k = 1, 2, ..., K$) underwent antialiasing filtering and sampling (usually combined with time multiplexing). The time-multiplexed pulse-amplitude-modulated samples were amplified and transmitted over a long-distance link to the system's Rx side, where the weakened group signal was amplified again and demultiplexed. Then, the signals of partial channels were interpolated separately for each recipient R_k. In this system, f_s was the same for all channels and a fixed time slot was allocated for each channel. Such systems utilized their equipment much better than single-channel ones because the most expensive parts were jointly used by all channels.

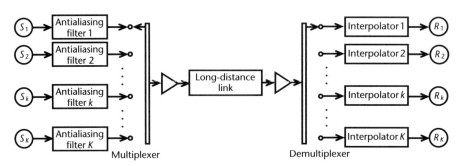

Figure 5.11 Early communication system with TDM.

The sampling structure used in the considered system is shown in Figure 5.12. For voice transmission, antialiasing filters suppressed the spectral components above ~3.0 kHz due to their negligible influence on the speech intelligibility and the possibility to reduce f_s. Sampling at the antialiasing filters' outputs was executed by an electronic switch controlled by a uniform train of short pulses $u_{s.p}(t)$. Although the sampling was lossy for the partial channels, the system's overall energy efficiency was high because most of the energy was used for transmission of group signals with high duty cycles over the long-distance link.

In the next generation of TDM systems, PAM was replaced with PWM and PPM due to their higher noise immunity, simpler pulse regeneration because of their constant amplitude, and availability of sufficiently accurate synchronization required for PWM and PPM. Finally, PWM and PPM in TDM systems were replaced with PCM where pulses were digitized and transmitted using binary code. High noise immunity of PCM, further enhanced by error-correcting coding, allows attaining virtually any required reliability of communications. On top of that, it provides unprecedented flexibility of selecting modulation/demodulation and encoding/decoding techniques.

The history of PCM is another example demonstrating that all three factors (practical demand, technological level, and theoretical basis) are needed for implementation of any innovation. PCM was invented by A. H. Reeves in 1937 [56, 57]. The inventor and his colleagues clearly understood its theory and advantages. The demand for this technique already existed. Still, despite many additional PCM-related inventions, its wide implementation started in the late 1950s when digital circuits, A/Ds, and D/As based on the transistor technology became mature enough. Thus, PCM implementation was delayed by about 20 years due to the absence of proper technology, whereas TDM implementation was delayed by insufficient theoretical knowledge. As shown in this book, insufficient theoretical knowledge also has significantly delayed the implementation of novel D&R techniques.

5.3.2.4 Derivation of the Sampling Theorem's Hybrid Interpretation

Formal derivation of the hybrid interpretation of the sampling theorem, which was mentioned above, requires representing the sampling function $\varphi_0(t)$ as a convolution of two functions:

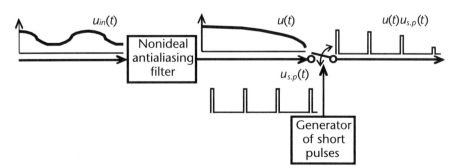

Figure 5.12 Sampling in early communication systems with TDM.

$$\varphi_0(t) = \tilde{\varphi}_0(t) * \hat{\varphi}_0(t) = \int_{-\infty}^{\infty} \tilde{\varphi}_0(\tau)\hat{\varphi}_0(t-\tau)d\tau = \int_{-\infty}^{\infty} \tilde{\varphi}_0(t-\tau)\hat{\varphi}_0(\tau)d\tau \quad (5.19)$$

In (5.19) and below, the sampling function designation $\varphi_0(t)$ means that the results of the subsequent derivation are applicable to both baseband and bandpass sampling functions: $\varphi_{0BB}(t)$ and $\varphi_{0BP}(t)$. Using translational equivalence of convolution, $\varphi_n(t)$ can be expressed as

$$\varphi_n(t) = \varphi_0(t - nT_s) = \tilde{\varphi}_0(t) * \hat{\varphi}_0(t - nT_s) = \int_{-\infty}^{\infty} \tilde{\varphi}_0(t-\tau)\hat{\varphi}_0(\tau - nT_s)d\tau \quad (5.20)$$

With regard to (5.20), (5.13) can be rewritten as:

$$\begin{aligned}
u(nT_s) &= c\int_{-\infty}^{\infty} u_{in}(t)\varphi_n(t)\,dt = c\int_{-\infty}^{\infty} u_{in}(t)\left[\tilde{\varphi}_0(t) * \hat{\varphi}_0(t - nT_s)\right]dt \\
&= c\int_{-\infty}^{\infty} u_{in}(t)\left[\int_{-\infty}^{\infty} \tilde{\varphi}_0(t-\tau)\hat{\varphi}_0(\tau - nT_s)d\tau\right]dt \\
&= c\int_{-\infty}^{\infty}\left[\int_{-\infty}^{\infty} u_{in}(t)\tilde{\varphi}_0(t-\tau)dt\right]\hat{\varphi}_0(\tau - nT_s)d\tau \\
&= c\int_{-\infty}^{\infty}\left[u_{in}(\tau) * \tilde{\varphi}_0(-\tau)\right]\hat{\varphi}_0(\tau - nT_s)d\tau \\
&= c\int_{-\infty}^{\infty}\left[u_{in}(t) * \tilde{\varphi}_0(-t)\right]\hat{\varphi}_0(t - nT_s)dt
\end{aligned} \quad (5.21)$$

The last part of (5.21) reflects the two-stage sampling procedure corresponding to the hybrid interpretation.

In practice, the physical unrealizability of $\varphi_0(t)$ requires replacing it with an appropriate physically realizable weight function $w_0(t)$ represented by convolution $w_0(t) = \tilde{w}_0(t) * \hat{w}_0(t)$:

$$u(nT_s) \approx c \int_{nT_s - 0.5T_{\hat{w}}}^{nT_s + 0.5T_{\hat{w}}} \left[u_{in}(t) * \tilde{w}_0(-t)\right]\hat{w}_0(t - nT_s)dt \quad (5.22)$$

where $T_{\hat{w}}$ is the length of $\hat{w}_0(t)$. According to (5.22), a prefilter with impulse response $\breve{h}(t) = c\tilde{w}_0(-t)$ starts antialiasing filtering of $u_{in}(t)$ at the first stage, as indicated by the convolution in brackets. Then, at the second stage, the prefilter output is integrated with weight $\hat{w}_0(t - nT_s)$, completing this filtering, accumulating the signal energy, and producing $u(nT_s)$.

To derive the sampling theorem's hybrid interpretation for interpolation, it is convenient to use the translational equivalence of convolution for expressing $\varphi_0(t)$ differently from (5.20):

$$\varphi_n(t) = \varphi_0(t - nT_s) = \tilde{\varphi}_0(t - nT_s) * \hat{\varphi}_0(t) = \int_{-\infty}^{\infty} \tilde{\varphi}_0(\tau - nT_s)\hat{\varphi}_0(t - \tau)d\tau \quad (5.23)$$

Rewriting (5.3) with regard to (5.23) yields:

$$\begin{aligned} u(t) &= \sum_{n=-\infty}^{\infty} u(nT_s)\varphi_n(t) = \sum_{n=-\infty}^{\infty} u(nT_s)\varphi_0(t - nT_s) \\ &= \sum_{n=-\infty}^{\infty}\left[u(nT_s) \int_{-\infty}^{\infty} \tilde{\varphi}_0(\tau - nT_s)\hat{\varphi}_0(t - \tau)d\tau \right] \\ &= \int_{-\infty}^{\infty}\left[\sum_{n=-\infty}^{\infty} u(nT_s)\tilde{\varphi}_0(\tau - nT_s) \right]\hat{\varphi}_0(t - \tau)d\tau \end{aligned} \quad (5.24)$$

After replacing physically unrealizable $\varphi_0(t)$ with an appropriate $w_0(t)$ represented again as $w_0(t) = \tilde{w}_0(t) * \hat{w}_0(t)$, (5.24) can be replaced with:

$$u(t) \approx \int_{-\infty}^{\infty}\left[\sum_{n=-\infty}^{\infty} u(nT_s)\tilde{w}_0(\tau - nT_s) \right]\hat{w}_0(t - \tau)d\tau \quad (5.25)$$

According to (5.25), the interpolation, corresponding to the hybrid interpretation, sums the products of the D/A output samples $u(nT_s)$ and $\tilde{w}_0(t - nT_s)$ at the first stage, concentrating most of the energy within the $u(t)$ bandwidth and starting the interpolating filtering. At the second stage, a postfilter with impulse response $\hat{h}(t) = \hat{w}_0(t)$ completes this filtering.

5.3.2.5 Initial Comparison of the Interpretations

This section confirms the constructive nature of the sampling theorem and presents its interpretations, corresponding to different forms of the theorem equations, which reflect specific S&I algorithms. While realization of these algorithms is considered in Chapter 6, the interpretations are concisely compared below. Since the direct and indirect interpretations represent the extreme cases, each of them has only one version (still allowing different realizations). In contrast, the hybrid interpretation has, in principle, an infinite number of versions because of the infinite number of $\{\tilde{w}_0(t), \hat{w}_0(t)\}$ pairs satisfying $w_0(t) = \tilde{w}_0(t) * \hat{w}_0(t)$ for the same weight function $w_0(t)$.

The implementation of PCM and DSP in the late 1950s and the emergence of digital radios in the 1970s radically changed the requirements for S&I techniques. Therefore, the interpretations are compared according to their capabilities to meet these requirements. First, the need to hold samples for the duration of their quantization, which stimulated the replacement of simple electronic switches by SHAs and THAs, is met by all the interpretations. Second, the loss of most of the signal energy during sampling, detrimental to the new applications, is prevented in the circuits based on the direct interpretation, but cannot be avoided in those based on the indirect one. Third, efficient interpolation requires concentrating most of the D/A output samples' energy within the reconstructed analog signal's bandwidth, and again the direct interpretation allows achieving this goal, but the indirect one does not. Fourth, IC implementation of S&I circuits in the new applications is very desirable, and the direct interpretation allows it, whereas the indirect one often prevents it, due to the incompatibility of the best bandpass antialiasing and interpolating filters with the IC technology. Fifth, high flexibility of S&I circuits, which is also very desirable in the new applications, is provided by the direct interpretation, but is not always attainable by the indirect one due to the inflexibility of the bandpass filters mentioned above.

Despite the advantages of S&I circuits based on the direct interpretation, these circuits are still in the R&D phase, although this interpretation was initially described a long time ago [46–50]. The delay of their wide implementation is partly caused by the need to solve some technological problems, but mostly by incomplete understanding of their substance and benefits. As to the hybrid interpretation, it is advantageous over the indirect interpretation but potentially inferior to the direct one. It was proposed in the 1980s [43–45], and, despite the absence of technological obstacles even at that time, its practical implementation was initially slow because its concept deviated from the entrenched paradigm.

The emergence of DSP and digital radios created not only new challenges but also new opportunities that simplify S&I realization. For instance, one such opportunity is the possibility to compensate in the digital domain some distortions produced in analog and mixed-signal circuits (see Chapters 3 and 4).

5.3.3 Baseband S&I Corresponding to Indirect Interpretation

The S&I in all D&R procedures, described in Chapters 3 and 4, are based on the indirect interpretation of the sampling theorem. Therefore, nonideal baseband sampling, corresponding to this interpretation, can be illustrated by the block and spectral diagrams shown, respectively, in Figures 3.4 and 3.5 (see Section 3.3.1). Several general facts presented in that section and related to sampling of baseband signals are emphasized below.

Nonideal sampling is closely connected not only to the subsequent quantization but also to the following DSP because some problems caused by it can be offset and/or compensated in the TDP. The spectrum $S_{in}(f)$ of an input signal $u_{in}(t)$ may contain spectra of ISs besides the spectrum $S(f)$ of a desired signal $u(t)$. In Txs, the ISs can be, for example, signals of neighboring channels not intended for transmission. As mentioned in Section 3.3.1 and explained in Section 5.3.1.1 (see (5.9) and (5.10)), sampling causes replication of the sampler's input signal spectrum, which is illustrated for the ideal sampling with $f_s = 2B$ in Figure 5.3 and for practical sampling with $f_{s1} = 1/T_{s1} > 2B$ in Figure 3.5. Such f_{s1} is selected due to a wide transition band of an antialiasing filter and the possibility of its passband being wider than B. As a result, some ISs pass through the antialiasing filter and are rejected by digital filters in the TDP, as shown in Figure 3.5(a, b).

As explained in Section 3.3.1, an antialiasing filter must suppress the $u_{in}(t)$ spectral components corresponding to $k \neq 0$ within intervals (3.14) where the $S(f)$ replicas appear after sampling. It does not have to suppress the $u_{in}(t)$ spectral components within the gaps between these intervals ("don't care" bands) because these components can be rejected in the TDP. Although traditional antialiasing filters do not utilize the existence of "don't care" bands, these bands allow increasing the efficiency of antialiasing and interpolating filtering based on the direct and hybrid interpretations of the sampling theorem, as explained in Chapter 6. Recall that the spectrum $S_{q1}(f)$ of the quantized (i.e., digital) signal $u_{q1}(nT_{s1})$ in Figure 3.5(b) is virtually identical to the spectrum $S_{d1}(f)$ of the discrete-time signal $u_1(nT_{s1})$ when the quantization is accurate (see also Figure 3.4). The downsampling with digital decimating filtering, performed after this quantization, increases the DSP efficiency in the TDP. The decimating filter also rejects ISs not completely suppressed by the antialiasing filter, as shown in Figure 3.5(b, c).

As mentioned above, many problems caused by nonideal sampling can be solved in the TDP. Indeed, downsampling with digital decimating filtering in the TDP offsets the negative effect of a nonideal antialiasing filter's wide transition band. Linear and nonlinear distortions within this filter's passband can be at least partly compensated in the TDP. However, insufficient suppression within this filter's stopbands cannot be compensated there. Therefore, this suppression determines the possibility of effective antialiasing filtering.

Sampling described in Section 3.3.1 and based on the indirect interpretation of the sampling theorem is widely used in practice. The employment of conventional analog antialiasing filters and THAs is typical for it. Although this interpretation is inferior to the direct and hybrid ones in real-world conditions, its inferiority is less significant in the case of baseband S&I than in the case of bandpass S&I.

The ideal interpolation of baseband signals, corresponding to the sampling theorem's indirect interpretation, is illustrated by the block diagram in Figure 5.10(b) and spectral diagrams in Figure 5.13.

Such interpolation is physically unrealizable, and practical interpolation is always nonideal. The block diagram in Figure 5.14 reflects the latter interpolation as a part of baseband signal reconstruction. The reconstruction of this type is often used at the outputs of digital Rxs. Prior to entering the D/A, digital signal $u_{q1}(nT_{s1})$ usually undergoes upsampling with digital interpolating filtering (see Appendix B).

This upsampling is needed because most of the RDP signal processing is performed at the minimum possible $f_s = f_{s1} = 1/T_{s1}$ to efficiently utilize the RDP hardware,

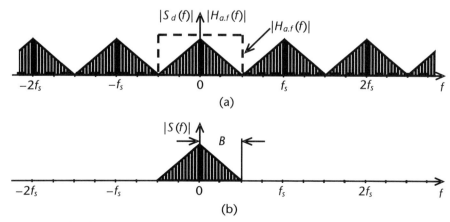

Figure 5.13 Amplitude spectra of $u_d(t)$ and $u(t)$ and AFR of ideal baseband interpolating filter (indirect interpretation): (a) $|S_d(t)|$ and $|H_{a,f}(f)|$ (dashed line) and (b) $|S(f)|$.

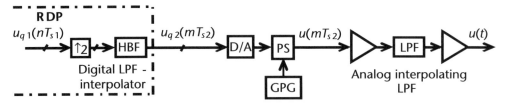

Figure 5.14 Reconstruction of baseband signals (indirect interpretation).

but a wide transition band B_t of the analog interpolating LPF requires increasing f_s at the D/A's input. Since an even AFR and a linear PFR of the digital interpolating filter are desirable, a FIR LPF is typically used for this interpolation. When the upsampling factor is equal to 2 (as shown in Figure 5.14) or a power of 2, the LPF is, respectively, an HBF or a cascade structure of HBFs (see Section B.4). The D/A converts digital samples $u_{q2}(mT_{s2})$ into the analog ones, but the transitions between the adjacent analog samples contain glitches caused by switching time disparities among the D/A bits and between on and off switching. The PS, controlled by the GPG, selects the undistorted segments of the D/A output samples as illustrated by the timing diagrams in Figure 3.7. There, Δt_s is the gating pulse length, whereas Δt_d is the time delay of gating pulses relative to the fronts of the D/A output samples, which must be equal to or longer than the length of the sample's distorted portion. The selected parts of the samples are interpolated by the analog LPFs. This interpolation transforms the sequence of samples $u(mT_{s2})$ into analog signal $u(t)$.

Figure 5.15 shows the signal spectrum transformations during this reconstruction and the required AFRs of the interpolating filters. The amplitude spectrum

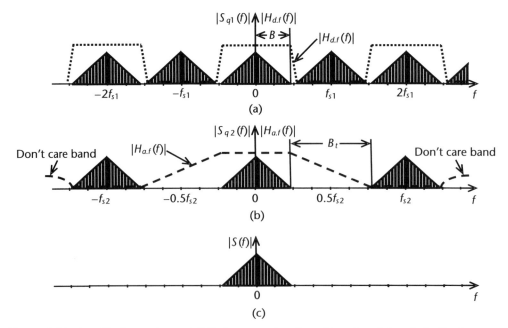

Figure 5.15 Amplitude spectra and AFRs for reconstruction of baseband signals (indirect interpretation): (a) $|S_{q1}(f)|$ and $|H_{d.f}(f)|$ (dotted line), (b) $|S_{q2}(f)|$ and $|H_{a.f}(f)|$ (dashed line), and (c) $|S(f)|$.

$|S_{q1}(f)|$ of $u_{q1}(nT_{s1})$ and AFR $|H_{d.f}(f)|$ of the digital interpolating LPF are depicted in Figure 5.15(a). Spectrum $S_{q1}(f)$ comprises the replicas of the spectrum $S(f)$ of $u(t)$ centered at kf_{s1} where k is any integer. Figure 5.15(b) shows the amplitude spectrum $|S_{q2}(f)|$ of $u_{q2}(nT_{s2})$, which is virtually identical to the amplitude spectrum $|S_{d2}(f)|$ of the discrete-time complex-valued signal $u(nT_{s2})$ when the D/A is accurate. The required AFR $|H_{a.f}(f)|$ of the analog interpolating filter is also displayed in Figure 5.15(b). The upsampling reflected by the spectral diagrams in Figure 5.15(a, b) doubles f_s. The analog interpolating filter reconstructs analog $u(t)$ by rejecting all the replicas of $S(f)$ in $S_{d2}(f)$ except the baseband one. As in the case of the antialiasing and interpolating filters in Chapters 3 and 4, the "don't care" bands of the analog interpolating filter with the AFR depicted in Figure 5.15(b) are not utilized by traditional filtering techniques, but they allow increasing the efficiency of interpolating filters based on the direct and hybrid interpretations of the sampling theorem. Figure 5.15(c) displays the amplitude spectrum $S(f)$ of $u(t)$.

Similar to antialiasing filtering, the problem caused by wide B_t of a nonideal analog interpolating LPF is offset by upsampling in the RDP. Linear and nonlinear distortions within the LPF passband can be at least partly compensated there. However, insufficient suppression within the LPF stopbands cannot be compensated in the RDP.

5.4 Uniform Sampling Theorem for Bandpass Signals

5.4.1 Baseband S&I of Bandpass Signals

The block diagram in Figure 5.16 and the spectral diagrams in Figure 5.17 illustrate the ideal baseband sampling of a bandpass real-valued signal $u_{in}(t)$ (see Section D.2.1) whose spectrum $S_{in}(f)$ may contain unwanted out-of-band components in addition to the spectrum $S(f)$ of the desired signal $u(t)$ (see Figure 5.17(a)). Signal $u_{in}(t)$ is first converted into its baseband complex-valued equivalent $Z_{in}(t)$. After this conversion, the ideal LPFs with AFR $|H_{a.f}(f)|$ perform antialiasing filtering, rejecting the out-of-band spectral components of $Z_{in}(t)$ (see Figure 5.17(b)). As a result, $Z(t)$, represented by $I(t)$ and $Q(t)$, as well as the discrete-time baseband complex-valued signal $Z(nT_s)$, represented by $I(nT_s)$ and $Q(nT_s)$, correspond only to $u(t)$, as illustrated

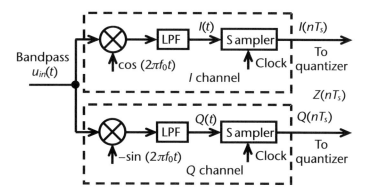

Figure 5.16 Ideal baseband sampling of bandpass signals.

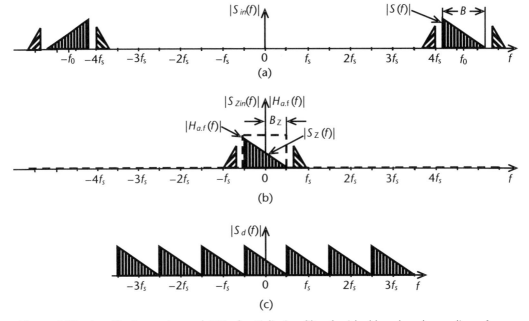

Figure 5.17 Amplitude spectra and AFR of antialiasing filter for ideal baseband sampling of bandpass signals: (a) $|S_{in}(f)|$, (b) $|S_{Zin}(f)|$ and $|H_{a.f}(f)|$ (dashed line), and (c) $|S_d(f)|$.

by the spectral diagrams in Figure 5.17(b, c). The ideal antialiasing filtering allows selecting $f_s = 2B_Z = B$.

Since the ideal antialiasing filtering is physically unrealizable, practical baseband sampling of bandpass signals is always nonideal. It is described as a part of baseband digitization of the Rx input signals in Section 4.4.1 and illustrated by the block and spectral diagrams in Figures 4.9 and 4.10, respectively. This sampling is based on the sampling theorem's indirect interpretation. Wide transition bands of nonideal antialiasing LPFs allow some out-of-band ISs to penetrate into the RDP and require an excessive sampling rate $f_s = f_{s1} > 2B_Z$. The ISs should be rejected in the RDP, and downsampling is needed after quantization to increase the RDP processing efficiency.

Similar to the described above nonideal S&I of baseband signals, many problems of nonideal baseband S&I of bandpass signals can be solved in the radios' RDPs and TDPs. For instance, the negative effect of wide transition bands of antialiasing and interpolating filters is usually offset by appropriate sampling rate conversions in the digital domain; also, linear and nonlinear distortions within the passbands of these filters can be at least partly compensated in the RDPs and TDPs. However, insufficient suppression within the stopbands of antialiasing and interpolating filters cannot be compensated in the digital domain. Therefore, adequate suppression within these stopbands is the necessary condition of effective antialiasing and interpolating filtering.

Condition $f_s > 2B_Z$ also creates "don't care" bands between the antialiasing filter's stopbands. Although the "don't care" bands are not utilized by traditional antialiasing filters, they allow increasing the efficiency of antialiasing filtering based on the direct and hybrid interpretations of the sampling theorem, as explained in Chapter 6.

As shown in Section 1.4.2, baseband complex-valued equivalent $Z(t)$ of a bandpass $u(t)$ can be represented, besides its I and Q components, by its envelope $U(t)$ and phase $\theta(t)$. The latter representation is used only when prior filtering sufficiently suppresses all undesired spectral components of $u_{in}(t)$ and the purpose of signal reception is extracting information from $U(t)$ and/or $\theta(t)$. A way to obtain the values of $U(t)$ and $\theta(t)$ is shown in Figure 5.18 where antialiasing filtering of $u_{in}(t)$ is performed by a BPF (see the sampling theorem for bandpass signals represented by $U(t)$ and $\theta(t)$ in Section D.2.2). Since the BPF and demodulators cannot be ideally realized in practice, some linear and/or nonlinear distortions may require compensation in the RDP.

Ideal baseband interpolation of a bandpass $u(t)$ is performed after separate D/A conversions of the I and Q components of its digital complex-valued equivalent $Z_q(nT_s)$ by two ideal LPFs, each with one-sided bandwidth B_Z, and is followed by forming $u(t)$ from $I(t)$ and $Q(t)$ obtained as a result of this interpolation. Practical interpolation is nonideal. As a part of baseband reconstruction of Tx output signals, it is described in Section 3.3.2 and illustrated by the block, timing, and spectral diagrams shown, respectively, in Figures 3.6, 3.7, and 3.8. Similar to the baseband sampling of bandpass signals discussed above, most of the negative effects of nonideal interpolation, such as a high f_s required due to wide transition bands of the analog interpolating LPFs as well as linear and nonlinear distortions within the LPFs' passband, can be offset or compensated in the TDP. Since insufficient suppression within the LPFs' stopbands cannot be compensated in the TDP, providing adequate suppression there is crucial for the interpolating filtering.

The interpolation described in Section 3.3.2 is based on the sampling theorem's indirect interpretation. The direct and hybrid interpretations of the sampling theorem not only allow utilizing the "don't care" bands for increasing the efficiency of interpolating filtering, but also change the approach to pulse shaping at the D/As' outputs, concentrating most of the energy within the output analog signal bandwidth.

5.4.2 Bandpass S&I of Bandpass Signals

This section describes ideal bandpass S&I of bandpass signals first. Then realization of nonideal bandpass S&I is discussed using the material of Chapters 3 and 4.

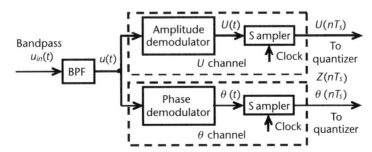

Figure 5.18 Baseband sampling of bandpass signals represented by $U(t)$ and $\theta(t)$.

5.4.2.1 Ideal Bandpass S&I

While baseband sampling of bandpass $u(t)$ does not impose any restrictions on $u(t)$ center frequency f_0, its bandpass sampling allows representing $u(t)$ by its instantaneous values $u(nT_s)$ with $f_s = 2B$ only if

$$f_0 = |m \pm 0.5|B \qquad (5.26)$$

where m is any integer (note that (3.16) turns into (5.26) when $f_s = 2B$). In this case (see Section D.2.3), reconstruction of $u(t)$ from $\{u(nT_s)\}$ is performed as follows:

$$u(t) = \sum_{n=-\infty}^{\infty} u(nT_s)\varphi_{nBP}(t) = \sum_{n=-\infty}^{\infty} u(nT_s)\varphi_{0BP}(t - nT_s) \qquad (5.27)$$

Equation (5.27) differs from (5.3) only by replacing the baseband sampling functions $\{\varphi_{nBB}(t)\}$ with the bandpass ones $\{\varphi_{nBP}(t)\}$. Here,

$$\varphi_{nBP}(t) = \operatorname{sinc}\left[\pi B(t - nT_s)\right]\cos\left[2\pi f_0(t - nT_s)\right] \qquad (5.28)$$

From (5.28),

$$\varphi_{0BP}(t) = \operatorname{sinc}(\pi Bt)\cos(2\pi f_0 t) \qquad (5.29)$$

Functions $\varphi_{nBP}(t)$ are mutually orthogonal if condition (5.26) is satisfied. Their squared norm, that is, the energy of $\varphi_{nBP}(t)$ dissipated in the 1Ω resistor, is

$$\|\varphi_{nBP}(t)\|^2 = \int_{-\infty}^{\infty} \frac{\sin^2\left[\pi B(t - nT_s)\right]}{\left[\pi B(t - nT_s)\right]^2}\cos^2\left[2\pi f_0(t - nT_s)\right]dt = T_s \qquad (5.30)$$

The spectral density $S_{\varphi nBP}(f)$ of $\varphi_{nBP}(t)$ is

$$S_{\varphi nBP}(f) = \begin{cases} T_s \exp(-j2\pi fnT_s) & \text{for } f \in \left[-(f_0 + 0.5B), -(f_0 - 0.5B)\right] \cup \left[(f_0 - 0.5B), (f_0 + 0.5B)\right] \\ 0 & \text{for } f \notin \left[-(f_0 + 0.5B), -(f_0 - 0.5B)\right] \cup \left[(f_0 - 0.5B), (f_0 + 0.5B)\right] \end{cases} \qquad (5.31)$$

In particular,

$$S_{\varphi 0BP}(f) = |S_{\varphi nBP}(f)|$$

$$= \begin{cases} T_s & \text{for } f \in \left[-(f_0 + 0.5B), -(f_0 - 0.5B)\right] \cup \left[(f_0 - 0.5B), (f_0 + 0.5B)\right] \\ 0 & \text{for } f \notin \left[-(f_0 + 0.5B), -(f_0 - 0.5B)\right] \cup \left[(f_0 - 0.5B), (f_0 + 0.5B)\right] \end{cases} \qquad (5.32)$$

The plots of $\varphi_{0BP}(t)$ and $S_{\varphi 0BP}(f)$ are shown in Figure 5.19.

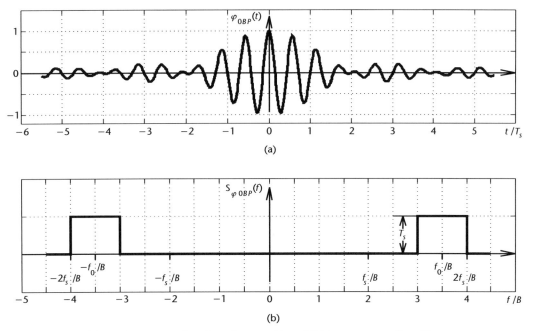

Figure 5.19 (a) Bandpass sampling function $\varphi_{0BP}(t)$ and (b) its spectrum $S_{\varphi 0BP}(f)$.

By following the line of reasoning used to derive (5.13), it easy to show that calculating samples $u(nT_s)$ of bandpass real-valued $u(t)$ according to the equation

$$u(nT_s) = c \int_{-\infty}^{\infty} u_{in}(t)\varphi_{nBP}(t)dt = c \int_{-\infty}^{\infty} u_{in}(t)\varphi_{0BP}(t - nT_s)dt \qquad (5.33)$$

minimizes the rms error of sampling. S&I of bandpass signals according to (5.33) and (5.27), respectively, correspond to the direct interpretation of the sampling theorem.

The methodology used to derive (5.16) to (5.18) allows for proving that the ideal sampling, corresponding to the sampling theorem's indirect interpretation, requires sending a bandpass real-valued signal $u_{in}(t)$ through an antialiasing BPF with the impulse response

$$h(t) = c\varphi_{0BP}(-t) = c\varphi_{0BP}(t) \qquad (5.34)$$

and applying the sifting property of the train of delta functions to the filter output. This methodology also proves that it is sufficient to send a discrete-time bandpass signal $u_d(t)$ through an interpolating BPF with the impulse response $h(t) = \varphi_{0BP}(t)$ for its ideal interpolation corresponding to the indirect interpretation. The block diagrams of the structures carrying out these operations are shown in Figure 5.20. The structures differ from those depicted in Figure 5.10 only in the types of the antialiasing and interpolating filters.

Transformation of the amplitude spectrum $|S_{in}(f)|$ of an input bandpass real-valued signal $u_{in}(t)$ in the case of ideal bandpass sampling is illustrated in Figure 5.21.

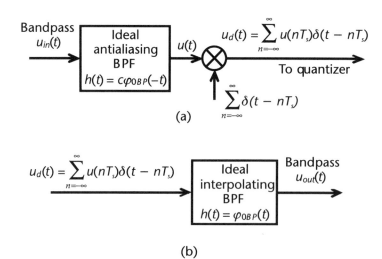

Figure 5.20 Ideal bandpass S&I of bandpass signals (indirect interpretation): (a) sampling structure and (b) interpolating structure.

Figure 5.21(a) presents $|S_{in}(f)|$ and the AFR $|H_{a.f}(f)|$ of the ideal antialiasing BPF. This BPF rejects all out-of-band spectral components of $u_{in}(t)$. As a result, only the desired bandpass signal $u(t)$ undergoes sampling. The amplitude spectrum $|S_d(f)|$ of the discrete-time signal $u_d(t)$ corresponding to $u(t)$ is shown in Figure 5.21(b). As follows from Figure 5.21(a), $u(t)$ meets condition (5.26), and Figure 5.21(b) shows that (5.26) guarantees nonoverlapping of the $S(f)$ replicas within $S_d(f)$ at the lowest sampling rate $f_s = 2B$.

5.4.2.2 Nonideal Bandpass S&I

Since ideal S&I are physically unrealizable, only nonideal bandpass S&I of bandpass signals can be implemented. Nonideal bandpass sampling is described as a part of bandpass digitization of Rx input signals in Section 4.4.2 where the block diagram

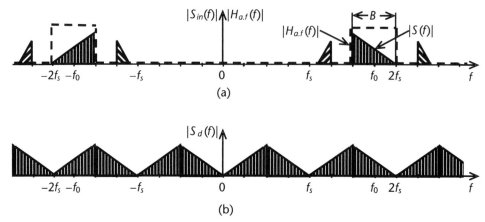

Figure 5.21 Amplitude spectra and AFR of antialiasing filter for ideal bandpass sampling: (a) $|S_{in}(f)|$ and $|H_{a.f}(f)|$ (dashed line) and (b) $|S_d(f)|$.

in Figure 4.11 corresponds to the general case, and the block diagram in Figure 4.13 to the digitization with optimal f_s that meets (3.16). Spectral transformations during this digitization are illustrated in Figure 4.12. Nonideal bandpass interpolation is described as a part of bandpass reconstruction of Tx output signals in Section 3.3.2 where its block, timing, and spectral diagrams are shown, respectively, in Figures 3.9, 3.10, and 3.11. The S&I presented in Sections 4.4.2 and 3.3.2 are based on the sampling theorem's indirect interpretation.

Nonideal bandpass S&I often suffer from wide transition bands of analog antialiasing and interpolating BPFs that require $f_s > 2B$, linear and nonlinear distortions within the BPFs' passbands, as well as insufficient suppression within their stopbands. While the BPFs' wide transition bands can be offset by sampling rate conversions in the digital domain, and distortions within the BPFs' passbands can be at least partly compensated in that domain, any required suppression within the BPFs' stopbands must be provided in the analog domain. As to the "don't care" bands caused by $f_s > 2B$, their presence is not used by conventional filters but can be utilized by S&I techniques based on the sampling theorem's direct and hybrid interpretations.

5.4.3 Comparison of Baseband and Bandpass S&I of Bandpass Signals

Baseband and bandpass sampling of bandpass signals are compared in Section 4.4.3 as a part of the comparison of baseband and bandpass digitization techniques. Similarly, baseband and bandpass interpolations of bandpass signals are compared in Section 3.3.3 as a part of the comparison of baseband and bandpass reconstructions. The results presented in those sections are outlined here, before explaining the contribution of the direct and hybrid interpretations to the advantages of bandpass S&I in Chapter 6.

Baseband S&I of bandpass signals have two fundamental disadvantages compared to bandpass ones. First, the baseband position of the analog signal spectrum in the case of baseband S&I (compare the spectral diagrams in Figures 3.8 and 4.10 to those in Figures 3.11 and 4.12) causes dc offset, flicker noise, and much larger power and number of IMPs within the signal bandwidth. Second, baseband S&I, in contrast with bandpass ones, require separate I and Q channels in the analog and mixed-signal domains (compare the block diagrams in Figures 3.6 and 4.9 to those in Figures 3.9, 4.11, and 4.13). The latter situation makes IQ imbalance unavoidable.

Some drawbacks of baseband S&I can be reduced using adaptive equalization and predistortion or postdistortion in the digital domain. However, these measures increase the complexity and cost of digital radios. Comparison of the block diagram in Figure 3.6 to that in Figure 3.9 and the block diagram in Figure 4.9 to those in Figures 4.11 and 4.13 show that bandpass S&I are also consistent with the major trend in the digital radio development: increasing the number of functions performed in the digital domain and reducing this number for the analog domain.

Still, bandpass S&I based on the indirect interpretation of the sampling theorem have two weaknesses. The first one is the inflexibility of the best traditional (e.g., SAW, BAW, crystal, electromechanical, and ceramic) BPFs and their incompatibility with the IC technology. This limits the adaptivity, reconfigurability, and scale of integration of digital radios. S&I based on the sampling theorem's direct interpretation allow overcoming these limitations, at least in principle. The second weakness is poor utilization of signal energy during S&I. It reduces the dynamic range and attainable bandwidth of digital radios. It also inhibits signal D&R close to the antennas. S&I based on the sampling theorem's direct and hybrid interpretations allow overcoming that weakness as well. The ways to overcome both weaknesses are discussed in Chapter 6.

5.5 Summary

The uniform sampling theorem was originated in 1915 as a part of interpolation theory. In communications, practical demand for this theorem existed from the first attempts to develop TDM telephony in the early 1900s. Despite the demand, sufficient technological level, and the fact that it was published in various forms in the 1930s, this theorem became common knowledge among engineers only in the late 1940s. Thus, the gap between theorists and practical engineers seriously delayed the implementation of TDM and S&I techniques.

The history of TDM and PCM demonstrates that three factors determine the fate of new technical ideas: practical demand, technological level, and theoretical basis. The absence of at least one of these factors precludes the implementation. For TDM, the missing factor was the theoretical basis (since the initial publications on the sampling theorem were overlooked). The implementation of PCM was delayed by the insufficient technological level.

The greatest contribution to the origination of the uniform sampling theorem for bandlimited signals was made by E. Whittaker, J. Whittaker, V. Kotelnikov, and C. Shannon during the period from 1915 to 1948. Therefore, it is often referred to as the WKS sampling theorem.

The sampling theorem explicitly demonstrates two components of its constructive nature, providing the ways of representing a bandlimited signal $u(t)$ by its samples $u(nT_s)$ or by the samples $y(nT_s)$ of its function $y(t) = y[u(t)]$, and the optimal, in the least-squares sense, interpolation algorithms. As shown above, the optimal sampling algorithms that also follow from the sampling theorem are the third component of its constructive nature.

The equations of the sampling theorem can be transformed and interpreted in various ways, and practical realization of S&I depends on the interpretation. The interpretation based on the theorem's original equations is direct. It does not require conventional analog filters for antialiasing and interpolating filtering. All interpretations based on transformed equations could, in principle, be called indirect. In this book, however, indirect interpretation refers only to the one where antialiasing and interpolating filtering are performed exclusively by conventional analog filters, and any interpretation combining the direct and indirect interpretations' methods is called a hybrid one.

The ideal S&I, based on any interpretation, are physically unrealizable because the sampling theorem uses many assumptions that could not be met in actuality. Therefore, only nonideal S&I are possible. Although all the interpretations are equally optimal in the ideal case, they provide different performance in real-world situations.

In addition to physical realizability, the implementation of S&I circuits requires taking into account linear and nonlinear distortions, jitter, and interference. Hence, the theoretical basis of S&I algorithms and circuits should include, besides the sampling theory, the theories of linear and nonlinear circuits, optimal filtering, and so forth.

The sampling theorem's indirect interpretation was the only one supported by the technology of the early 1950s, and it did not demonstrate its drawbacks then. Later, wide implementation of DSP and digital radios exposed its drawbacks, and the technological progress has made realization of S&I based on other interpretations feasible. Still, incomplete understanding of the substance and benefits of those interpretations delayed their implementation.

Practical realization of S&I often suffers from wide transition bands of analog antialiasing and interpolating filters that require $f_s > 2B$, linear and nonlinear distortions within the filters' passbands, and insufficient suppression within their stopbands. While the filters' wide transition bands can be offset by sampling rate conversions in the digital domain, and the distortions within the passbands can be at least partly compensated there, any required suppression within the stopbands must be provided in the analog domain.

Bandpass S&I of bandpass signals are advantageous over baseband ones. Still, the indirect interpretation of the sampling theorem does not allow complete utilization of the advantages. For this reason, implementation of bandpass S&I based on the hybrid and direct interpretations is practically important.

References

[1] Cattermole, K. W., *Principles of Pulse Code Modulation*, London, U.K.: Iliffe Books, 1969.

[2] Lüke, H. D., "The Origins of the Sampling Theorem," *IEEE Commun. Mag.*, Vol. 37, No. 4, 1999, pp. 106–108.

[3] Nyquist, H., "Certain Factors Affecting Telegraph Speed," *Bell Syst. Tech. J.*, No. 3, 1924, pp. 324–345.

[4] Nyquist, H., "Certain Topics in Telegraph Transmission Theory," *AIEE Trans.*, Vol. 47, April 1928, pp. 617–644.

[5] Küpfmüller, K., "Über die Dynamik der Selbsttätigen Verstärkungsregler," ("On the Dynamics of Automatic Gain Controllers"), *Elektrische Nachrichtentechnik*, Vol. 5, No. 11, 1928, pp. 459–467.

[6] Küpfmüller, K., "Utjämningsförloppinom Telegrafoch Telefontekniken," ("Transients in Telegraph and Telephone Engineering"), *Teknisk Tidskrift*, No. 9, 1931, pp. 153–160 and No. 10, 1931, pp. 178–182.

[7] Hartley, R. V. L., "Transmission of Information," *Bell Syst. Tech. J.*, Vol. 7, No. 3, 1928, pp. 535–563.

[8] Miner, W. M., "Multiplex Telephony," U.S. Patent 745743, filed February 26, 1903.

[9] Whittaker, E. T., "On the Functions Which Are Represented by the Expansions of the Interpolation Theory," *Proc. Roy. Soc. Edinburgh*, Vol. 35, 1915, pp. 181–194.

[10] Ferrar, W. L., "On the Consistency of Cardinal Function Interpolation," *Proc. Roy. Soc. Edinburgh*, Vol. 47, 1927, pp. 230–242.

[11] Whittaker, J. M., "The Fourier Theory of the Cardinal Functions," *Proc. Math. Soc. Edinburgh*, Vol. 1, 1929, pp. 169–175.

[12] Whittaker, J. M., *Interpolatory Function Theory*, Cambridge, U.K.: Cambridge University Press (Tracts in Mathematics and Mathematical Physics), No. 33, 1935.

[13] Kotelnikov, V. A., "On the Transmission Capacity of 'Ether' and Wire in Electrocommunications," *Proc. First All-Union Conf. Commun. Problems*, Moscow, January 14, 1933.

[14] Raabe, H., "Untersuchungen an der Wechselzeitigen Mehrfachübertragung (Multiplexübertragung)," *Elektrische Nachrichtentechnic*, Vol. 16, No. 8, 1939, pp. 213–228.

[15] Bennett, W. R., "Time Division Multiplex System," *Bell Syst. Tech. J.*, Vol. 20, 1941, pp. 199–221.

[16] Shannon, C. E., "A Mathematical Theory of Communication," *Bell Syst. Tech. J.*, Vol. 27, No. 3, pp. 379–423, and No. 4, 1948, pp. 623–655.

[17] Shannon, C. E., "Communications in the Presence of Noise," *Proc. IRE*, Vol. 37, No. 1, January 1949, pp. 10–21.

[18] Gabor, D., "Theory of Communication," *JIEE*, Vol. 93, Part 3, 1946, pp. 429–457.

[19] Someya, I., *Signal Transmission* (in Japanese), Tokyo: Shukyo, 1949.

[20] Weston, J. D., "A Note on the Theory of Communication," *London, Edinburgh, Dublin Philos. Mag. J. Sci.*, Ser. 7, Vol. 40, No. 303, 1949, pp. 449–453.

[21] Jerry, A. J., "The Shannon Sampling Theorem—Its Various Extensions and Applications: A Tutorial Review," *Proc. IEEE*, Vol. 65, No. 11, 1977, pp. 1565–1595.

[22] Papoulis, A., *Signal Analysis*, New York: McGraw-Hill, 1977.

[23] Marks II, R. J., *Introduction to Shannon Sampling and Interpolation Theory*, New York: Springer-Verlag, 1991.

[24] Higgins, J. R., *Sampling Theory in Fourier and Signal Analysis*, Oxford, U.K.: Clarendon Press, 1995.

[25] Unser, M., "Sampling—50 Years after Shannon," *Proc. IEEE*, Vol. 88, No. 4, 2000, pp. 569–587.

[26] Meijering, E., "A Chronology of Interpolation: from Ancient Astronomy to Modern Signal and Image Processing," *Proc. IEEE*, Vol. 90, No. 3, 2002, pp. 319–342.

[27] Landau, H. J., "Necessary Density Conditions for Sampling and Interpolation of Certain Entire Functions," *Acta Math.*, Vol. 117, February 1967, pp. 37–52.

[28] Mishali, M., and Y. C. Eldar, "Blind Multiband Signal Reconstruction: Compressed Sensing for Analog Signals," *IEEE Trans. Signal Process.*, Vol. 57, No. 3, 2009, pp. 993–1009.

[29] Marvasti, F., "Random Topics in Nonuniform Sampling," in *Nonuniform Sampling: Theory and Practice*, F. Marvasti (ed.), New York: Springer, 2001, pp. 169–234.

[30] Bilinskis, I., *Digital Alias-Free Signal Processing*, New York: John Wiley & Sons, 2007.

[31] Kozmin, K., J. Johansson, and J. Delsing, "Level-Crossing ADC Performance Evaluation toward Ultrasound Application," *IEEE Trans. Circuits Syst. I*, Vol. 56, No. 8, 2009, pp. 1708–1719.

[32] Tang, W., et al., "Continuous Time Level Crossing Sampling ADC for Bio-Potential Recording Systems," *IEEE Trans. Circuits Syst. I*, Vol. 60, No. 6, 2013, pp. 1407–1418.

[33] Wu, T.-F., C.-R. Ho, and M. Chen, "A Flash-Based Nonuniform Sampling ADC Enabling Digital Anti-Aliasing Filter in 65nm CMOS," *Proc. IEEE Custom Integrated Circuits Conf.*, San Jose, CA, September 28–30, 2015, pp. 1–4.

[34] Candès, E. J., J. Romberg, and T. Tao, "Signal Recovery from Incomplete and Inaccurate Measurements." *Comm. Pure Appl. Math.*, Vol. 59, No. 8, 2005, pp. 1207–1223.

[35] Candès, E. J., "Compressive Sampling," *Proc. Int. Cong. Mathematicians*, Madrid, Spain, Vol. 3, August 1–20, 2006, pp. 1433–1452.

[36] Baraniuk, R., "Compressive Sensing," *IEEE Signal Process. Mag.*, Vol. 24, No. 4, 2007, pp. 118–120, 124.

[37] Romberg, J., "Imaging Via Compressive Sampling," *IEEE Signal Process. Mag.*, Vol. 25, No. 2, 2008, pp. 14–20.

[38] Candes, E. J., and M. B. Wakin, "An Introduction to Compressive Sampling," *IEEE Signal Process. Mag.*, Vol. 25, No. 2, 2008, pp. 21–30.

[39] Jiang, X., "Linear Subspace Learning-Based Dimensionality Reduction," *IEEE Signal Process. Mag.*, Vol. 25, No. 2, 2011, pp. 16–25.

[40] Tosic, I., and P. Frossard, "Dictionary Learning," *IEEE Signal Process. Mag.*, Vol. 25, No. 2, 2011, pp. 27–38.

[41] Eldar, Y. C., *Sampling Theory: Beyond Bandlimited Systems*, Cambridge, U.K.: Cambridge University Press, 2015.

[42] Unser, M., and J. Zerubia, "A Generalized Sampling without Bandlimiting Constraints," *IEEE Trans. Circuits Syst. I*, Vol. 45, No. 8, 1998, pp. 959–969.

[43] Poberezhskiy, Y. S., *Digital Radio Receivers* (in Russian), Moscow, Russia: Radio & Communications, 1987.

[44] Poberezhskiy, Y. S., and M. V. Zarubinskiy, "Sample-and-Hold Devices Employing Weighted Integration in Digital Receivers," *Telecommun. and Radio Engineering*, Vol. 44, No. 8, 1989, pp. 75–79.

[45] Poberezhskiy, Y. S., and G. Y. Poberezhskiy, "Optimizing the Three-Level Weighting Function in Integrating Sample-and-Hold Amplifiers for Digital Radio Receivers," *Radio and Commun. Technol.*, Vol. 2, No. 3, 1997, pp. 56–59.

[46] Poberezhskiy, Y. S., and G. Y. Poberezhskiy, "Sampling with Weighted Integration for Digital Receivers," *Dig. IEEE MTT-S Symp. Technol. Wireless Appl.*, Vancouver, Canada, February 21–24, 1999, pp. 163–168.

[47] Poberezhskiy, Y. S., and G. Y. Poberezhskiy, "Sampling Technique Allowing Exclusion of Antialiasing Filter," *Electronics Lett.*, Vol. 36, No. 4, 2000, pp. 297–298.

[48] Poberezhskiy, Y. S., and G. Y. Poberezhskiy, "Sample-and-Hold Amplifiers Performing Internal Antialiasing Filtering and Their Applications in Digital Receivers," *Proc. IEEE ISCAS*, Geneva, Switzerland, May 28–31, 2000, pp. 439–442.

[49] Poberezhskiy, Y. S., and G. Y. Poberezhskiy, "Sampling Algorithm Simplifying VLSI Implementation of Digital Radio Receivers," *IEEE Signal Process. Lett.*, Vol. 8, No. 3, 2001, pp. 90–92.

[50] Poberezhskiy, Y. S., and G. Y. Poberezhskiy, "Signal Reconstruction Technique Allowing Exclusion of Antialiasing Filter," *Electronics Lett.*, Vol. 37, No. 3, 2001, pp. 199–200.

[51] Poberezhskiy, Y. S., and G. Y. Poberezhskiy, "Sampling and Signal Reconstruction Structures Performing Internal Antialiasing Filtering and Their Influence on the Design of Digital Receivers and Transmitters," *IEEE Trans. Circuits Syst. I*, Vol. 51, No. 1, 2004, pp. 118–129.

[52] Poberezhskiy, Y. S., and G. Y. Poberezhskiy, "Implementation of Novel Sampling and Reconstruction Circuits in Digital Radios," *Proc. IEEE ISCAS*, Vol. IV, Vancouver, Canada, May 23–26, 2004, pp. 201–204.

[53] Poberezhskiy, Y. S., and G. Y. Poberezhskiy, "Flexible Analog Front-Ends of Reconfigurable Radios Based on Sampling and Reconstruction with Internal Filtering," *EURASIP J. Wireless Commun. Netw.*, No. 3, 2005, pp. 364–381.

[54] Poberezhskiy, Y. S., and G. Y. Poberezhskiy, "Impact of the Sampling Theorem Interpretations on Digitization and Reconstruction in SDRs and CRs," *Proc. IEEE Aerosp. Conf.*, Big Sky, MT, March 1–8, 2014, pp. 1–20.

[55] Poberezhskiy, Y. S., and G. Y. Poberezhskiy, "Influence of Constructive Sampling Theory on the Front Ends and Back Ends of SDRs and CRs," *Proc. IEEE COMCAS*, Tel Aviv, Israel, November 2–4, 2015, pp. 1–5.

[56] Reeves, A. H., French Patent 852,183, 1938; British Patent 535,860, 1939; and US Patent 2,272,070, 1942.

[57] Reeves, A. H., "The Past, Present, and Future of PCM," *IEEE Spectrum*, May 1965, pp. 58–63.

CHAPTER 6
Realization of S&I in Digital Radios

6.1 Overview

This chapter describes and analyzes S&I techniques in digital radios with focus on conceptual problems of their design. Therefore, many technicalities are intentionally ignored or simplified to clearly explain the essence and potential capabilities of fundamentally different approaches to the development of S&I algorithms and circuits. For instance, although differential structures of S&I circuits provide higher performance than single-ended ones, only single-ended circuits are discussed for conciseness. For the same reason, many technical aspects (e.g., presence or absence of feedback, number of stages, types of amplifiers and/or switches) important for practical realization of S&I circuits are omitted because they are extensively described in other books (see, for example, [1–13]) and numerous papers. As noted in the previous chapters, bandpass S&I impose much higher requirements on the corresponding circuits than baseband S&I. Also, requirements for sampling circuits are usually much higher than those for interpolating ones. Therefore, this chapter is centered on bandpass S&I, and most attention is paid to sampling.

The role played by S&I techniques is often unclear to the designers of digital radios for many reasons. One of them is that samplers and quantizers are placed in the same package and often integrated on the same chip in modern A/Ds. This practice, justifiable from a technological perspective, makes it difficult to determine which of these devices (sampler or quantizer) contributes to or limits any given parameter of an A/D. Although these contributions and limitations depend on the types and technology of A/Ds, their analog bandwidths and input characteristics are usually determined by the samplers, which also significantly influence the A/Ds' dynamic ranges. The interpolation circuits similarly affect the reconstruction of analog signals. This chapter shows that S&I techniques have a great potential for improvement.

Section 6.2 analyzes S&I based on the indirect interpretation of the sampling theorem, shows their intrinsic drawbacks, and explains why the termination of development and production of SHAs was incorrect. The opportunities provided by the hybrid interpretation for bandpass S&I are described in Section 6.3. S&I based on the direct interpretation, their potential capabilities, and challenges of their realization and implementation are discussed in Section 6.4. Since all S&I circuits based on the direct interpretation and most of those based on the hybrid interpretation are multichannel, methods of channel mismatch mitigation in these circuits are concisely examined in Section 6.5. Selection of weight functions that determine most properties of S&I circuits based on the direct interpretation and many properties of S&I circuits based on the hybrid interpretation is explored in Section 6.6. Section 6.7 evaluates S&I circuits based on the hybrid and direct interpretations and explains the need for them, illustrating it by two examples of ISs' spatial suppression.

6.2 S&I Based on the Sampling Theorem's Indirect Interpretation

6.2.1 Sampling Based on the Indirect Interpretation

As mentioned in Chapter 5, the technology at the dawn of DSP implementation allowed practical realization of S&I circuits based only on the sampling theorem's indirect interpretation. Despite their drawbacks and the emerged feasibility of S&I based on the alternative interpretations, these circuits (albeit significantly evolved) are still widely used in digital radios and other applications. Therefore, their fundamental properties are analyzed in this section.

A conventional sampling structure based on the indirect interpretation (see Figure 6.1(a)) includes an analog antialiasing filter and a sampler. Two types of samplers were initially used: THAs and SHAs. However, SHAs were gradually phased out because it was decided in the 1970s that they are inefficient for bandpass sampling. The concise analysis of THAs and SHAs below is intended to demonstrate that this decision was incorrect, and SHAs are actually advantageous over THAs. Moreover, SHAs are a convenient starting point for the transition from the indirect to other interpretations of the sampling theorem.

6.2.1.1 THAs

There are many versions of THAs. They can use conventional operational amplifiers or transconductance stages, have or not have feedback, and contain various numbers of stages, types of switches, and so forth. Still, they all operate in two modes: track and hold, alternating with period T_s. Figure 6.1(b) shows a basic THA in the track mode, in which switch S_1 is closed, switch S_2 is open, and the voltage across capacitor C follows the THA input voltage $G_b u(t)$. In the subsequent hold mode, switch S_1 is open, switch S_2 is closed, and the voltage across C has to stay constant while it is quantized (usually the THA output is buffered). After the quantization completion, the THA goes into the track mode again and so on, as illustrated in Figure 6.1(c) where $G_b u(t)$ is depicted by a solid line and the voltage across C is depicted by a dashed line.

The mathematical model of a THA in Figure 6.2(a), corresponding to the track mode, is simplified to demonstrate fundamental drawbacks of THAs. The antialiasing filter with impulse response $h(t)$ is nonideal but acceptable if it provides sufficient suppression in the stopbands and tolerable distortion in the passband. In this case, $u_1(t) \approx u(t)$. The BA decouples the antialiasing filter from the THA and compensates the energy losses in both. Being nonideal, it introduces additive noise and nonlinear distortion. In properly designed radios, the spectral density of this noise within the bandwidth B is much lower than the spectral density of the noise

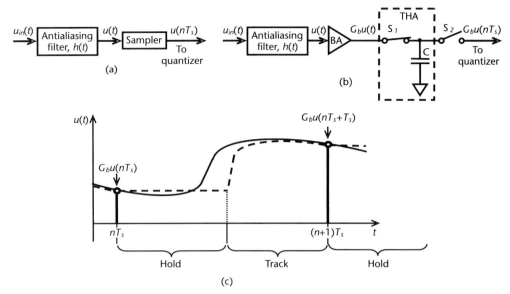

Figure 6.1 Sampling circuits: (a) general structure, (b) sampling circuit with THA, and (c) its timing diagram.

generated and amplified by the preceding stages. Consequently, it can be neglected. However, while the out-of-band noise of the preceding stages is suppressed by the antialiasing filter, the out-of-band noise of the BA is not suppressed, and its significant part falls within the signal spectrum after sampling. Therefore, the BA output signal, instead of being equal to $u_2(t) \approx G_b u_1(t) \approx G_b u(t)$, is $u_2(t) \approx f\{G_b[u(t) + n_{b.i}(t)]\} \approx G_b u(t)$ where G_b reflects the BA gain, function $f(\cdot)$ reflects its nonlinearity, and $n_{b.i}(t)$ reflects its noise. The BA is acceptable only if $u_2(t) \approx G_b u_1(t) \approx G_b u(t)$. This condition usually requires increasing the power consumption and gain of the preceding stages.

For accurate tracking, the time constant τ_t of the circuit comprising R_c and C should be sufficiently short. Here, R_c is the sum of the BA's output resistance and the resistance of closed S_1. For bandpass signals (see Figure 6.2(b)), it should meet the condition

$$\tau_t = R_c C \ll \frac{1}{f_0} \tag{6.1}$$

As follows from (6.1),

$$R_c \ll \frac{1}{2\pi f_0 C}, \tag{6.2}$$

that is, the THA impedance in the track mode is determined by C. Since C is sufficient to keep the voltage across it constant in the hold mode, the BA current is large. The voltage across C is also large to utilize the quantizer resolution. Thus, the

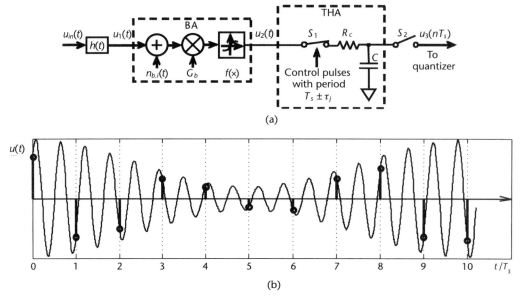

Figure 6.2 (a) Mathematical model of sampling circuit with THA and (b) sampled bandpass signal.

necessity of fast signal energy accumulation in the track mode, expressed by (6.1), requires significant total gain and high power consumption of all circuits preceding the THA. This also increases nonlinear distortion and, consequently, reduces the dynamic range.

Sampling jitter is deviation ($\pm\tau_j$) of samples from their correct positions. The jitter-induced error is proportional to the sampled signal's derivative and the absolute value of the clock error. It has regular and random components. Usually, it is easier to compensate the regular component than to limit the random one. THAs are very susceptible to jitter.

6.2.1.2 Integrating SHAs

In the 1960s and 1970s, SHAs were used as an alternative to THAs. The block and timing diagrams of a sampling circuit with an SHA are shown in Figure 6.3. An SHA has three modes: sample, hold, and clear (in Figure 6.3, it is in the hold mode). In the timing diagram (Figure 6.3(b)), $G_b u(t)$ is depicted by a solid line and the integrator output by a dashed line. During the sample (or integration) mode with length T_i (switch S_1 is closed, switches S_2 and S_3 are open), the SHA integrates $G_b u(t)$ to produce a sample. Throughout the subsequent hold mode with length T_h (S_2 is closed, S_1 and S_3 are open), the SHA keeps $u(nT_s)$ constant for its quantization. In the clear mode with length T_c (S_3 is closed, S_1 and S_2 are open), the integrator is discharged. The length of the SHA operational cycle is $T_i + T_h + T_c = T_s$. Since T_i in an SHA is longer than τ_t in the corresponding THA, the SHA charging current is lower. This reduces the IMPs' power and the required gain of the preceding stages. Simultaneously, the integration reduces jitter-induced error as well as out-of-band noise and IMPs.

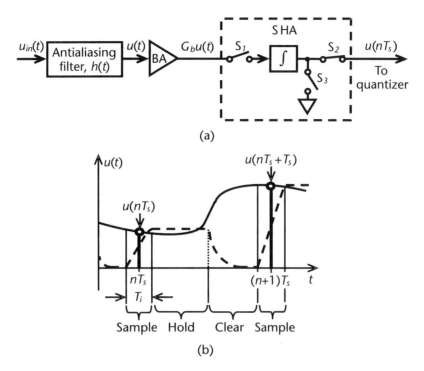

Figure 6.3 Sampling circuit with SHA: (a) block diagram and (b) timing diagram.

Longer T_i strengthens these advantages. However, the increase of T_i is limited by the required accuracy of determining $u(nT_s)$. Indeed, when $G_b u(t)$ is a straight line during T_i (assumption 1),

$$u(nT_s) = \frac{1}{T_i} \int_{nT_s - 0.5T_i}^{nT_s + 0.5T_i} G_b u(t)\, dt \tag{6.3}$$

For baseband signals, assumption 1 allows selecting a relatively long T_i. Still, increasing T_i beyond a certain point reduces the sampling accuracy due to unacceptable deviation of $G_b u(t)$ from a straight line (see Figure 6.3(b)). While this deviation reflects the sampling inaccuracy in the time domain, the nonuniformity of the integrator AFR $|H_i(f)|$ within the signal bandwidth B reflects it in the frequency domain. Here,

$$H_i(f) = \frac{T_i}{\tau_i}\operatorname{sinc}(\pi f T_i) = \frac{\sin(\pi f T_i)}{\pi f \tau_i} \tag{6.4}$$

where τ_i is the integrator time constant. The maximum acceptable T_i is determined by the maximum tolerable relative difference between the $|H_i(f)|$ values at $f = 0$ and $f = B$:

$$\frac{H_i(0) - H_i(B)}{H_i(0)} = 1 - \operatorname{sinc}(\pi B T_i) \tag{6.5}$$

The digital-domain compensation of $|H_i(f)|$ nonuniformity allows expanding T_i compared to that limited by (6.5).

The development of digital radios has made bandpass sampling desirable. However, it was assumed that T_i in an SHA should still be limited to the time interval on which a bandpass signal $u(t)$ is close to a straight line, that is, the same assumption 1 was applied to both baseband and bandpass signals. In this case,

$$T_i \ll \frac{1}{f_0} \qquad (6.6)$$

Such a short T_i eliminated all the advantages of SHAs over THAs and caused the phase-out of SHAs.

Actually, as was shown 1974 and later presented in [14, 15], assumption 1 is not adequate for bandpass signals. Indeed, a bandpass signal is much closer to a sinusoid than to a straight line (see Figure 6.2(b)), while the integral of $u(t)$ over T_i is proportional (with a time-independent coefficient) to the value of $u(t)$ at the midpoint of T_i-long interval for both straight line and sinusoid. Therefore, assumption 1 should be replaced by assumption 2 that a bandpass signal $u(t)$ is a sinusoid during T_i. In this case, it is reasonable to select

$$T_i \leq 0.5 T_0 \qquad (6.7)$$

The transition from assumption 1 to assumption 2 does not require changing the SHA design; it just allows increasing T_i, easily regaining the advantages of SHAs over THAs for bandpass sampling. The improvement of the bandpass sampling quality is especially significant at a relatively low or moderate IF, that is, when $T_0 = 1/f_0$ is only a few times shorter than $T_s = 1/f_s$. SHAs based on assumption 2 were developed and implemented in several high-dynamic-range HF Rxs in Russia in the mid-1980s. The measurements, later described in [16–18], confirmed the capability to radically increase Rx dynamic range (see Section 4.2.1) and showed that $T_i = (1/3)T_0$ provides the highest dynamic range for optimal f_s because in this case SHAs reject the sum-frequency third-order IMPs of the preceding BAs.

6.2.1.3 Problems of Conventional Bandpass Sampling

After the phase-out of SHAs, THAs became indispensable parts of conventional sampling structures (see Figure 6.1) for a long time. The root causes of the major drawbacks of such structures used for bandpass sampling are the shortcomings of the best traditional antialiasing BPFs (SAW, BAW, crystal, etc.), which are inflexible and incompatible with the IC technology, and extremely short time of the signal energy accumulation in THAs. The flowchart in Figure 6.4 illustrates the chain of relations between these root causes and their consequences for digital Rxs, namely, limited adaptivity and reconfigurability, dynamic range, attainable bandwidth, and scale of integration, as well as high power consumption and impossibility of close-to-the-antenna digitization.

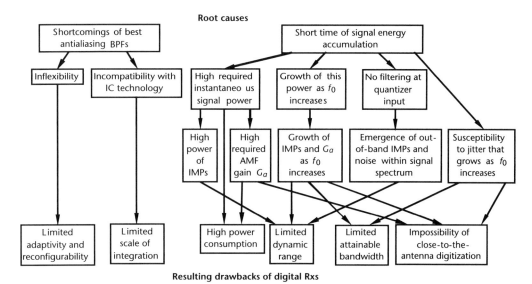

Figure 6.4 Drawbacks of conventional bandpass sampling and their consequences for digital Rxs.

6.2.2 Interpolation Based on the Indirect Interpretation

As explained in Section 1.2.1 and shown in Figure 1.3(b), reconstruction includes analog decoding of digital signals, which converts them into discrete-time signals and is performed by a D/A, and analog interpolation that converts the discrete-time signals into the analog ones. The interpolation, in turn, comprises pulse shaping and interpolating filtering, which are performed separately when this interpolation is based on the sampling theorem's indirect interpretation. Although D&R are opposite procedures, S&I are similar and have different requirements mainly due to different conditions of their realization. Therefore, most theoretical results and technical solutions related to sampling are applicable to interpolation and vice versa.

Signal interpolation is carried out in both Txs and Rxs. Below, however, attention is focused on bandpass interpolation in Txs due to its higher complexity. Baseband and bandpass interpolations based on the sampling theorem's indirect interpretation are described and compared in Sections 3.3.2 and 3.3.3, as parts of the corresponding reconstruction procedures. As follows from that material, the advantages of bandpass reconstruction (such as the absence of IQ imbalance, dc offset, and flicker noise, as well as a low IMP level) are fundamental because they are caused by the bandpass location of the analog signal spectrum. Simultaneously, its drawbacks (such as inefficient utilization of the energy of D/As' output samples, inflexibility of the best traditional analog BPFs, and their incompatibility with IC technology) are provisional because they are related to the sampling theorem's indirect interpretation and limitations of contemporary technology. The alternative interpretations of the sampling theorem allow partial or complete elimination of these drawbacks.

Bandpass interpolation is illustrated by the block, timing, and spectral diagrams in Figures 3.9, 3.10, and 3.11, respectively, as a part of reconstruction in digital Txs. The block diagram in Figure 6.5(a) is actually a part of that in Figure 3.9, and the timing diagram in Figure 6.5(b) is a copy of the diagram in Figure 3.10. The AFRs of PSs in Figures 6.5(c, d), shown in linear scale, correspond to different lengths of gating pulses. They show that if Δt_s meets (3.15), its reduction improves the uniformity of the PS AFR $|H_{p.s}(f)|$ within the Nyquist zone bandwidth $B_{NZ} = 0.5f_s$, but reduces the utilization of the energy of D/A's output samples. Therefore, gating pulse length is usually determined as a result of a trade-off that takes into account ratio f_0/B, acceptable nonuniformity of $|H_{p.s}(f)|$, and availability of its compensation by predistortion in TDP.

In most digital Txs, bandpass interpolation is performed at IF using optimal f_s that meet (3.16). Therefore, $f_0 = 2.25f_s$ in Figures 6.5(c, d). Besides the optimal relation between f_0 and f_s, there are many other factors that influence the selection of f_0, including the availability of effective BPFs. Pulse shaping should not necessarily be performed the way shown in Figure 6.5(a), and rectangular gating pulses are not always optimal. They are used here as a convenient starting point for the development of interpolating techniques based on the alternative interpretations of the sampling theorem.

The inflexibility of the best traditional interpolating BPFs and their incompatibility with the IC technology limit the adaptivity, reconfigurability, and the scale of integration of Tx drives. Taking into account the inefficient utilization of the energy of D/As' output samples, bandpass interpolation that is based on the sampling theorem's indirect interpretation limits the performance of Tx drives

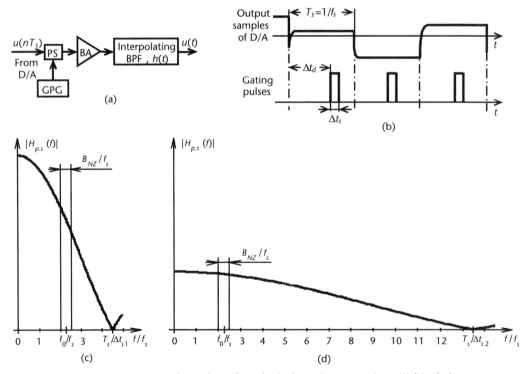

Figure 6.5 Bandpass interpolation based on the indirect interpretation: (a) block diagram, (b) timing diagram, (c) AFR of PS with $\Delta t_{s1} = 0.5T_0 = 1/(2f_0)$, and (d) AFR of PS with $\Delta t_{s2} = (1/6)T_0 = 1/(6f_0)$.

similarly to the way bandpass sampling based on the same interpretation limits the performance of Rxs.

6.3 S&I Based on the Sampling Theorem's Hybrid Interpretation

6.3.1 Sampling Based on the Hybrid Interpretation

As shown above, SHAs based on assumption 2 allow significant improvement of Rx performance compared to THAs when f_0 is only a few times higher than f_s because T_i restricted by (6.7) is a substantial fraction of $T_s = 1/f_s$. Such a situation corresponds to sampling at low or moderate IF. Sampling at high IF or RF makes $f_0 \gg f_s$ and $T_i \ll T_s$, preventing slow accumulation of the signal energy in SHAs.

6.3.1.1 SHAs with Weighted Integration

Elimination of the dependence of T_i on f_0 allows overcoming this obstacle in SHAs with weighted integration (SHAWIs) [19–21]. The SHAWI in Figure 6.6(a) has the same modes as the simplest SHA in Figure 6.3, but $G_b u(t)$ is multiplied by weight functions $w_n(t) = w_0(t - nT_s)$ (generated by the WFG) at the integrator input during the sampling mode. Therefore,

$$u(nT_s) = \int_{nT_s - 0.5T_w}^{nT_s + 0.5T_w} G_b u(t) w_n(t)\, dt \tag{6.8}$$

For bandpass sampling,

$$w_0(t) = W_0(t) c_0(t) \text{ for } t \in \left[-0.5T_w,\ 0.5T_w\right] \tag{6.9}$$

where $W_0(t)$ is the envelope of $w_0(t)$, $c_0(t)$ is its periodic carrier with period $T_0 = 1/f_0$, and $T_w = T_i$ is its length.

The same center frequency of $G_b u(t)$ and $w_n(t)$ makes T_i independent of f_0 but does not create a baseband spectral replica of $G_b u(t)$ (unless f_0 is a multiple of f_s) since the phases of $w_n(t)$ are tied to sampling instants nT_s. The independence between T_i and f_0 allows increasing T_i compared to (6.7). This reduces the BA output current, lowers the susceptibility to jitter, and provides filtering right at the quantizer input. The reduction of the BA output current lowers the IMPs and the filtering rejects a part of out-of-band noise and IMPs that can fall within the signal spectrum. All these factors increase the dynamic range and attainable bandwidth of Rxs.

The SHAWI operational cycle is shown Figure 6.6(b). Here, $w_0(t)$ has rectangular $W_0(t)$ and square-wave $c_0(t)$ with $f_0 = 6.25 f_s$. The necessity of allocating all the modes within T_s restricts T_i to

$$T_i = T_s - T_h - T_c \leq 0.5 T_s \tag{6.10}$$

Since the spectrum of $w_0(t)$ determines the SHAWI transfer function $H_w(f)$, (6.10) limits its spectral resolution (see Figure 6.6(c)), preventing effective antialiasing

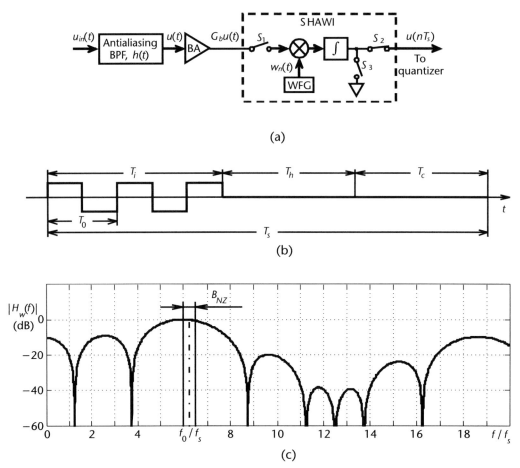

Figure 6.6 Sampling circuit with SHAWI: (a) block diagram, (b) timing diagram, and (c) AFR of the SHAWI.

filtering. Indeed, all the spectral diagrams in Chapters 3 to 5 show that the distances between the center frequencies of the neighboring passbands and stopbands of antialiasing and interpolating filters should be equal to f_s for baseband S&I and to $0.5f_s$ for bandpass S&I with optimal f_s.

6.3.1.2 Time-Interleaved SHAWIs

In the case of baseband S&I, antialiasing and interpolating filters must suppress spectral components within frequency intervals (4.43). In the case of bandpass S&I with optimal f_s, suppression must be provided within frequency intervals

$$\left[-\left(f_0 + 0.5B + 0.5kf_s\right), -\left(f_0 - 0.5B + 0.5kf_s\right)\right] \cup \left[f_0 - 0.5B + 0.5kf_s, f_0 + 0.5B + 0.5kf_s\right]$$
(6.11)

where an integer $k \in [(0.5 - 2f_0/f_s), \infty[$; $k \neq 0$. SHAWIs perform mixed-signal FIR filtering. To provide suppression within all intervals (4.43) or (6.11), the AFR of a

FIR filter should have at least one zero within each stopband. The necessary conditions for that are $T_i = T_w \geq 1/f_s = T_s$ for baseband S&I and $T_i = T_w \geq 2/f_s = 2T_s$ for bandpass S&I. The hold and clear modes make the operational cycles longer. When

$$T_h + T_c = T_s \qquad (6.12)$$

the minimum lengths of the operational cycles for baseband and bandpass sampling are, correspondingly, $2T_s$ and $3T_s$. Therefore, only time-interleaved SHAWIs with the minimum numbers of channels $L = 2$ and $L = 3$ for baseband and bandpass sampling, respectively, can perform genuine internal antialiasing filtering [22–33].

While single-channel SHAWIs merely correspond to the sampling theorem's hybrid interpretation, their time-interleaved structures can correspond to either hybrid or direct interpretation. When they require prefiltering, they correspond to the hybrid interpretation; when they provide sufficient internal antialiasing filtering, they correspond to the direct interpretation (see Section 5.3.2). Increasing L enables improving the filtering quality. Requirements for antialiasing filtering and properties of $w_0(t)$ determine the minimum L at which transition from the hybrid to the direct interpretation is possible. The variety of $w_0(t)$ that can be used in time-interleaved SHAWIs grows fast as L increases, but a small L seriously restricts the selection of $w_0(t)$. For instance, baseband S&I with $L = 2$ allow only rectangular $w_0(t)$ with $T_w = T_s$, and bandpass S&I with $L = 3$ allow only $w_0(t)$ with $T_w = 2T_s$ and rectangular $W_0(t)$. No other $w_0(t)$ provide suppression within all intervals (4.43) or (6.11).

The block and timing diagrams of a three-channel structure of time-interleaved SHAWIs are shown in Figures 6.7(a, b), respectively. Figure 6.7(c) reflects the relative positions of samples $u(nT_s)$ and corresponding $w_n(t)$. In Figure 6.7(a), the initial antialiasing filtering is still performed by a traditional BPF, but the three-channel structure greatly improves suppression within the stopbands. The multiplexer (Mx) cyclically connects the channels' outputs to the quantizer. Channel l ($l = 1, 2, 3$) forms all samples with numbers $l + 3k$ where k is an integer. The $3T_s$-long operational cycle of each channel is delayed by T_s relative to that of the prior channel. The WFG generates $w_n(t) = w_0(t - nT_s)$ for all the channels. Since the transfer function $H_w(f)$ of the ideal three-channel structure is the Fourier transform of $w_0(-t)$, the latter is usually selected even, that is, $w_0(-t) = w_0(t)$, to provide a linear PFR and simplify the WFG. In Figure 6.7(c), $w_0(t)$ has rectangular $W_0(t)$ and square-wave $c_0(t)$ with $f_0 = 2.25f_s$. During the sample mode with length $T_i = T_w = 2T_s$, a sample $u(nT_s)$ is formed according to (6.8). Throughout the hold mode, the channel is connected to the quantizer, and $u(nT_s)$ is quantized. In the clear mode, the channel integrator is disconnected from the quantizer and discharged. The lengths of the last two modes meet (6.12).

Although only rectangular $W_0(t)$ is allowed for bandpass $w_0(t)$ with $T_w = 2T_s$, the variety of carriers $c_0(t)$ is much larger and limited mostly by the requirement to be periodic with period $T_0 = 1/f_0$. Replacing multipliers in SHAWIs with a small number of switches increases the dynamic range of sampling circuits and simplifies WFGs. This possibility depends on $w_0(t)$. When $W_0(t)$ is rectangular, it is sufficient to properly select $c_0(t)$ for this replacement. Reducing the number of levels in $c_0(t)$ simplifies SHAWI realization. The major obstacle to reducing this number is the resulting deterioration of filtering properties.

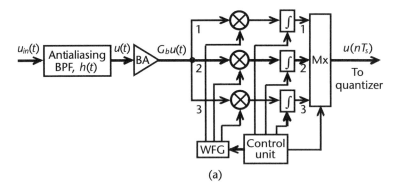

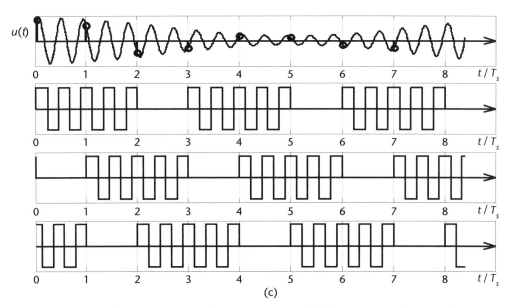

Figure 6.7 Three-channel structure of time-interleaved SHAWIs: (a) block diagram, (b) timing diagram (C stands for the clear mode), and (c) relative positions of $u(nT_s)$ and corresponding $w_n(t)$.

Figure 6.8 presents single-period timing diagrams of four carriers $c_{0i}(t)$: cosine carrier $c_{01}(t)$, square-wave carrier $c_{02}(t)$, three-level carrier $c_{03}(t)$, and four-level carrier $c_{04}(t)$ ($c_{04}(t)$ was suggested in [34]). It is obvious that $c_{02}(t)$, $c_{03}(t)$, and $c_{04}(t)$ are approximations of $c_{01}(t)$ with accuracy that depends on the number of levels.

Weight functions $w_0(t)$ with different $c_{0i}(t)$ and corresponding AFRs $|H_w(f)|$ of the structure in Figure 6.7(a) are shown in Figures 6.9 and 6.10, respectively, for $f_0 = 1.25 f_s$ and $f_0 = 2.75 f_s$.

These figures and results of simulations show the following. First, an increase of f_0/f_s reduces the influence of the $c_0(t)$ type on $H_w(f)$, and this influence can be neglected when $f_0/f_s > 10$ because the harmonics of $c_0(t)$ can be easily filtered out.

6.3 S&I Based on the Sampling Theorem's Hybrid Interpretation

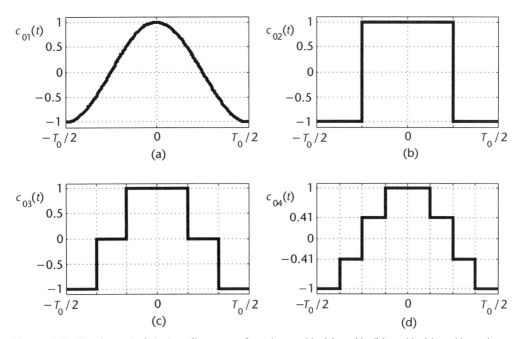

Figure 6.8 Single-period timing diagrams of carriers $c_{0i}(t)$: (a) $c_{01}(t)$, (b) $c_{02}(t)$, (c) $c_{03}(t)$, and (d) $c_{04}(t)$.

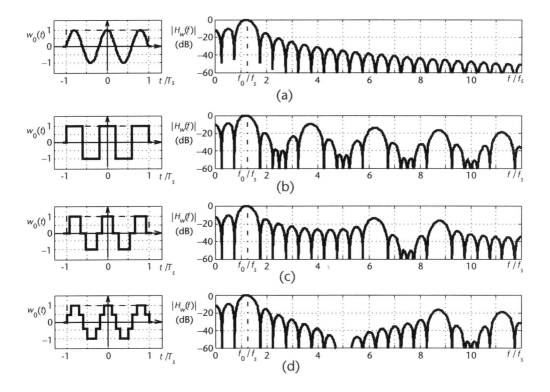

Figure 6.9 Weight functions $w_0(t)$ with rectangular $W_0(t)$ (dashed line) and corresponding AFRs $|H_w(f)|$ for $f_0 = 1.25 f_s$ and various $c_{0i}(t)$: (a) $c_{01}(t)$, (b) $c_{02}(t)$, (c) $c_{03}(t)$, and (d) $c_{04}(t)$.

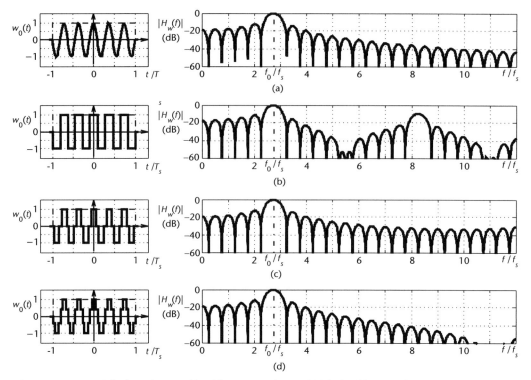

Figure 6.10 Weight functions $w_0(t)$ with rectangular $W_0(t)$ (dashed line) and corresponding AFRs $|H_w(f)|$ for $f_0 = 2.75 f_s$ and various $c_{0i}(t)$: (a) $c_{01}(t)$, (b) $c_{02}(t)$, (c) $c_{03}(t)$, and (d) $c_{04}(t)$.

Second, the troughs of $|H_w(f)|$ are located within intervals (6.11), increasing suppression there, while its sidelobes correspond to "don't care" bands between intervals (6.11) where suppression is unimportant. Thus, multichannel sampling circuits based on the hybrid interpretation utilize the existence of "don't care" bands. Third, antialiasing filtering provided by three-channel structures of time-interleaved SHAWIs is usually insufficient, and, therefore, traditional antialiasing BPFs (although with relaxed requirements) are still needed.

Sampling circuits based on the hybrid interpretation accumulate the signal energy much slower than circuits based on the indirect interpretation. Consequently, much smaller instantaneous signal power is required at their inputs, and the power is independent of f_0. This lowers the IMP level and the needed AMF gain. These sampling circuits perform filtering right at the quantizer input, suppressing all out-of-band IMPs, noise, and jitter. Therefore, they radically increase the dynamic range and attainable bandwidth of Rxs. They also bring digitization closer to the antenna and somewhat increase Rx adaptivity and scale of integration. Still, the necessity to retain traditional antialiasing filters limits the adaptivity and scale of integration and prevents digitization close enough to the antenna.

6.3.2 Interpolation Based on the Hybrid Interpretation

The similarity between sampling and interpolation, mentioned in Section 6.2.2, makes it clear that weighted pulse shapers (WPSs) should improve bandpass interpolation quality similarly to the way SHAWIs improve bandpass sampling quality. The block and timing diagrams of a WPS are shown, respectively, in Figures 6.11(a, b), while Figures 6.11(c, d) demonstrate its AFR, correspondingly, in linear and logarithmic scales. As follows from the previous section, $w_0(t)$ should have rectangular $W_0(t)$ and periodic $c_0(t)$ with period $T_0 = 1/f_0$. A comparison of the AFRs in Figures 6.5(c) and 6.11(c) shows that the WPS concentrates the energy at the D/A's output within the analog signal bandwidth better than the simplest PS.

Still, single-channel WPSs cannot perform genuine internal bandpass interpolating filtering because their $w_0(t)$ have lengths $T_w < T_s$. For the same reasons as sampling (see Section 6.3.1.2), baseband interpolating filtering requires $w_0(t)$ with length $T_w \geq T_s$ and bandpass interpolating filtering requires $w_0(t)$ with length $T_w \geq T_s$. Thus, only time-interleaved WPSs can perform internal interpolating filtering, and their minimum numbers of channels are $L = 2$ and $L = 3$, respectively, for baseband and bandpass interpolation. Also, baseband interpolation with $L = 2$

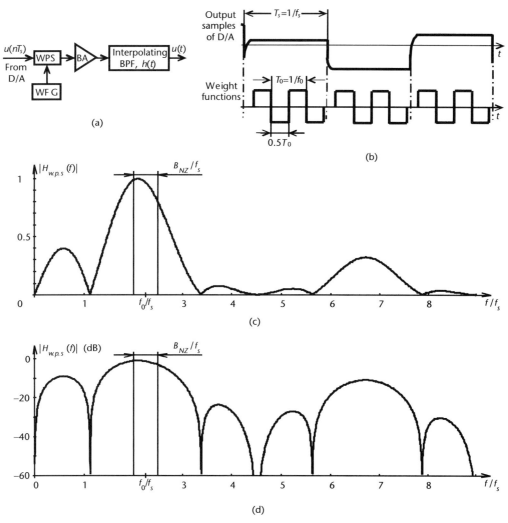

Figure 6.11 Weighted bandpass single-channel interpolation: (a) block diagram, (b) timing diagram, (c) AFR of WPS (linear scale), and (d) AFR of WPS (logarithmic scale).

allows only rectangular $w_0(t)$ with $T_w = T_s$, and bandpass interpolation with $L = 3$ allows only $w_0(t)$ with $T_w = 2T_s$ and rectangular $W_0(t)$, leaving significant freedom for selecting $c_0(t)$.

The block and timing diagrams of a three-channel structure of time-interleaved WPSs are shown in Figures 6.12(a, b), respectively. The Dmx periodically (with period $3T_s$) connects the D/A output to each channel. A channel's operational cycle has length $3T_s$, is delayed by T_s relative to that of the previous channel, and comprises three modes: clear, sample, and multiply. During the clear mode at the beginning of each operational cycle, the SHA's capacitor is discharged. This mode's duration T_c should be longer than the D/A glitches and sufficient for the capacitor discharge. Usually, $T_c \approx T_s/3$ meets both requirements. During the subsequent sample mode with length $T_i = T_s - T_c$, the SHA is connected to the D/A by the Dmx, and its capacitor is charged by the undistorted part of the D/A's output sample $u(nT_s)$. Throughout the multiply mode with length $T_w = 2T_s$, $u(nT_s)$ is multiplied by $w_n(t) = w_0(t - nT_s)$ generated by the WFG, and the product is summed with those of all other channels. Channel l ($l = 1, 2, 3$) processes samples with numbers $l + 3k$ where k is an integer.

To clarify the similarity between sampling and interpolation, note that the sample mode in time-interleaved structures of SHAWIs corresponds to the multiply mode in time-interleaved structures of WPSs, the hold mode in time-interleaved structures of SHAWIs corresponds to the sample mode in time-interleaved structures of WPSs, and functions performed during the clear mode are the same in the structures of both types.

Since the same $w_0(t)$ are used in the sampling and interpolation structures shown in Figures 6.7(a) and 6.12(a), respectively, the AFRs $|H_w(f)|$ in Figures 6.9 and 6.10 characterize the filtering quality in both structures. When the interpolating filtering performed by time-interleaved WPSs is insufficient, traditional interpolating BPFs must provide postfiltering, although requirements for them can be relaxed. Similar to sampling, an increase in the number of channels L in time-interleaved structures of WPSs improves the filtering quality, allowing transition from the circuits based

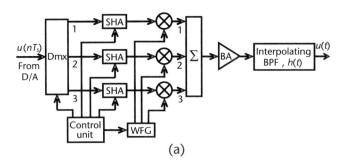

Figure 6.12 Three-channel structure of time-interleaved WPSs: (a) block diagram and (b) timing diagram (C stands for the clear mode).

on the hybrid interpretation to those based on the direct interpretation of the sampling theorem.

Time-interleaved structures of WPSs based on the hybrid interpretation enhance the energy utilization at D/As' outputs, reduce the influence of jitter, lower the required AMB gain, and allow higher accuracy of modulation in Txs compared to interpolation circuits based on the sampling theorem's indirect interpretation. They also bring reconstruction closer to the Tx antenna and somewhat increase adaptivity and scale of integration of Tx drives. Still, the necessity to retain traditional interpolating filters limits the adaptivity and scale of integration and prevents reconstruction close enough to the antenna.

It is important to emphasize that currently there are no serious technological or technical problems precluding the implementation of S&I circuits based on the sampling theorem's hybrid interpretation.

6.4 S&I Based on the Sampling Theorem's Direct Interpretation

6.4.1 Sampling Based on the Direct Interpretation

As shown in the previous section, increasing the number of channels L in time-interleaved structures of SHAWIs and WPSs improves the quality of filtering and allows transitioning from the hybrid to the direct interpretation of the sampling theorem as L gets sufficiently large. For convenience, S&I circuits based on the direct interpretation are considered below together with the corresponding quantizers and D/As. For that reason, they are called, respectively, novel digitization circuits (NDCs) and novel reconstruction circuits (NRCs).

6.4.1.1 Conceptual Structures of NDCs

Two conceptual structures of NDCs are depicted in Figures 6.13(a, b). The NDC in Figure 6.13(b) requires L quantizers but allows reducing their speed by factor L compared to the quantizer of the NDC in Figure 6.13(a). It also allows replacing an analog Mx with a digital one. The timing diagram in Figure 6.13(c) and the relative positions of samples $u(nT_s)$ and corresponding $w_n(t)$ in Figure 6.13(d) reflect the operation of both NDCs for $L = 5$. The comparison of Figures 6.13(c, d) with Figures 6.7(b, c) shows that both NDCs have the same operational modes as the structure in Figure 6.7(a), but, unlike the latter, they perform the entire antialiasing filtering internally and, therefore, no traditional filter is required at their inputs.

Thus, $u_{in}(t)$ directly enters time-interleaved SHAWIs in NDCs. The nth sample at the NDC output is

$$u(nT_s) = \int_{nT_s - 0.5T_w}^{nT_s + 0.5T_w} u_{in}(t) w_n(t) \, dt \qquad (6.13)$$

Taking into account (6.12), the length T_i of the sample mode in NDCs is

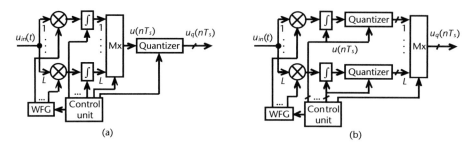
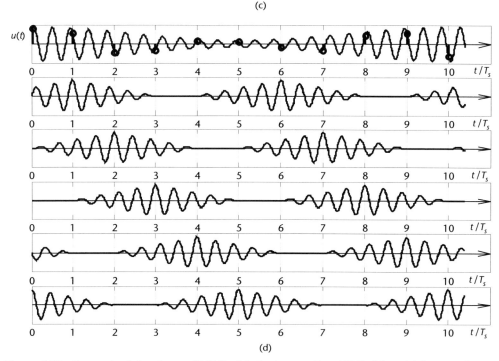

Figure 6.13 Conceptual structures of NDCs: (a) single-quantizer NDC, (b) multiple-quantizer NDC, (c) timing diagrams (C stands for the clear mode), and (d) relative positions of $u(nT_s)$ and corresponding $w_n(t)$.

$$T_i = T_w = (L-1)T_s \qquad (6.14)$$

Longer T_i enables better antialiasing filtering and larger freedom in selecting $w_0(t)$. The examples of longer baseband and bandpass $w_0(t)$ and the corresponding AFRs $|H_w(t)|$ of the NDCs are presented in Figures 6.14 and 6.15. The baseband $w_0(t)$ in Figure 6.14(a) has $T_w = 4T_s$ and is a fourth-order B-spline (see Section A.3). The bandpass $w_0(t)$ in Figure 6.15(a) has $T_w = 8T_s$ and cosine carrier $c_0(t)$. Its envelope $W_0(t)$ is also a fourth-order B-spline.

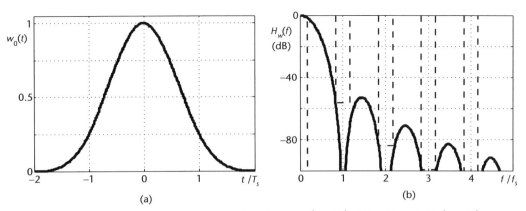

Figure 6.14 Baseband weight function $w_0(t)$ and its AFR $|H_w(f)|$: (a) $w_0(t)$ and (b) $|H_w(f)|$.

The reason for the different lengths of baseband and bandpass $w_0(t)$ is the different lengths of their original rectangles: T_s for baseband $w_0(t)$ and $2T_s$ for bandpass $w_0(t)$. These lengths guarantee proper positions of the AFRs' zeros (see Section 6.3.1.2). As a result, the troughs of $|H_w(t)|$ in Figure 6.14(b) are located within intervals (4.43), and the troughs of $|H_w(t)|$ in Figure 6.15(b) are located within intervals (6.11), while their sidelobes correspond to "don't care" bands between these intervals. Thus, B-spline-based $w_0(t)$ allow effective utilization of "don't care" bands. The rectangular envelopes $W_0(t)$ of $w_0(t)$ presented in Figures 6.7(c), 6.9, and 6.10 are first-order B-splines, and the triangular envelope $W_0(t)$ of $w_0(t)$ presented in Figure 6.13(d) is a second-order B-spline.

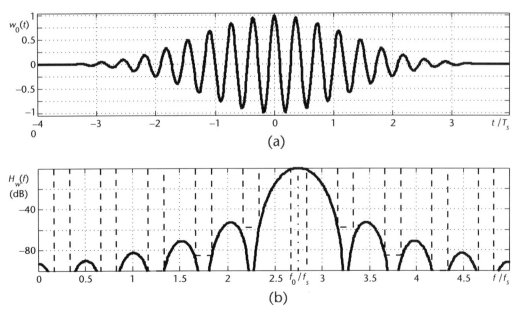

Figure 6.15 Bandpass weight function $w_0(t)$ and its AFR $|H_w(f)|$ for $f_0 = 2.75 f_s$: (a) $w_0(t)$ and (b) $|H_w(f)|$.

The NDCs with B-spline-based $w_0(t)$ provide different suppression in different stopbands (the lowest one is in the first stopband), and uneven suppression within each stopband. Their AFRs are nonuniform within their passbands (in Figures 6.14(b) and 6.15(b), the passbands and stopbands are marked with dashed lines). The stopband suppression and passband nonuniformity depend primarily on the B-spline order and ratio f_s/B. In the bandpass NDCs, they also depend on the ratio f_0/f_s, but this dependence is negligible for $f_0/f_s > 10$. When this dependence can be neglected, the bandpass and baseband NDCs with $w_0(t)$ based on fourth-order B-splines have minimum stopband suppression and passband nonuniformity, respectively, 58 dB and ±0.7 dB for $f_s/B = 6$ and 42 dB and ±1.5 dB for $f_s/B = 4$. B-spline-based $w_0(t)$ are extensively discussed in this chapter due to their useful properties and relative simplicity of their analysis.

6.4.1.2 Alternative Structures of NDCs

The block diagrams of NDCs in Figure 6.13 do not exhaust all their possible structures. Practical NDCs can significantly differ from them and among themselves. Their realization can be $w_0(t)$-specific and may depend on the Rx architecture. A single-quantizer NDC that differs from the conceptual one is shown in Figure 6.16(a). It uses voltage-controlled amplifiers (VCAs) to carry out multiplications. Generating $w_n(t)$ by a digital WFG (DWFG) and replacing the VCAs with digitally controlled amplifiers (DCAs), as shown in Figure 6.16(b), increases digitization accuracy. Selecting $w_0(t)$ that can be accurately represented by a small number of bits simplifies these NDCs. Note that significant increase in the control channel bandwidth is required for the VCAs and DCAs, especially in bandpass NDCs. Amplification of $u_{in}(t)$ in the VCAs or DCAs allows placing the NDCs closer to the antennas. In principle, DCAs can be replaced with multiplying D/As (MD/As). It is straightforward to develop multiple-quantizer NDCs with VCAs or DCAs, based on the structures shown in Figures 6.13 and 6.16. Note that, in principle, the number of quantizers L_q in NDCs can be $1 \leq L_q \leq L$.

6.4.1.3 Advantages of NDCs

The properties of sampling circuits based on the sampling theorem's hybrid interpretation, described in Section 6.3.1.2, clarify most advantages of NDCs. Indeed, longer weight functions $w_0(t)$ in NDCs allow performing the entire antialiasing filtering

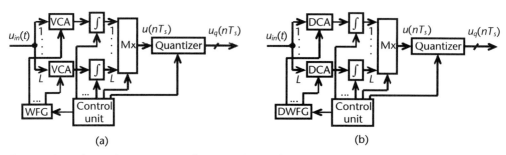

Figure 6.16 Alternative structures of NDCs: (a) VCA-based and (b) DCA-based.

internally and accumulating the signal energy during sampling even slower than in the circuits based on the hybrid interpretation. The NDCs' filtering properties are determined by $w_0(t)$, whose shapes and parameters can be easily changed. This and close-to-the-antenna digitization make Rxs flexible and reconfigurable. Exclusion of traditional antialiasing filters also increases the Rx scale of integration. Slower accumulation of signal energy results in the higher dynamic range, wider attainable bandwidth, close-to-the-antenna digitization, and lower power consumption. The flowchart in Figure 6.17 illustrates the chain of relations between the merits of NDCs and the resulting advantages of digital Rxs.

6.4.2 Interpolation Based on the Direct Interpretation

As follows from Section 6.3, increasing the number of channels L in time-interleaved WPS structures improves the quality of interpolating filtering and allows transitioning from the hybrid to the direct interpretation of the sampling theorem. Similar to the sampling circuits based on the direct interpretation, it is convenient to consider the interpolation circuits based on this interpretation together with the corresponding D/As. The combined circuits are named NRCs at the beginning of Section 6.4.1.

Two conceptual structures of NRCs are depicted in Figures 6.18(a, b). The NRC in Figure 6.18(b) requires L D/As but allows reducing their speed by factor L compared to the D/A of the NRC in Figure 6.18(a). It also allows replacing an analog Dmx with a digital one. The timing diagram in Figure 6.18(c) reflects the operation of both NRCs for $L = 5$. Comparison of Figure 6.18(c) with Figure 6.12(b) shows that both NRCs have the same operational modes as the structure in Figure 6.12(a), but, unlike the latter, they perform the entire interpolating filtering internally and, therefore, no traditional filter is required. Thus, the NRC output signal is

$$u(t) = \sum_{n=-\infty}^{n=\infty} u(nT_s) w_n(t) = \sum_{n=-\infty}^{n=\infty} u(nT_s) w_0(t - nT_s) \qquad (6.15)$$

As follows from Section 6.3.2, the multiply mode in NRCs corresponds to the sample mode in NDCs, the sample mode in NRCs corresponds to the hold mode in

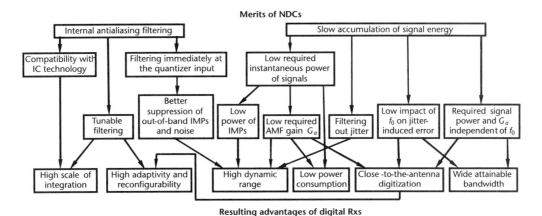

Figure 6.17 Advantages of NDCs and their influence on performance of digital radios.

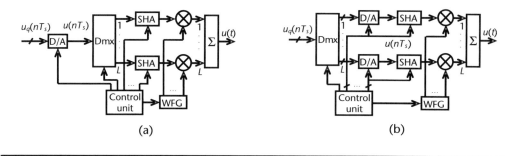

Channel 1	C	S	Multiply 1		C	S		Multiply 6		
Channel 2			C	S	Multiply 2			C	S	Multiply 7
Channel 3				C	S	Multiply 3			C	S	Mu
Channel 4					C	S	Multiply 4			C	S
Channel 5						C	S	Multiply 5			C	S

(c)

Figure 6.18 Conceptual structures of NRCs: (a) single-D/A NRC, (b) multiple-D/A NRC, and (c) timing diagrams (C and S stand, respectively, for the clear and sample modes).

NDCs, and the clear mode has the same purpose in both types of structures. Therefore, the length T_w of the multiply mode in NRCs is determined by (6.14). Longer T_w enables better interpolating filtering and larger freedom in selecting $w_0(t)$. The approaches to selecting $w_0(t)$ in NDCs and NRCs are similar, and the main difference is lower stopband suppression usually required in NRCs.

Similar to NDCs, the ways of realizing NRCs are not exhausted by their conceptual structures. The multiple-D/A NRC shown in Figure 6.19(a) uses MD/As that perform both digital-to-analog conversion of the input digital samples and subsequent multiplication by $w_n(t)$. In the NRC shown in Figure 6.19(b), these procedures are performed by the DCAs. In contrast with the NDC in Figure 6.16(b) where the DCAs amplify the input signal with the gains controlled by digitally generated $w_n(t)$, the DCAs of the NRC in Figure 6.19(b) amplify analog $w_n(t)$ with the gains controlled by digital samples (still, an increase in the control channel bandwidth can be required for the DCAs). Various single-D/A NRCs can also be suggested.

The benefits provided by NRCs to Txs are similar to those provided by NDCs to Rxs. NRCs make AMBs highly adaptive and easily reconfigurable because $w_0(t)$,

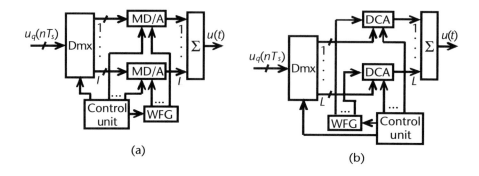

Figure 6.19 Alternative structures of multiple-D/A NRCs: (a) MD/A-based and (b) DCA-based.

determining NRCs' properties, can be dynamically changed. Additionally, NRCs increase the scale of integration due to the removal of traditional filters incompatible with the IC technology. Interpolating filtering, combined with pulse shaping, reduces the jitter-induced error and required AMB gain. It also concentrates the NRCs' output energy within the signal bandwidth. This increases the AMB dynamic range, improves modulation accuracy, and allows closer-to-the-antenna reconstruction. Many of these and other advantages depend on the specific AMB architecture. An example of the AMB in Figure 6.20 illustrates this. Here, the NRC performs signal reconstruction at the RF (although it can be performed at the IF), and the output signals of the NRC channels are summed in the air (alternatively, they can be summed before their common antenna). The filtering by LPFs (or BPFs) is optional and used only if the NRC filtering is insufficient. Similarly, the PAs are used only if the levels of the DCA output signals are inadequate.

The most important feature of this AMB is that NRC uses the samples for changing not only the DCA gains but also the rail voltages of the DCAs and PAs. For high-quality reconstruction, the DCA gains should be controlled with fairly high resolution, while the rail voltages can be controlled with much lower resolution. Therefore, the code transformers (CTs) take only a few most significant bits of the samples' absolute values to control the rail voltages. The rough proportionality of the rail voltages to the absolute values of the samples makes the DCAs and PAs operate as Class-G amplifiers and enables both high dynamic range and low power consumption at once. The changing of the rail voltages is possible because the time intervals LT_s between neighboring samples in each channel are L times longer than T_s, and an increase in L simplifies this method's realization. Still, sufficient guard intervals between consecutive $w_n(t)$ are needed to exclude the influence of the transients caused by the rail voltage changes. Summing the NRC channels' output signals in the air or before the common antenna requires solving several technical problems but also creates some opportunities.

All the structures presented in this section are not intended for immediate implementation; rather, they illustrate new possibilities provided by the sampling theorem's direct interpretation.

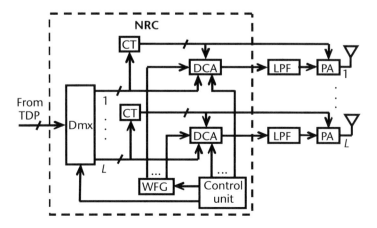

Figure 6.20 AMB with multiple-D/A NRC.

6.5 Channel Mismatch Mitigation

6.5.1 Approaches to the Problem

Since all NDCs and NRCs based on the sampling theorem's direct interpretation and some of those based on the hybrid interpretation are inherently multichannel, the influence of channel mismatch on their performance must be mitigated. Channel mismatch is especially dangerous for NDCs because $u_{in}(t)$ there is a sum of a desired signal $u(t)$, ISs, and noise. When the average power of ISs and noise is much larger than that of $u(t)$, the error caused by channel mismatch can be comparable with or even higher than $u(t)$. Therefore, the attention below is focused on channel mismatch mitigation in NDCs.

There are three approaches to the problem. The first one includes technical and technological measures that reduce the mismatch: placing all the channels on the same die, digital generation of $w_n(t)$, and proper realization of multiplications. The second approach is based on preventing spectral overlap of signal and mismatch error. This allows suppressing the error spectrum in the RDP [26, 28, 29]. The third approach is adaptive compensation of the channel mismatch in the RDP [28, 29, 35]. The first approach alone is sufficient in many types of Txs and in Rxs with small dynamic range. In high-quality Rxs, it is useful but insufficient and should be combined with other measures. Therefore, the second and third approaches are concisely analyzed below.

6.5.2 Separation of Signal and Error Spectra

Let us determine the conditions that prevent overlapping of the spectra $S_{d.u}(t)$ of $u_d(t)$ and $S_{d.e}(f)$ of $e_d(t)$ for bandpass sampling at optimal f_s (3.16). Here, $u_d(t)$ is the discrete-time signal produced by sampling of $u(t)$, and $e_d(t)$ is the discrete-time error caused by channel mismatch. Without loss of generality, we assume the $u(t)$ center frequency $f_0 = 0.25 f_s$ for better visualization of the spectral diagrams.

The delay mismatch among L channels is usually very small because all clock impulses are generated using the same reference oscillator, and proper design minimizes the timing skew. Therefore, the amplitude mismatch caused by the differences among the channel gains $g_1, g_2, \ldots, g_L$ is considered first. The average gain is $g_0 = (g_1 + g_2 + \ldots + g_L)/L$, and the gain deviation in the lth channel is $\gamma_l = g_l - g_0$. Since samples $u(nT_s)$ are generated in turn by all channels, the deviations $\gamma_1, \gamma_2, \ldots, \gamma_L$ appear at sampling instants $t = nT_s$ as a discrete-time periodic function $\gamma_d(t)$ with period LT_s:

$$\gamma_d(t) = \sum_{k=-\infty}^{\infty} \sum_{l=1}^{L} \left\{ \gamma_l \delta\left[t - (kL + l)T_s\right] \right\} \tag{6.16}$$

where $\delta(t)$ is the delta function. The spectrum of $\gamma_d(t)$ (see Section A.2) is

$$S_{d.\gamma}(f) = \sum_{m=-\infty}^{\infty} \left[C_m \delta\left(f - \frac{m}{L} f_s\right) \right] \tag{6.17}$$

with coefficients

$$C_m = \frac{1}{LT_s} \sum_{l=1}^{L} \left[\gamma_l \exp\left(\frac{-j2\pi ml}{L}\right) \right] \qquad (6.18)$$

As reflected by (6.17) and (6.18), $S_{d.\gamma}(f)$ is a periodic function of frequency with period $f_s = 1/T_s$ due to the discrete-time nature of $\gamma_d(t)$. Therefore, it is sufficient to consider $S_{d.\gamma}(f)$ only within the interval $[-0.5f_s, 0.5f_s[$. Since $\gamma_d(t)$ is real-valued and periodic with period LT_s, $S_{d.\gamma}(f)$ is even and discrete with the harmonics located at frequencies $\pm mf_s/L$ where $m = 1, 2, \ldots,$ floor $(L/2)$ within the interval $[-0.5f_s, 0.5f_s[$. The spectral components of $\gamma_d(t)$ are shown in Figure 6.21(a) for $L = 5$.

Since $f_0 = 0.25f_s$, the spectrum $S_u(f)$ of $u(t)$ occupies the bands

$$\left[-(0.25f_s + 0.5B), -(0.25f_s - 0.5B)\right] \cup \left[0.25f_s - 0.5B, 0.25f_s + 0.5B\right] \qquad (6.19)$$

within the interval $[-0.5f_s, 0.5f_s[$. Here, B is the $u(t)$ bandwidth. Figure 6.21(b) shows $|S_u(f)|$ and the AFR $|H_{a.f}(f)|$ of the antialiasing filtering performed by the NDC. Spectrum $S_{d.e}(f)$ is a convolution of $S_u(f)$ and $S_{d.\gamma}(f)$:

$$S_{d.e}(f) = \sum_{m=-\infty}^{\infty} \left\{ C_m \left\{ S_u \left[f - f_s\left(\frac{m}{L} - 0.25\right) \right] + S_u \left[f - f_s\left(\frac{m}{L} + 0.25\right) \right] \right\} \right\} \qquad (6.20)$$

Since $e_d(t)$ is a real-valued discrete-time function with sampling period T_s, $|S_{d.e}(f)|$ is an even periodic function with period f_s, unique within the interval $[-0.5f_s, 0.5f_s[$.

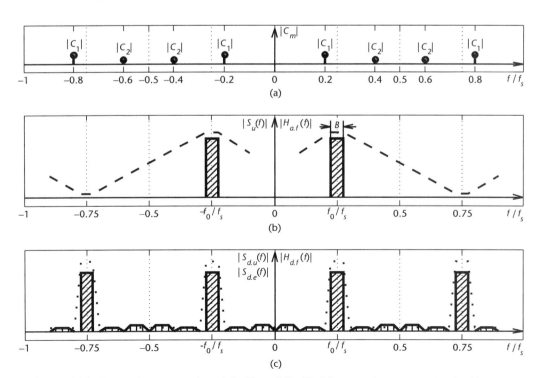

Figure 6.21 Preventing an overlap of $S_{d.u}(f)$ and $S_{d.e}(f)$: (a) spectral components of $\gamma_d(t)$, (b) $|S_u(f)|$ and $|H_{a.f}(f)|$ (dashed line), and (c) $|S_{d.u}(f)|$, $|S_{d.e}(f)|$, and $|H_{d.f}(f)|$ (dotted line).

As follows from (6.20), the mismatch error's spectral terms corresponding to $m = \pm 0.5L$ fall within the signal spectrum if L is even. Therefore, $S_{d.u}(f)$ and $S_{d.e}(f)$ cannot be separated. If L is odd, the center frequencies of the mismatch error's spectral terms are $\pm(r + 0.5)f_s/(2L)$ where $r = 0, 1, ..., 0.5(L-1) - 1, 0.5(L-1) + 1, ..., L - 1$, within the interval $[-0.5f_s, 0.5f_s[$, that is, they differ from the center frequencies of the signal spectral replicas. This is a necessary condition for avoiding any overlap of $S_{d.u}(f)$ and $S_{d.e}(f)$, which becomes sufficient when the distances between the center frequencies of the signal spectral replicas and the mismatch error terms exceed a certain minimum distance, as shown in Figure 6.21(c) for $L = 5$. Calculations in [28, 29] show that the minimum distance should only slightly exceed B when the power within B is higher than the power of the error caused by the channel mismatch. In this case, the relation among f_s, L, and B, which prevents an overlap of $S_{d.u}(f)$ and $S_{d.e}(f)$ for bandpass sampling with optimum f_s and odd L, is quite simple:

$$f_s > 2LB \tag{6.21}$$

Relation (6.21) allows rejecting the mismatch error by a digital filter with AFR $|H_{d.f}(f)|$ in the RDP (see Figure 6.21(c)).

As mentioned above, delay mismatch among channels can usually be made very small. If the resulting phase mismatch is also small, the error it causes has spectral distribution similar to (6.20). Consequently, relation (6.21) also separates the signal and phase mismatch error spectra, allowing the rejection of the latter by the same digital filter.

When the minimum distance between the center frequencies of the signal spectral replicas and the mismatch error terms is smaller than B, $S_{d.u}(f)$ and $S_{d.e}(f)$ overlap. However, this overlap can be lowered by increasing f_s and, for $L \geq 5$, by reducing the C_m neighboring $\pm 0.5 f_s$ since they create the spectral replicas of $S_{d.e}(f)$ closest to the signal. Changing the channel switching sequence can reduce these harmonics. The sequence that makes $\gamma_d(t)$ close to a sampled sinusoid minimizes the overlap.

In contrast with bandpass sampling, overlapping of signal and mismatch error spectra in the baseband case can be avoided when L is even and

$$f_s > LB \tag{6.22}$$

Channel mismatch mitigation by separating the signal and error spectra and rejecting the latter in the RDP does not interrupt signal reception but requires increasing f_s proportionally to L, as follows from (6.21) and (6.22). Therefore, this approach is adequate when L is relatively small or a high ratio f_s/B is needed anyway (for instance, in sigma-delta A/Ds). Under these conditions, it is not restricted to NDCs and can be used in other time-interleaved data converters.

6.5.3 Channel Mismatch Compensation

Adaptive channel mismatch compensation in the RDP is the most universal approach, widely used in time-interleaved and parallel structures (see, for example, [36]). This compensation is simplified when all channels identically process time-interleaved portions of the same signals. For proper compensation, the mismatch must be estimated

6.5 Channel Mismatch Mitigation

with adequate accuracy. The estimation and compensation can be performed either simultaneously with signal reception or in a separate calibration mode. The latter method is faster and more accurate, but it interrupts signal reception. Therefore, they are often combined in practice. Mismatch estimation during operation can be performed using a calibration signal $u_c(t)$ or a received signal $u_{in}(t)$ (blind estimation).

To avoid the influence of channel mismatch estimation on signal reception, $u_c(t)$ should be orthogonal to $u(t)$. The orthogonality can be achieved, for example, by choosing $u_c(t)$ whose spectrum $S_{u,c}(f)$ does not overlap with the spectrum $S_u(f)$ of $u(t)$ but is sufficiently close to it, as shown in Figure 6.22(a) where $S_{u,c}(f)$ is

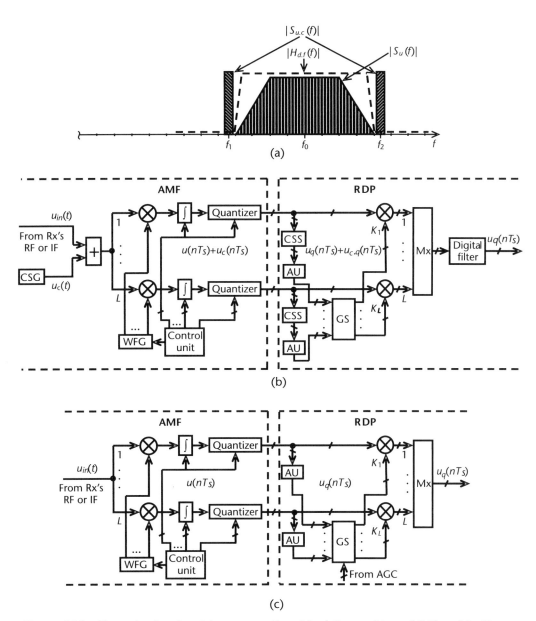

Figure 6.22 Channel gain mismatch compensation: (a) relative positions of $S_u(f)$ and $S_{u,c}(f)$, (b) compensator with calibration signal, and (c) compensator with blind mismatch estimation.

concentrated around frequencies f_1 and f_2. The block diagram of a channel gain mismatch compensator with such $u_c(t)$ for a multiple-quantizer NDC is depicted in Figure 6.22(b).

Here, $u_c(t)$ from the calibration signal generator (CSG) and $u_{in}(t)$ from the Rx's RF or IF strip are fed into the NDC. At the output of each NDC channel, $u_c(t)$ is extracted by the calibration signal selector (CSS) and sent to the averaging unit (AU), which calculates the average magnitude of the $u_c(t)$ samples. The magnitudes from all the channels are processed in the gain scaler (GS) that generates coefficients K_l ($l = 1, 2, ..., L$) compensating the channel mismatch. The Mx combines the scaled output signals of all the channels and sends them to the digital filter that rejects $u_c(t)$. Selection of an elementary deterministic $u_c(t)$ simplifies the compensation and reduces the averaging time. A couple of sinusoids with frequencies f_1 and f_2 is an example of such $u_c(t)$. Correlating $u_c(t)$ samples in the first channel with those in other channels allows estimating and compensating the delay mismatch.

Blind channel mismatch estimation is schematically simpler than that with $u_c(t)$, as shown in Figure 6.22(c). Its major problem is a long estimation time T_{est} caused by the fact that $u(t)$ is a stochastic process. This time [35] is

$$T_{est} = 1.5\left(2^{N_b-1} - 1\right)^2 D^2_{|u|} r_e L T_s \tag{6.23}$$

where N_b is the number of bits in the NDC quantizers, r_e is an acceptable ratio of the quantization noise power to that of the estimation error, and $D^2_{|u|}$ depends on the one-dimensional distribution of $u(t)$. For most signals, $D^2_{|u|} \in [0.2, 0.6]$ where 0.2 corresponds to a sinusoid and 0.6 to Gaussian noise. Blind estimation of the delay mismatch is based on correlating the signal samples in different channels [36].

6.6 Selection and Implementation of Weight Functions

6.6.1 Theoretical Basis

Since weight functions $w_0(t)$ determine many properties of S&I circuits based on the sampling theorem's hybrid interpretation and most properties of S&I circuits based on its direct interpretation, proper selection of $w_0(t)$ is critical for implementing these circuits. Selection of $w_0(t)$ for S&I circuits based on the hybrid interpretation is relatively simple because their short T_w limit the variety of $w_0(t)$ suitable for baseband S&I and the variety of envelopes $W_0(t)$ suitable for bandpass S&I. Carriers $c_0(t)$ of $w_0(t)$ should allow simple generation of $w_0(t)$, possibly replacing multipliers with a small number of switches, and effective suppression of IMPs. In differential bandpass S&I circuits with optimal f_s, third-order sum-frequency IMPs require the strongest suppression. Among the carriers $c_0(t)$ considered in Section 6.3.1.2, $c_{03}(t)$ and $c_{04}(t)$ meet these conditions better than others. As noted in that section, increasing f_0/f_s reduces the influence of $c_0(t)$. Therefore, square-wave carrier $c_{02}(t)$ is a logical choice when $f_0/f_s > 10$ due to its simplicity. For these reasons, the discussion below is focused on the selection of $W_0(t)$ for bandpass NDCs and NRCs, with the emphasis on NDCs operating in harsh RF envirohments.

6.6.1.1 Approaches

As mentioned above, $w_0(t)$ and $W_0(t)$ should allow simple generation and replacing multipliers in NDCs and NRCs with a small number of switches. They also should provide adequate filtering that automatically slows the accumulation of signal energy in NDCs and concentrates the NRCs' output energy within the signal bandwidth. It is easier to formalize the criteria for antialiasing and interpolating filtering in the frequency domain.

The least squares (LS) and Chebyshev criteria are most suitable for selecting $w_0(t)$ with optimal spectrum that determines the transfer function $H_w(f)$ of an NDC or NRC. The first criterion minimizes the weighted rms deviation of $H_w(f)$ from the ideal transfer function $H_{\text{ideal}}(f)$:

$$\sigma_e = \left\{ \int_{f \in F} q(f) \left[H_w(f) - H_{\text{ideal}}(f) \right]^2 df \right\}^{0.5} \to \min \quad (6.24)$$

where $H_{\text{ideal}}(f) = 1$ in the passbands, $H_{\text{ideal}}(f) = 0$ in the stopbands, and is undefined in transition bands; $q(f)$ is the error weight; and set F includes only passbands and stopbands. This criterion often allows closed-form solutions, and well-developed numeric algorithms of multistopband FIR filter design can be used in other cases. The disadvantage of this criterion is that it does not limit the maximum deviation of $H_w(f)$ from $H_{\text{ideal}}(f)$. The constrained LS criterion [37] fixes the problem.

The Chebyshev criterion minimizes the maximum weighted deviation of $H_w(f)$ from $H_{\text{ideal}}(f)$:

$$e = \max_{f \in F} \left\{ q(f) \left| H_w(f) - H_{\text{ideal}}(f) \right| \right\} \to \min \quad (6.25)$$

It most adequately reflects the filtering quality, but usually does not lead to closed-form solutions. The Parks-McLellan algorithm is typically used for the design of multi-stopband FIR filters based on this criterion.

The LS and Chebyshev criteria require similar lengths T_w of $w_0(t)$ to achieve similar quality of filtering. The main drawback of both criteria is a high required accuracy of optimal $w_0(t)$ generation that, in particular, prevents replacing the multipliers with a small number of switches.

In the time domain, assessing the complexity of generating a selected $w_0(t)$ is easier than evaluating its filtering properties. The following heuristic procedure resolves the problem. First, a class of easily generated $w_0(t)$ with supposedly good filtering properties is selected in the time domain, based on prior experience and/or educated guesses. Second, it is determined if this class satisfies the theoretical constraints that assure proper distribution of the AFR nulls over the frequency axis. Third, the filtering properties of this class are verified by computing $H_w(f)$. If the properties are inadequate, another class of $w_0(t)$ is tested. This trial-and-error process is efficient if the theoretical constraints significantly reduce the number of candidates at the second step.

6.6.1.2 Theoretical Constraints

Before explaining the theoretical constraints mentioned above, the notion of frame should be introduced. All versions of the WKS sampling theorem use signal expansions with respect to orthogonal bases (see Appendix D), but the functions constituting these bases are physically unrealizable. As explained in Section 1.3.1, a set of functions $\{w_n(t)\}$ forms a basis in a certain function space if it spans this space, and $w_n(t)$ are linearly independent. A frame generalizes the notion of basis to the function sets that still span the space but may be linearly dependent. Thus, a basis is a special case of a frame. Here, terms "function" and "signal" are used interchangeably. Signal expansion with respect to a frame allows redundancy that provide larger freedom in selecting $\{w_n(t)\}$.

From the theoretical standpoint, S&I of bandlimited signals can be viewed as follows. The signals within the passband of an antialiasing filter belong to the desired function space F_0 of bandlimited signals, while the signals within its stopbands and "don't care" bands belong, respectively, to the undesired function spaces F_k (with $k \neq 0$) and to the irrelevant function spaces. S&I corresponding to the sampling theorem's direct interpretation require choosing a set $\{w_n(t) = w_0(t - nT_s)\}$ of physically realizable functions, which approximately meets two requirements: it is orthogonal to spaces F_k ($k \neq 0$) and its projection onto F_0 can be considered a frame in F_0 ("approximately" is the key word because there are no physically realizable functions that strictly meet these requirements). Imperfect suppression in the stopbands and nonuniform AFR in the passband (the PFR is usually linear) reflect the approximation inaccuracy. For a given set $\{w_n(t)\}$, an increase in the ratio $f_s/(2B) > 1$, which reflects its redundancy, reduces this inaccuracy. Therefore, $f_s/(2B)$ is selected sufficiently high for S&I. Since the redundancy lowers the DSP efficiency, signals are downsampled right after digitization and upsampled right before reconstruction in Txs and Rxs (see Chapters 3 and 4). Any decrease or increase of $f_s/(2B)$ always means, respectively, narrowing or widening of "don't care" bands. Note that selection of antialiasing and interpolating filters' characteristics in S&I circuits based on the indirect interpretation has the same theoretical foundation.

Physical realizability of selected $w_0(t)$ is easily noticeable in the time domain, and all sets $\{w_n(t) = w_0(t - nT_s)\}$ that meet the theoretical constrains described below are reasonable candidates for further consideration.

A finite-length baseband $w_0(t)$ performs FIR filtering by suppressing the unwanted signals in stopbands (4.43) with its spectral nulls. To provide regular spacing of these nulls in the stopbands and finite nonzero gain in the passband, such $w_0(t)$ should meet the partition of unity condition [38]:

$$\varsigma \sum_{n=-\infty}^{\infty} w_0(t - nT_s) = 1 \qquad (6.26)$$

where ς is a scaling factor. Bandpass $w_0(t)$ with optimal f_s must suppress unwanted signals within stopbands (6.11), and therefore (6.26) should be replaced with the partition of cosine condition [30]:

$$\varsigma \sum_{n=-\infty}^{\infty} (-1)^n w_0(t - 2nT_s) = \cos(2\pi f_0 t) \qquad (6.27)$$

As follows from (6.27), envelope $W_0(t)$ of a bandpass $w_0(t)$ must meet its own partition of unity condition:

$$\varsigma \sum_{n=-\infty}^{\infty} W_0(t - 2nT_s) = 1 \qquad (6.28)$$

6.6.2 B-Spline-Based Weight Functions

B-spline-based $w_0(t)$, suggested for NDCs and NRCs in [28, 31], were identified using heuristic procedure described above. When the maximum permissible T_w are $T_w = T_s$ for baseband $w_0(t)$ and $T_w = 2T_s$ for bandpass $w_0(t)$ (see Section 6.3.1.2), they are optimal according to any reasonable criteria. For longer T_w, they are not optimal but may still remain an acceptable choice even for relatively long T_w. This section analyzes filtering properties of bandpass B-spline-based $w_0(t)$ and some ways of their implementation in NDCs.

6.6.2.1 Filtering Properties

A baseband $w_0(t)$, which is a K-order B-spline (see Section A.3), has length $T_w = KT_s$ and spectrum

$$H_w(f) = A_K \left[\mathrm{sinc}(\pi f T_s) \right]^K \qquad (6.29)$$

where A_K is a scaling factor. Compare (6.29) with (A.27), where k is a B-spline degree.

The envelope $W_0(t)$ of a bandpass $w_0(t) = W_0(t)c_0(t)$ based on a K-order B-spline is a convolution of K rectangles, each of duration $2T_s$. This $w_0(t)$ has length $T_w = 2KT_s$ and spectrum

$$H_w(f) = 0.5 A_K \left\{ \left\{ \mathrm{sinc}\left[2\pi(f+f_0)T_s\right] \right\}^K + \left\{ \mathrm{sinc}\left[2\pi(f-f_0)T_s\right] \right\}^K \right\} \qquad (6.30)$$

when $c_0(t) = c_{01}(t) = \cos(2\pi f_0 t)$.

B-spline-based $w_0(t)$ meet the constraints outlined in the previous section. For all K, the nulls of $H_w(f)$ defined by (6.29) are located within stopbands (4.43), and the nulls of $H_w(f)$ defined by (6.30) are located within stopbands (6.11). In both cases, the sidelobes of $H_w(f)$ correspond to "don't care" bands between the stopbands. These properties are illustrated by baseband and bandpass $w_0(t)$ shown, respectively, in Figures 6.14 and 6.15.

An increase of K improves filtering at the cost of NDC complexity and larger L. Since B-spline-based $w_0(t)$ are not optimal for $K > 1$, they require longer T_w than those of optimal $w_0(t)$ to achieve the same quality of filtering. As mentioned in Section 6.4.1.1, their filtering properties are determined exclusively by the B-spline order when $f_0/f_s > 10$. Thus, the comparison of baseband B-spline-based and optimal $w_0(t)$ performed in [39] can be applied to both baseband and bandpass $w_0(t)$, taking into account that bandpass $w_0(t)$ are two times longer than the corresponding baseband ones. This comparison is complicated by the difference in the filtering patterns. Indeed, as the distance from the passband increases, stopband suppression grows fast for B-spline-based $w_0(t)$ and slower for optimal $w_0(t)$. Within each stopband, B-spline-based $w_0(t)$ provide the highest suppression at its midpoint and the lowest one at its edges for any T_w. The optimal $w_0(t)$ provide more uniform suppression within the stopbands when $T_w > T_s$ for baseband $w_0(t)$ and $T_w > 2T_s$ for bandpass $w_0(t)$. Figure 6.23 [39] shows the minimum and rms stopband suppressions theoretically attainable for three equal-length baseband $w_0(t)$: Chebyshev-optimal, LS-optimal, and B-spline-based. The rms suppression is computed for three closest stopbands on each side of the passband. In practice, these theoretical results are usually limited by hardware imperfections when suppression exceeds 80 dB.

To achieve a given stopband suppression, the lengths T_w of B-spline-based $w_0(t)$ must be increased by a certain factor η compared to those of optimal $w_0(t)$. When $f_s/B = 6$, $\eta \leq 1.3$ for rms suppression and $\eta \leq 1.7$ for minimum suppression. When $f_s/B = 4$, $\eta \leq 1.5$ for rms suppression and $\eta \leq 2$ for minimum suppression.

The capability of B-spline-based $w_0(t)$ to provide the highest suppression at the midpoints of frequency intervals (6.11) makes them a good choice for NDCs with sigma-delta A/Ds where ratios f_s/B are very high. Table 6.1 presents the minimum stopband suppression provided by B-spline-based $w_0(t)$ of different orders. It demonstrates that even low-order B-splines can provide adequate suppression when f_s/B is sufficiently high.

6.6.2.2 Implementation

As follows from the conceptual structures of NDCs in Figure 6.13, each NDC channel can be considered a correlator with reference signal $w_0(t)$. Ideally, this

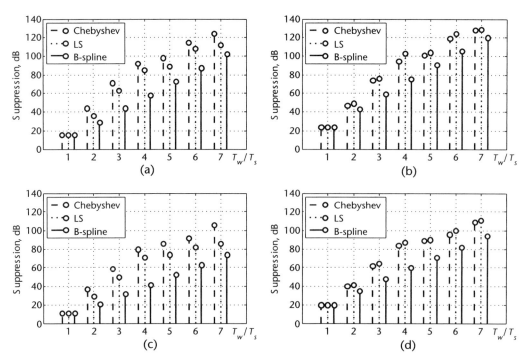

Figure 6.23 Stopband suppression for three equal-length baseband $w_0(t)$: (a) minimum suppression for $f_s/B = 6$, (b) rms suppression for $f_s/B = 6$, (c) minimum suppression for $f_s/B = 4$, and (d) rms suppression for $f_s/B = 4$.

6.6 Selection and Implementation of Weight Functions

Table 6.1 Minimum Suppression Provided by B-Spline-Based $w_0(t)$

f_s/B	4	6	8	16	32	64	128
Suppression, dB (first-order B-spline)	10	14	17	24	30	36	42
Suppression, dB (second-order B-spline)	21	29	34	47	60	72	84
Suppression, dB (third-order B-spline)	31	43	51	71	90	108	126
Suppression, dB (fourth-order B-spline)	42	58	69	94	119	144	168

correlator should equally correspond to all signals' spectral components within the bandwidth B and reject out-of-band components. Considering NDCs a specific type of mixed-signal correlators prompts their implementation as mixed-signal matched filters. This possibility is demonstrated below for NDCs with B-spline-based $w_0(t)$, which, in principle, can also be realized the way shown in Figure 6.13. However, transforming (6.13) with regard to (A.25) and (A.26), a sample at the output of an NDC channel can be presented as:

$$u(nT_s) = \sum_{k=0}^{K} \left[C_k \int_{t'}^{nT_s + 0.5T_w - akT_s} v(\tau_{K-1}) d\tau_{K-1} \right] \qquad (6.31)$$

where

$$C_k = (-1)^k \binom{K}{k}, \; v(t) = \int_{t'}^{t} d\tau_{K-2} \ldots \int_{t'}^{\tau_2} d\tau_1 \int_{t'}^{\tau_1} \hat{c}(\tau) u_{in}(\tau) d\tau, \text{ and } t' = nT_s - 0.5T_w \qquad (6.32)$$

and $\tau, \tau_1, \ldots, \tau_{K-1}$ are integration variables. Here, $a = 1$ and $\hat{c}(t) = 1$ for baseband $w_n(t)$, while $a = 2$ and $\hat{c}(t) = c_n(t)$ for bandpass $w_n(t)$.

A general structure of a bandpass NDC channel and the table of switches' states for this structure, which follow from (6.31) and (6.32), are shown in Figure 6.24. The structure of a baseband NDC channel differs from that in Figure 6.24(a) only by the absence of the input multiplier. The table in Figure 6.24(b) shows dynamically changing states of switches throughout the channel sample mode and is true for both bandpass and baseband structures. Here, 0 and 1 correspond to the open and closed states, respectively. The switches are controlled by the WFG. The time interval length is equal to T_s for baseband $w_0(t)$ and to $2T_s$ for bandpass $w_0(t)$. The cosine carriers of bandpass $w_n(t)$ have zero phases at the midpoints of $w_n(t)$. These carriers can be replaced with stepwise carriers often without degrading the sampling quality. This allows replacing the multiplier at the channel input with a small number of switches. Properties of alternating-sign binomial coefficients C_K allow simplifing the structure in Figure 6.24(a). As a result, the number of subchannels can be reduced to $(K + 1)/2$ for odd K and to $(K/2) + 1$ for even K. Since low values of K are of the highest interest, the simplified channel structures for $K = 2, \ldots, 5$ and the states of their switches are shown in Figures 6.25 through 6.28.

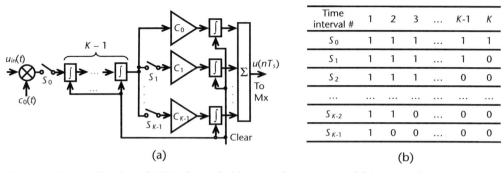

Figure 6.24 B-spline-based NDC channel: (a) general structure and (b) states of switches.

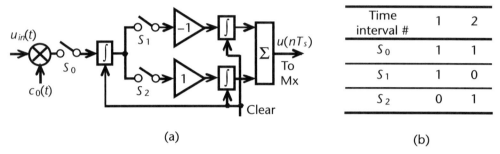

Figure 6.25 B-spline-based NDC channel for $K = 2$: (a) general structure and (b) states of switches.

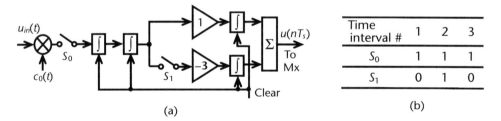

Figure 6.26 B-spline-based NDC channel for $K = 3$: (a) general structure and (b) states of switches.

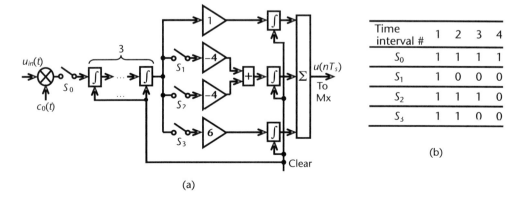

Figure 6.27 B-spline-based NDC channel for $K = 4$: (a) general structure and (b) states of switches.

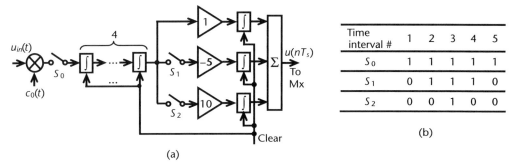

Figure 6.28 B-spline-based NDC channel for $K = 5$: (a) general structure and (b) states of switches.

In bandpass NDCs with the channel structures described above, the WFGs are very simple because their only tasks are generating the couriers with proper phases and controlling the switches. The main challenge of these NDCs' realization is mitigation of the mismatch among the subchannels of the NDC channels.

6.6.3 Additional Remarks on Weight Function Implementation

Besides B-spline-based $w_0(t)$, there are many other classes of $w_0(t)$ that satisfy the theoretical constraints (6.26) or (6.27) and (6.28), and some of them can be implemented using both correlator and matched-filter approaches. Still, the correlator approach is probably more widely applicable to S&I circuits based on the hybrid and direct interpretations of the sampling theorem. The main challenges of that approach are improving and simplifying generation of $w_0(t)$ and multiplications in NDCs and NRCs.

Digital generation of $w_0(t)$ is most accurate. To simplify it and increase its speed, the techniques used in nonrecursive DDSs can be employed. Specifically, the six-step procedure, described at the end of Section 3.2.3, can be directly applied to the $w_0(t)$ generation. Converting a digitally generated $w_0(t)$ into the analog domain before multiplying it by $u_{in}(t)$ would reduce the dynamic range and accuracy of processing. The preferable way is performing such multiplications by MD/As or DCAs with properly selected parameters. From this standpoint, the selection of $w_0(t)$ that can be accurately represented by a low number of bits is highly desirable, and when a few bits are sufficient, the MD/As or DCAs can be replaced with a small number of switches. Some techniques introduced for the development of multiplierless FIR filters can be used to that end.

Independently of the domain where $w_0(t)$ is selected, its implementation in the frequency domain can be attractive. For instance, if $w_0(t)$ is represented by a few frequency components and ratio f_0/f_s is high, substituting cosines with simple stepwise functions eases the WFG realization, and division of each NDC channel into a few subchannels allows replacing multipliers with switches. Minimizing the subchannel mismatch is critical in this case.

The use of MD/As, DCAs, or a small number of switches for multiplication in the channels may require some prefiltering at the inputs of NDCs and postfiltering at the outputs of NRCs to prevent or reject the unwanted spectral components caused by such simplifications. Both prefiltering and postfiltering can often be done by low-quality analog filters compatible with the IC technology. Some kinds of prefiltering in Rxs and postfiltering in Txs always exist. Therefore, $w_0(t)$ selection requires maximum attention to the stopbands closest to the passbands.

Bandpass S&I can be performed at IF or RF. An optimal f_s is usually selected for IF S&I. In this case, signal spectrum is centered in the middle of a Nyquist zone (see Section 3.3.2), samples of different parities belong to different signal components (I and Q), and, therefore, the carriers of neighboring $w_n(t)$ are shifted by ±90° relative to each other. RF S&I require precautions to prevent leakage of WFG signals through the antenna.

The authors' papers on SHAs with extended integration time, SHAWIs, and S&I with internal antialiasing and interpolating filtering based on the sampling theorem's hybrid and direct interpretations started to appear in English-language publications in the early 1980s [14–33]. Although the papers confirmed that the implementation of even the initial theoretical results had radically increased the dynamic ranges of mass-produced radios, they were initially overlooked by the scientific and engineering community. First publications of other authors on the topic appeared only in the early 2000s, and their number continues to grow (see, for instance, [40–57]). Analysis of those publications is out of this book's scope, but this example shows that acceptance of new concepts is a long process.

6.7 Need for Hybrid and Direct Interpretations

6.7.1 Evaluation of Hybrid and Direct Interpretations' Advantages

As shown in this chapter, sampling techniques based on the sampling theorem's hybrid and direct interpretations enable radical increase in Rx dynamic range, attainable bandwidth, adaptivity, reconfigurability, and scale of integration. They also allow close-to-the-antenna digitization and lower power consumption. These advantages are interconnected, and their importance depends on the Rx purpose and required parameters.

The dynamic range increase is probably the most important advantage provided by the hybrid and direct interpretations. As shown in Chapter 4, dynamic range, which reflects Rx capability to pick up a weak desired signal in the presence of strong in-band ISs, determines reception reliability in the frequency bands where Rxs can be subject to interference. For instance, rejection of ISs in the frequency and spatial domains is ineffective when the Rx dynamic range is inadequate (see Figures 4.4 and 4.5 for the frequency domain).

High dynamic range is needed even if signals in the RDP undergo nonlinear transformations, intended, for example, for compensating nonlinear distortion in the AMF or realizing robust AJ algorithms [58]. Widening the Rx bandwidth also requires increasing dynamic range due to the higher probability and level of interference, selective fading, and diversity of simultaneously received signals. The dynamic range increase required for HF Rx is illustrated by Figure 4.8(a). Widening the radios' bandwidths is beneficial for many applications because it improves the throughput of communications, range resolution of radar, processing gain of

SS systems, and spectrum utilization by CRs. The hybrid and direct interpretations allow increasing both bandwidth and dynamic range.

As explained throughout this chapter, the hybrid and direct interpretations make the integration time T_i in sampling circuits independent of f_0 and f_s, enabling its significant increase that reduces the integrator's charging current and, consequently, the IMPs and the required AMF gain G_a. Internal antialiasing filtering reduces jitter-induced error and, being performed immediately at the quantizer input, suppresses all out-of-band IMPs and noise. These factors together radically increase dynamic range. A long T_i, independent of f_0 and f_s, makes the required G_a and signal power at the sampling circuit input independent of f_0. It also reduces the impact of f_0 on jitter-induced error. The last two factors substantially extend the analog bandwidth of A/Ds with sampling circuits based on the hybrid or direct interpretation. The extended analog bandwidth increases attainable Rx bandwidth and, together with reduced G_a, allows close-to-the-antenna digitization (see Figure 6.17).

Reducing the required G_a and signal power at the sampling circuit's input lowers the Rx power consumption. Close-to-the-antenna digitization means that some functions, previously performed in the AMF, are now performed in the RDP. This enables higher adaptivity and reconfigurability of processing. The removal of traditional antialiasing filters, allowed by the direct interpretation, further improves the adaptivity and reconfigurability because the NDCs' filtering properties can be varied by changing $w_0(t)$. It also increases the scale of integration. The interpolation techniques based on the hybrid and direct interpretations provide similar advantages to Txs.

Accurate quantitative assessment of performance advantages provided by S&I techniques based on the hybrid and direct interpretations can be done only on a case-by-case basis. However, the theory and experiments allow us to expect orders of magnitude improvement in the dynamic range as well as multiple increase in the A/D analog bandwidth and, consequently, in the maximally acceptable frequency of digitized signals.

Although all the novel S&I techniques discussed above follow from the sampling theorem, they were derived taking into account factors that traditionally are not considered by the sampling theory (signal energy accumulation, IMPs, noise, etc.). Thus, as noted in Section 5.3.2, the theoretical basis of S&I should include, besides the sampling theory, the theories of linear and nonlinear circuits, optimal filtering, and the like. Again, presently there are no serious technological or technical obstacles to the implementation of S&I circuits based on the hybrid interpretation, but there are still some challenges to the implementation of the circuits based on the direct interpretation. The challenges depend on the purpose and required parameters of the radios.

Currently, structures consisting of L conventional time-interleaved A/Ds are widely used to provide L-fold increase in f_s for the same effective number of bits (ENOB) $N_{b.e}$. Since they resemble the L-quantizer NDC in Figure 6.13(b), it is logical to compare their capabilities. The L-quantizer NDC allows the same increase in f_s compared to a single quantizer, but it also increases $N_{b.e}$ (for the same nominal number of bits), analog bandwidth, and flexibility of digitization. The increased analog bandwidth allows higher f_0 and broader B of input signals.

The importance of the dynamic range increase, provided by the hybrid and direct interpretations, for frequency-domain IS rejection is clear from Section 4.3. Its importance for spatial-domain rejection is demonstrated in the next two sections together with outlining two unconventional IS rejection techniques that utilize this increase.

6.7.2 Two-Stage Spatial Suppression of ISs

Currently, adaptive antenna arrays are widely used for IS suppression and beamforming. In these arrays, signals from different antenna elements (AE_k) are summed with coefficients w_k^* that maximize the sum's SNR. Here $k \in [1, K]$ and K is the number of AEs in the array. The accuracy of digital calculation of w_k^* and summation of weighted signals has made adaptive IS suppression and beamforming in RDPs predominant. This accuracy must be supported by sufficient dynamic range of the Rx digitization circuits. The stronger ISs are, the higher dynamic range is needed. Thus, the dynamic range increase, provided by the hybrid and direct interpretations, is beneficial. Still, many Rxs can experience extremely strong ISs capable of desensitizing or even damaging their input circuits. Such ISs can be intentional (i.e., EW) or result from poor regulation or accidents. For instance, aircraft navigation Rxs may suffer from such ISs during landing near high-power broadcast stations, and Rxs of various vehicles can come dangerously close to Txs operating at the same or adjacent frequencies.

In such cases, an increase in the digitization circuits' dynamic range alone is insufficient, but combining it with two-stage spatial suppression [59, 60] improves the situation. The two-stage suppression first weakens extremely strong ISs (α-type ISs) at the AMF input, and then suppresses the residual α-type ISs and moderate ISs (β-type ISs) in the RDP. While the first-stage suppression in the analog domain cannot increase SNR as much as the second-stage suppression in the digital domain, it protects the AMF from desensitization or even damage.

The method divides the Rx array into subarrays, executes the first stage within the subarrays, and uses the subarrays' output signals at the second stage. The subarrays allow various arrangements and can be nonoverlapping (i.e., each AE belongs to only one subarray) or overlapping. The arrays can have different K, geometries, and AE spacing. The example below explains the method's basic principles.

The block diagram of a Rx with a 4-AE uniform linear array (ULA) divided into three 2-AE overlapping subarrays is shown in Figure 6.29. Denoting the number of AEs in each subarray by M and the number of subarrays in the array by N, it can be written that $K = 4$, $M = 2$, and $N = 3$ in this example. In the AMF, each AE is connected to the input of its primary channel through a guard attenuator (GAt) that can be enabled and disabled by the attenuator control circuit (AtC). Besides the GAt and AtC, each AE is connected to an analog phase and amplitude tuner (PAT) controlled by the digital complex-valued coefficient w_k^* formed in the RDP. The PATs' outputs within the same subarray are summed by an adder connected to a secondary channel through a switch. The indices $k \in [1, K]$ of the primary channels, GAts, AtCs, and PATs are the same as those of the corresponding AEs. The indices of the adders, switches, and secondary channels consist of the AEs' indices of the corresponding subarrays.

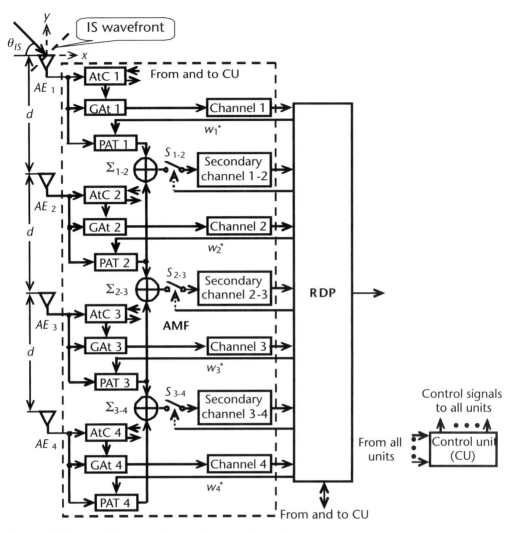

Figure 6.29 Rx with a 4-AE ULA and overlapping subarrays.

The maximum number $N_{\max}$ of overlapping subarrays is

$$N_{\max} = \binom{K}{M} = \frac{K!}{M!(K-M)!} \qquad (6.33)$$

Since $K = 4$ and $M = 2$ in Figure 6.29, the array there can be divided into six overlapping subarrays: AE_1 and AE_2, AE_2 and AE_3, AE_3 and AE_4, AE_1 and AE_3, AE_1 and AE_4, and AE_2 and AE_4 with spacing between the AEs equal to d, $2d$, or $3d$. Although subarrays with different distances between AEs have different and, in some cases, problematic gain patterns, they can still be used in the discussed method. In Figure 6.29, however, only the first three subarrays are depicted to simplify the explanation.

The Rx in Figure 6.29 can operate in two modes: standard or extreme. The standard mode corresponds to the absence of α-type ISs, whereas the extreme one corresponds to their presence. In either mode, β-type ISs may be present or absent. The AEs' signals are monitored to select a proper mode. By default, the Rx is in the standard mode, in which the signal reception is performed only by the primary channels because the GAts are disabled and the secondary channels are also disabled and disconnected from the adders. The array has $K - 1 = 3$ degrees of freedom, and, consequently, can null up to three β-type ISs. When the number of β-type ISs $L_\beta <$ $K - 1$, the unutilized degrees of freedom are used for beamforming that shapes the array gain pattern towards the desired signal source.

Exceeding a certain threshold by the signal level at AEs indicates the presence of α-type ISs, and changes the Rx mode to the extreme one, enabling the GAts and the secondary channels. As a result, the primary channels are protected but their sensitivity is significantly reduced. Due to this reduction, the desired signal and smaller β-type ISs cannot be sensed, but α-type and stronger β-type ISs are still observable. Immediately after transitioning to the extreme mode, the number and power range of ISs is estimated in the RDP using the weakened input signals from the primary channels. The estimation (usually based on the eigendecomposition of the signals' spatial covariance matrix) determines the number L_α of α-type ISs and thus the minimum $M = L_\alpha + 1$ required for their nulling at the first stage. The secondary channels' structures are configured based on this information.

In the RDP, the weakened α-type ISs from the primary channels are also used for calculating $\{w_k^*\}$ that tune the phase shifters and attenuators in the PATs to null the α-type ISs at the outputs of the adders in each subarray. After the nulling, these outputs are connected to the corresponding secondary channels. It is hard to expect that the α-type ISs can be weakened by more than 30 dB at the first stage due to low accuracy of analog phase and gain adjustment.

The weakening of α-type ISs at the first stage protects the AMF input circuits. However, the residual α-type and β-type ISs are still substantial, and the second stage of suppression is required to prevent blocking the signal reception or reducing its quality. The second-stage nulling is performed in the RDP using signals from the secondary channels. It requires $N \geq M$.

Since the second-stage suppression in the extreme mode and the suppression in the standard mode are completely performed in the RDP, the digitization circuits' dynamic range is the main factor determining the depth of suppression in both modes. In the considered method, it also determines the boundary between α-type and β-type ISs. The higher this boundary, the stronger suppression of both types of ISs can be achieved.

6.7.3 Virtual-Antenna-Motion-Based Spatial Suppression of ISs

The use of antennas based on the intentionally generated Doppler effect for navigation and DF is outlined in Section 2.4.3. The Doppler effect produced by switching among the AEs of an array creates virtual antenna motion (VAM) that can be used for various purposes in Rxs and Txs. In this section, its use for spatial IS suppression is concisely discussed. This application was selected because, first, like conventional nulling, it requires high digitization quality, and, second, it was examined only in [61] despite its benefits.

6.7 Need for Hybrid and Direct Interpretations

When the AEs of a ULA are linked to the central digital processor (CDP) of a Rx through an ECS that sequentially connects each AE to the CDP (see Figure 6.30), switching of the AEs creates the effect of the virtual antenna moving leftward with speed v, which is the magnitude of the velocity vector $\mathbf{v}$, and returning back with a much higher speed. Since such motion of the virtual antenna changes its distance from the Tx antenna, a single-tone signal, transmitted at the frequency f_0, is received at the frequency

$$f_{0r} = f_0 + f_d = f_0 + \left(\frac{v_r}{c}\right)f_0 = f_0\left[1 + \left(\frac{v}{c}\right)\cos\theta\right] \qquad (6.34)$$

where f_d is the Doppler shift, c is the speed of light, v_r is the VAM radial speed, that is, the projection of $\mathbf{v}$ onto the line-of-sight (LOS) unit vector $\mathbf{l}$ pointing from the Rx towards the Tx, and θ is the angle between $\mathbf{v}$ and $\mathbf{l}$. Therefore,

$$v_r = \mathbf{v} \cdot \mathbf{l} = v\cos\theta \qquad (6.35)$$

Being a scalar, v_r can be positive or negative depending on the VAM direction.

Since VAM is produced by switching among spaced AEs, v can be many orders of magnitude higher than the speed of the fastest Rx platforms. Therefore, Rx and Tx can be assumed motionless for VAM analysis. Simultaneously, since still $v \ll c$, relativistic effects can be neglected and f_d calculated using the classical equations. The VAM with speed v and almost immediate return back is not the only possible

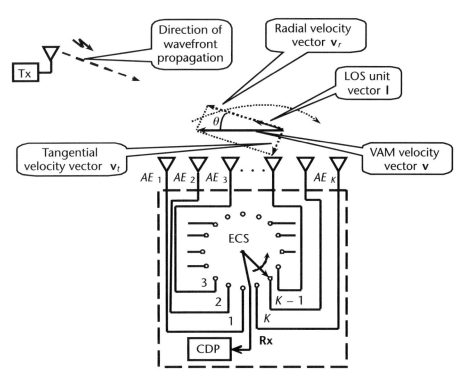

Figure 6.30 VAM in ULA.

type of VAM in ULAs. Changing the switching order in the ECS allows realizing many other types of VAM that can be performed not only in sequence but also simultaneously in the same ULA.

In general, the Doppler effect is time-frequency scaling of a transmitted signal rather than its frequency shift (see Section 1.2.3). However, the resulting spectrum compression or expansion is much less noticeable than the spectrum shift when $f_0/B \geq 100$. Therefore, only the Doppler shift (caused by the rightward VAM) is shown on the spectral diagrams in Figure 6.31. The figure illustrates the possibility of utilizing the dependence of this shift on the directions to Txs for separating $S(f)$ and $S_{IS}(f)$ that are the spectra of a desired signal $s(t)$ and an IS, respectively.

The spectra occupy the same frequency interval at the Rx's AEs (see Figure 6.31(a)), but their positions are changed by VAM (see Figures 6.31(b–e)) depending on the directions to the Txs. The separation of $S(f)$ and $S_{IS}(f)$ allows subsequent suppression of $S_{IS}(f)$ in the frequency domain.

Figure 6.31 corresponds to a hypothetical situation when $s(t)$ and an IS are the only signals in the air. In practice, Rxs usually operate in multisignal environments, and many out-of-band undesired signals from various directions can appear within the Rx passband due to VAM. To exclude the influence of out-of-band signals, AE switching should be executed after the main frequency selection in the RDP. An example of a digital Rx with such switching and a separate AMF for each AE is shown in Figure 6.32.

Here, the identical AMFs amplify, filter, and digitize input signals that, after digitization with the sampling rate f_{s1}, are independently processed by the digital filter-interpolators (DFIs). The DFIs initially perform digital filtering with downsampling, reducing the sampling rate to $f_{s2} = \gamma_2 B$ with $\gamma_2 > 2$, and then upsampling with digital filtering, increasing this rate to f_{s3}. The latter is needed prior to the ECS because the VAM-induced direction-dependent Doppler shifts expand the input signal spectrum from the initial bandwidth B to B_1 that can be as wide as

$$B_1 = 2\frac{v}{c}f_0 + B \tag{6.36}$$

Since $B_1 \gg B$,

$$B_1 \approx 2\frac{v}{c}f_0 \tag{6.37}$$

To prevent aliasing caused by widening the signal bandwidth from B to B_1, f_{s3} should meet the condition

$$f_{s3} = \frac{1}{T_{s3}} = \gamma_3 B_1 \approx 2\gamma_3 \frac{v}{c} f_0 \tag{6.38}$$

where $\gamma_3 > 2$. When the virtual antenna "travels" towards the Tx of interest,

$$v_r = v = \frac{d}{T_{s3}} = df_{s3} = d\gamma_3 B_1 \tag{6.39}$$

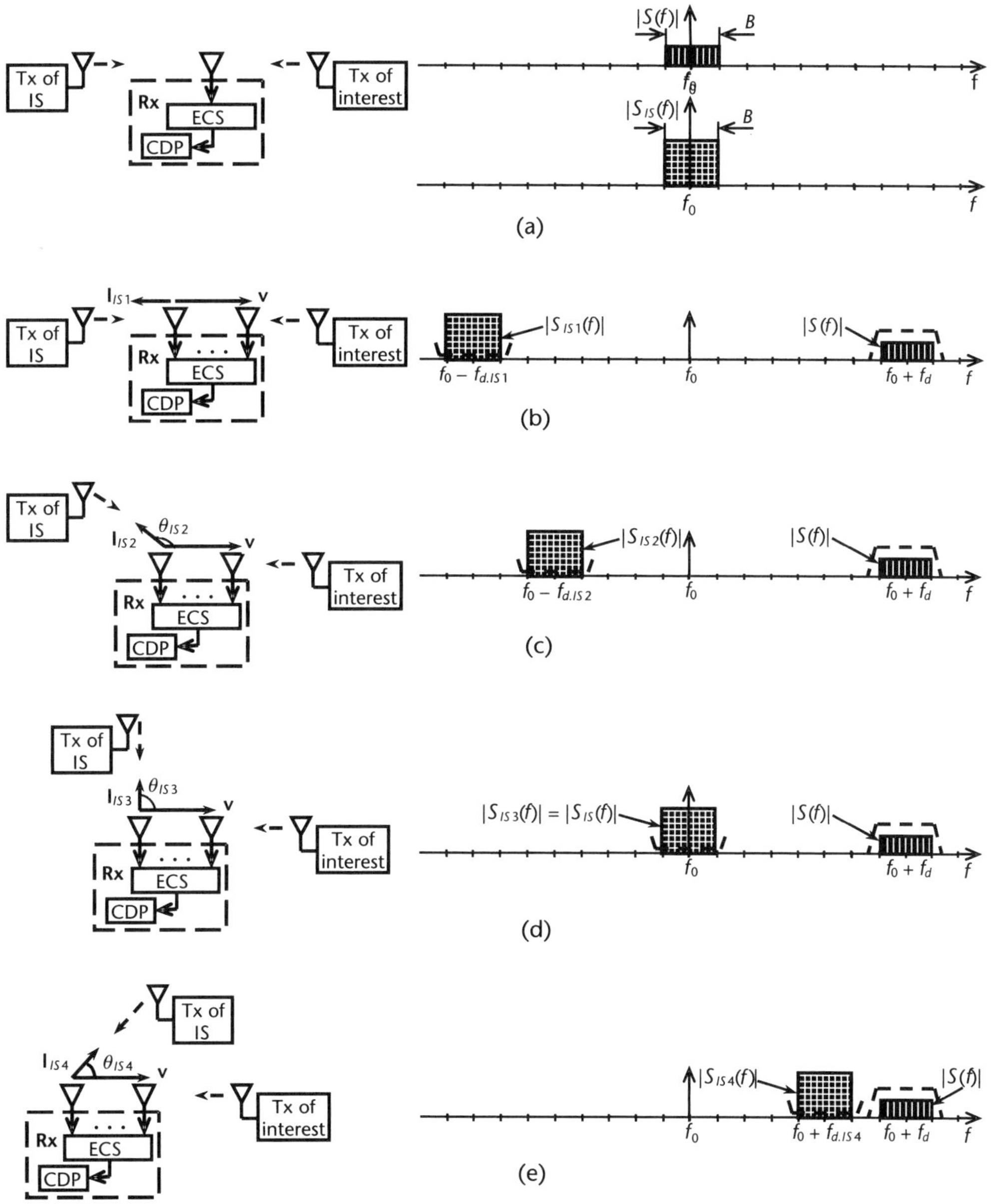

6.7 Need for Hybrid and Direct Interpretations

Figure 6.31 Locations of $|S(f)|$ and $|S_{IS}(f)|$ for various arrangements of positions of Rxs and Txs: (a) absence of VAM, (b) first arrangement, (c) second arrangement, (d) third arrangement, and (e) fourth arrangement.

where d is the distance between the neighboring AEs in the ULA. As follows from (6.38) and (6.39),

$$d \approx \frac{c}{2\gamma_3 f_0} = \frac{\lambda_0}{2\gamma_3} \qquad (6.40)$$

where λ_0 is the wavelength of f_0. Therefore, the maximum d and the maximum length D of a K-element ULA are, respectively,

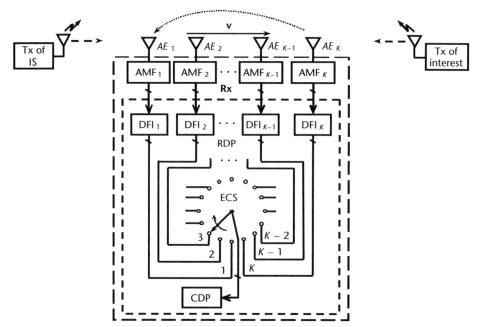

Figure 6.32 VAM-based Rx.

$$d_{max} < 0.25\lambda_0 \text{ and } D_{max} = (K-1)d_{max} < 0.25(K-1)\lambda_0 \qquad (6.41)$$

In most cases, $d \approx 0.2\lambda_0$ can be recommended. As follows from (6.38) and (6.41), an increase in f_0 reduces d and D but raises f_{s3} and, consequently, the required speed of AE switching and signal processing in the RDP.

The Rx operates in cycles. Within each cycle with duration KT_{s3}, ECS sequentially sends K samples from AEs with rate f_{s3} to the CDP while the virtual antenna "travels" towards the Tx of interest with speed v. Each sample contains all the signals within the Rx passband B arriving from different directions. Depending on the direction of arrival, the spectra of these signals are not only shifted but also expanded or compressed. When the VAM is directed towards the Tx of interest, $s(t)$ is the most time-compressed signal. Consequently, $S(f)$ is the most expanded spectrum and has the largest positive Doppler shift. The CDP selects $s(t)$ and suppresses all other signals in the frequency domain, reduces the sampling rate, and then demodulates and decodes $s(t)$.

Since $s(t)$ is time-compressed with factor $(K+1)/K$ by the VAM, one sample should be removed at the end of each cycle to seamlessly concatenate the $s(t)$ portions of sequential cycles. The emergence of one redundant sample at the end of each cycle means that

$$v = \frac{c}{K} \qquad (6.42)$$

At each instant, the VAM-based Rx in Figure 6.32 utilizes the signal energy only from one AE, that is, $(1/K)$th of the available energy. The Rx in Figure 6.33

utilizes the energy from all AEs. It contains K digital ECSs, and each ECS is offset by T_{s3} relative to the previous one. The output signals of all the ECSs are summed with appropriate offsets after initial processing. The proper summation forms a beam towards the Tx of interest.

Comparison of conventional nulling and beamforming with the VAM-based ones shows that both techniques require a high quality of digitization and each one has important advantages over the other.

Most advantages of the VAM-based technique are caused by converting angular separation into frequency separation and by the lower cost of the degrees of freedom in the frequency domain. These advantages are:

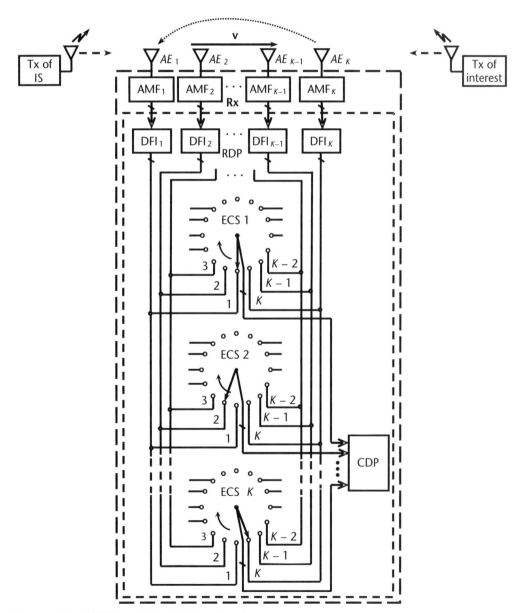

Figure 6.33 VAM-based Rx with beamforming.

1. VAM-based suppression can be stronger than that provided by conventional nulling, especially when the number of ISs is $N_{IS} > K - 1$.
2. For $N_{IS} > K - 1$, VAM-based suppression can be effective within much larger spatial sectors (at least 240° in a two-dimensional case) and with a much smaller angular separation from the desired signal.
3. A growth of N_{IS} without decreasing the angular separation between the desired signal and ISs does not influence the complexity of VAM-based processing, whereas conventional nulling requires increasing K (approximately proportionally to N_{IS}) and the complexity of calculations (faster than K^2).

Conventional nulling and beamforming have the following advantages:

1. They require lower K for $N_{IS} < 10$, and simpler signal processing for $N_{IS} < 5$.
2. They can work with a larger variety of array geometries and are less sensitive to the array orientation.
3. Knowledge of the direction to the Tx of interest is not critical for conventional adaptive nulling, whereas VAM-based suppression requires this knowledge and proper array orientation.
4. The optimal distance between the neighboring AEs in the arrays intended for conventional nulling and beamforming is $0.5\lambda_0$. In the VAM-based technique, it is $\sim 0.2\lambda_0$, and mutual coupling among the AEs presents a larger problem.

Thus, the choice between the discussed techniques is application-specific, and they can be combined in some cases.

6.8 Summary

At the dawn of DSP implementation, the existing technology allowed realization of S&I circuits based only on the sampling theorem's indirect interpretation. This interpretation separates antialiasing filtering and sampling as well as pulse shaping and interpolating filtering. Despite its drawbacks and the emerged feasibility of S&I based on other interpretations, these circuits (albeit evolved) are still widely used in digital radios and other applications.

Two types of samplers were initially used: THAs and SHAs. Due to the incorrect assumption that T_i in an SHA is limited to the time interval on which a bandpass signal $u_{in}(t)$ is close to a straight line, it was decided that T_i should meet (6.6). Such a short T_i eliminates all the advantages of SHAs over THAs, so SHAs were phased out.

Improvements of THAs did not remove their main drawback: the necessity of fast signal accumulation in the track mode. As shown in Section 6.2.1, this drawback limits the dynamic range and attainable bandwidth of Rxs, increases their power consumption, and makes close-to-the-antenna digitization impossible. Traditional BPFs used for bandpass sampling based on the indirect interpretation limit the adaptivity, reconfigurability, and scale of integration of Rxs. Bandpass interpolation based on the indirect interpretation creates similar disadvantages for Txs.

6.8 Summary

Understanding that restriction (6.6) on T_i can be replaced with (6.7) in the simplest SHAs and the development of SHAWIs where T_i is independent of f_0 were the first steps towards sampling circuits based on the hybrid and direct interpretations. Similar steps were made for interpolation circuits.

In sampling circuits based on the hybrid and direct interpretations, the lengths $T_w = T_i$ of weight functions $w_0(t)$ are independent of both f_0 and f_s, and antialiasing filtering is performed internally (partly for the hybrid interpretation and completely for the direct one). In interpolation circuits based on the hybrid and direct interpretations, $w_0(t)$ and interpolating filtering have the same properties.

Sampling circuits based on the hybrid and direct interpretations substantially increase the dynamic range, attainable bandwidth, adaptivity, reconfigurability, and scale of integration of Rxs. They also reduce Rx power consumption and enable close-to-the-antenna digitization. The advantages provided by interpolation based on the hybrid and direct interpretations are similar.

The direct interpretation provides broader advantages than the hybrid one (especially in terms of adaptivity, reconfigurability, scale of integration, and closeness to the antenna). However, while currently there are no serious technological or technical obstacles to the implementation of S&I circuits based on the hybrid interpretation, some challenges to the implementation of the circuits based on the direct interpretation still exist.

In multichannel S&I circuits based on the hybrid and direct interpretations, channel mismatch must be mitigated. Among three approaches to the problem, the first one, based on technical and technological measures, should always be used but is not always sufficient. The second approach, based on preventing any overlap of signal and mismatch error spectra and digital suppression of the error spectrum, is adequate when L is relatively small or f_s/B is relatively high. The third approach, adaptive channel mismatch compensation, can be accurately realized because all channels identically process time-interleaved portions of the same signals.

Selection of $w_0(t)$, which determine many properties of S&I based on the hybrid interpretation and most properties of S&I based on the direct interpretation, can be performed in the frequency or time domain. Frequency-domain selection is well formalized for choosing $w_0(t)$ with the best filtering properties, but these $w_0(t)$ do not allow simple realization of S&I. Time-domain selection simplifies choosing $w_0(t)$ that allow simple realization of S&I, but their filtering properties must be examined afterwards for each $w_0(t)$. The heuristic procedure described in Section 6.6.1 resolves the problem.

Realization of S&I circuits based on the direct interpretation can significantly differ from their conceptual structures. These circuits can be implemented as specific types of mixed-signal correlators or matched filters. The latter implementation is demonstrated for B-spline-based $w_0(t)$.

The dynamic range increase is the most important advantage provided by the hybrid and direct interpretations. IS rejection in the frequency and spatial domains requires especially high dynamic range. For frequency-domain IS rejection, the need for high dynamic range is explained in Section 4.3. For spatial-domain rejection, it is demonstrated in Sections 6.7.2 and 6.7.3 using examples of two unconventional IS rejection techniques.

References

[1] Hnatec, E. R., *A User's Handbook of D/A and A/D Converters*, New York: John Wiley & Sons, 1976.

[2] Dooley, D. J., *Data Conversion Integrated Circuits*, New York: IEEE Press, 1980.

[3] Sheingold, D. H. (ed.), *Analog-Digital Conversion Handbook*, 2nd ed., Englewood, NJ: Prentice Hall, 1986.

[4] Razavi, B., *Principles of Data Conversion System Design*, New York: Wiley-IEEE Press, 1995.

[5] Van de Plassche, R., *CMOS Integrated Analog-to-Digital and Digital-to-Analog Converters*, 2nd ed., Norwell, MA: Kluwer Academic Publishers, 2003.

[6] Kester, W. (ed.), *The Data Conversion Handbook*, Norwood, MA: Analog Devices and Newnes, 2005.

[7] Zumbahlen, H. (ed.), *Linear Circuit Design Handbook*, Boston, MA: Elsevier-Newnes, 2008.

[8] Baker, R. J., *CMOS: Mixed-Signal Circuit Design*, 2nd ed., New York: John Wiley & Sons, 2008.

[9] Cao, Z., and S. Yan, *Low-Power High-Speed ADCs for Nanometer CMOS Integration*, New York: Springer, 2008.

[10] Ahmed, I., *Pipelined ADC Design and Enhancement Techniques*, New York: Springer, 2010.

[11] Zjajo, A., and J. de Gyvez, *Low-Power High-Resolution Analog to Digital Converters*, New York: Springer, 2011.

[12] Ali, A., *High Speed Data Converters*, London, U.K.: IET, 2016.

[13] Pelgrom, M, *Analog-to-Digital Conversion*, New York: Springer, 2017.

[14] Poberezhskiy, Y. S., "Gating Time for Analog-Digital Conversion in Digital Reception Circuits," *Telecommunications and Radio Engineering*, Vol. 37/38, No. 10, 1983, pp. 52–54.

[15] Poberezhskiy, Y. S., "Digital Radio Receivers and the Problem of Analog-to-Digital Conversion of Narrow-Band Signals," *Telecommunications and Radio Engineering*, Vol. 38/39, No. 4, 1984, pp. 109–116.

[16] Poberezhskiy, Y. S., M. V. Zarubinskiy, and B. D. Zhenatov, "Large Dynamic Range Integrating Sampling and Storage Device," *Telecommun. and Radio Engineering*, Vol. 41/42, No. 4, 1987, pp. 63–66.

[17] Poberezhskiy, Y. S., et al., "Design of Multichannel Sampler-Quantizers for Digital Radio Receivers," *Telecommun. and Radio Engineering*, Vol. 46, No. 9, 1991, pp. 133–136.

[18] Poberezhskiy, Y. S., et al., "Experimental Investigation of Integrating Sampling and Storage Devices for Digital Radio Receivers," *Telecommun. and Radio Engineering*, Vol. 49, No. 5, 1995, pp. 112–116.

[19] Poberezhskiy, Y. S., *Digital Radio Receivers* (in Russian), Moscow, Russia: Radio & Communications, 1987.

[20] Poberezhskiy, Y. S., and M. V. Zarubinskiy, "Sample-and-Hold Devices Employing Weighted Integration in Digital Receivers," *Telecommun. and Radio Engineering*, Vol. 44, No. 8, 1989, pp. 75–79.

[21] Poberezhskiy, Y. S., and G. Y. Poberezhskiy, "Optimizing the Three-Level Weighting Function in Integrating Sample-and-Hold Amplifiers for Digital Radio Receivers," *Radio and Commun. Technol.*, Vol. 2, No. 3, 1997, pp. 56–59.

[22] Poberezhskiy, Y. S., and G. Y. Poberezhskiy, "Sampling with Weighted Integration for Digital Receivers," *Dig. IEEE MTT-S Symp. Technol. Wireless Appl.*, Vancouver, Canada, February 21–24, 1999, pp. 163–168.

[23] Poberezhskiy, Y. S., and G. Y. Poberezhskiy, "Sampling Technique Allowing Exclusion of Antialiasing Filter," *Electronics Lett.*, Vol. 36, No. 4, 2000, pp. 297–298.

[24] Poberezhskiy, Y. S., and G. Y. Poberezhskiy, "Sample-and-Hold Amplifiers Performing Internal Antialiasing Filtering and Their Applications in Digital Receivers," *Proc. IEEE ISCAS*, Geneva, Switzerland, May 28–31, 2000, pp. 439–442.

[25] Poberezhskiy, Y. S., and G. Y. Poberezhskiy, "Signal Reconstruction Technique Allowing Exclusion of Antialiasing Filter," *Electronics Lett.*, Vol. 37, No. 3, 2001, pp. 199–200.

[26] Poberezhskiy, Y. S., and G. Y. Poberezhskiy, "Sampling Algorithm Simplifying VLSI Implementation of Digital Radio Receivers," *IEEE Signal Process. Lett.*, Vol. 8, No. 3, 2001, pp. 90–92.

[27] Poberezhskiy, Y. S., and G. Y. Poberezhskiy, "Sampling and Signal Reconstruction Structures Performing Internal Antialiasing Filtering and Their Influence on the Design of Digital Receivers and Transmitters," *IEEE Trans. Circuits Syst. I*, Vol. 51, No. 1, 2004, pp. 118–129.

[28] Poberezhskiy, Y. S., and G. Y. Poberezhskiy, "Implementation of Novel Sampling and Reconstruction Circuits in Digital Radios," *Proc. IEEE ISCAS*, Vol. IV, Vancouver, Canada, May 23–26, 2004, pp. 201–204.

[29] Poberezhskiy, Y. S., and G. Y. Poberezhskiy, "Flexible Analog Front-Ends of Reconfigurable Radios Based on Sampling and Reconstruction with Internal Filtering," *EURASIP J. Wireless Commun. Netw.*, No. 3, 2005, pp. 364–381.

[30] Poberezhskiy, Y. S., and G. Y. Poberezhskiy, "Signal Reconstruction in Digital Transmitter Drives," *Proc. IEEE Aerosp. Conf.*, Big Sky, MT, March 1–8, 2008, pp. 1–19.

[31] Poberezhskiy, Y. S., and G.Y. Poberezhskiy, "Some Aspects of the Design of Software Defined Receivers Based on Sampling with Internal Filtering," *Proc. IEEE Aerosp. Conf.*, Big Sky, MT, March 7–14, 2009, pp. 1–20.

[32] Poberezhskiy, Y. S., and G. Y. Poberezhskiy, "Impact of the Sampling Theorem Interpretations on Digitization and Reconstruction in SDRs and CRs," *Proc. IEEE Aerosp. Conf.*, Big Sky, MT, March 1–8, 2014, pp. 1–20.

[33] Poberezhskiy, Y. S., and G. Y. Poberezhskiy, "Influence of Constructive Sampling Theory on the Front Ends and Back Ends of SDRs and CRs," *Proc. IEEE COMCAS*, Tel Aviv, Israel, November 2–4, 2015, pp. 1–5.

[34] Jamin, O., *Broadband Direct RF Digitization Receivers*, New York: Springer, 2014.

[35] Poberezhskiy, G. Y., and W. C. Lindsey, "Channel Mismatch Compensation in Multichannel Sampling Circuits with Weighted Integration," *Proc. IEEE Aerosp. Conf.*, Big Sky, MT, March 7–14, 2009, pp. 1–15.

[36] El-Chammas, M., and B. Murmann, *Background Calibration of Time-Interleaved Data Converters*, New York: Springer, 2012.

[37] Selesnick, I. W., M. Lang, and C. S. Burrus, "Constrained Least Square Design of FIR Filters Without Specified Transition Bands," *IEEE Trans. Signal Process.*, Vol. 44, No. 8, 1996, pp. 1879–1892.

[38] Unser, M., "Sampling—50 Years After Shannon," *Proc. IEEE*, Vol. 88, No. 4, 2000, pp. 569–587.

[39] Poberezhskiy, G. Y., and W. C. Lindsey, "Weight Functions Based on B-Splines in Sampling Circuits with Internal Filtering," *Proc. IEEE Aerosp. Conf.*, Big Sky, MT, March 5–12, 2011, pp. 1–12.

[40] Yuan, J., "A Charge Sampling Mixer with Embedded Filter Function for Wireless Applications," *Proc. Int. Conf. Microw. Millimeter Wave Technol.*, Beijing, China, September 14–16, 2000, pp. 315–318.

[41] Karvonen, S., T. Riley, and J. Kostamovaara, "A Low Noise Quadrature Subsampling Mixer," *Proc. IEEE ISCAS*, Sydney, Australia, May 6–9, 2001, pp. 790–793.

[42] Karvonen, S., T. Riley, and J. Kostamovaara, "Charge Sampling Mixer with ΔΣ Quantized Impulse Response," *Proc. IEEE ISCAS*, Vol. 1, Phoenix-Scottsdale, AZ, May 26–29, 2002, pp. 129–132.

[43] Lindfors, S., A. Pärssinen, and K. Halonen, "A 3-V 230-MHz CMOS Decimation Subsampler," *IEEE Trans. Circuits Syst. II*, Vol. 50, No. 3, 2003, pp. 105–117.

[44] Xu, G., and J. Yuan, "Charge Sampling Analogue FIR Filter," *Electronics Letters*, Vol. 39, No. 3, 2003, pp. 261–262.

[45] Muhammad, K., and R. B. Staszewski, "Direct RF Sampling Mixer with Recursive Filtering in Charge Domain," *Proc. IEEE ISCAS*, Vol. 1, Dallas, TX, May 23–26, 2004, pp. 577–580.

[46] Xu, G., and J. Yuan, "Accurate Sample-and-Hold Circuit Model," *Electronics Lett.*, Vol. 41, No. 9, 2005, pp. 520–521.

[47] Muhammad, K., et al., "A Discrete-Time Quad-Band GSM/GPRS Receiver in a 90-nm Digital CMOS Process," *Proc. IEEE Custom Integr. Circuits Conf.*, San Jose, CA, September 18–21, 2005, pp. 809–812.

[48] Xu, G., and J. Yuan, "Performance Analysis of General Charge Sampling." *IEEE Trans. Circuits Syst. II*, Vol. 52, No. 2, 2005, pp. 107–111.

[49] Cenkeramaddi, L. R., and T. Ytterdal, "Jitter Analysis of General Charge Sampling Amplifiers," *Proc. IEEE ISCAS*, Kos, Greece, May 21–24, 2006, pp. 5267–5270.

[50] Mirzaei, A., et al., "Software-Defined Radio Receiver: Dream to Reality," *IEEE Commun. Mag.*, Vol. 44, No. 8, pp. 111–118.

[51] Bagheri, R., et al., "An 800-MHz-6-GHz Software-Defined Wireless Receiver in 90-nm CMOS," *IEEE J. Solid-State Circuits*, Vol. 41, No. 12, 2006, pp. 2860–2876.

[52] Cenkeramaddi, L. R., and T. Ytterdal, "Analysis and Design of a 1V Charge Sampling Readout Amplifier in 90-nm CMOS for Medical Imaging," *Proc. IEEE Int. Symp. VLSI Design, Autom. Test*, Hsinchu, Taiwan, April 25–27, 2007, pp. 1–4.

[53] Abidi, A., "The Path to the Software-Defined Radio Receiver," *IEEE J. Solid-State Circuits*, Vol. 42, No. 5, 2007, pp. 954–966.

[54] Mirzaei, A., et al., "Analysis of First-Order Anti-Aliasing Integration Sampler," *IEEE Trans. Circuits Syst. I*, Vol. 55, No. 10, 2008, pp. 2994–3005.

[55] Mirzaei, A., et al., "A Second-Order Antialiasing Prefilter for a Software-Defined Radio Receiver," *IEEE Trans. Circuits Syst. I*, Vol. 56, No. 7, 2009, pp. 1513–1524.

[56] Tohidian, M., I. Madadi, and R. B. Staszewski, "Analysis and Design of a High-Order Discrete-Time Passive IIR Low-Pass Filter," *IEEE J. Solid-State Circuits*, Vol. 49, No. 11, 2014, pp. 2575–2587.

[57] Bazrafshan, A., M. Taherzadeh-Sani, and F. Nabki, "A 0.8-4-GHz Software-Defined Radio Receiver with Improved Harmonic Rejection Through Non-Overlapped Clocking," *IEEE Trans. Circuits Syst. I*, 2018, Vol. 65, No. 10, pp. 3186–3195.

[58] Poberezhskiy, Y. S., and G. Y. Poberezhskiy, "On Adaptive Robustness Approach to Anti-Jam Signal Processing," *Proc. IEEE Aerosp. Conf.*, Big Sky, MT, March 2–9, 2013, pp. 1–20.

[59] Poberezhskiy, Y. S., and G. Y. Poberezhskiy, "Suppression of Multiple Jammers with Significantly Different Power Levels," *Proc. IEEE Aerosp. Conf.*, Big Sky, MT, March 3–10, 2012, pp. 1–12.

[60] Poberezhskiy, Y. S., and G. Y. Poberezhskiy, "Spatial Nulling and Beamforming in Presence of Very Strong Jammers," *Proc. IEEE Aerosp. Conf.*, Big Sky, MT, March 5–12, 2016, pp. 1–20.

[61] Poberezhskiy, Y. S., and G. Y. Poberezhskiy, "Efficient Utilization of Virtual Antenna Motion," *Proc. IEEE Aerosp. Conf.*, Big Sky, MT, March 5–12, 2011, pp. 1–17.

CHAPTER 7
Improving Resolution of Quantization

7.1 Overview

As mentioned in the previous chapters, an increase in the dynamic range and analog bandwidth of A/Ds, provided by sampling based on the sampling theorem's hybrid and direct interpretations, is beneficial even for a given resolution of quantizers, especially for bandpass signals with high f_0. It also stimulates further improvement of the quantizers' resolution, additionally increasing the dynamic range of digitization. In contrast with S&I, where the implementation of new concepts was very slow during the last three decades, the speed, accuracy, sensitivity, and resolution of quantizers significantly increased and their power consumption decreased throughout that time not only because of the IC technology development but also due to the introduction and realization of new ideas. This chapter shows that, despite the outstanding progress, new quantization concepts still can be suggested.

The highest resolution and speed of quantization are usually required at the inputs of Rxs. They also can be needed at the inputs of Txs, for instance, to quantize fast-changing multipixel images intended for transmission. Section 7.2 shows that these two cases usually require different approaches. The availability of many excellent publications on various quantization methods (see, for instance, [1–17]) allows reducing the section material to a concise analysis of the most effective techniques currently used in digital radios.

Section 7.3 demonstrates the possibility to increase the sensitivity and resolution of quantization based on joint mixed-signal processing of several samples.

Section 7.4 presents an image quantization technique that combines quantization with source encoding, effectively utilizing not only statistical dependences among pixels within and between images but also the sparsity of discontinuities in most images.

7.2 Conventional Quantization

7.2.1 Quantization of Rx Input Signals

PCM quantizers performing uniform quantization (constant sampling rate f_s, quantization step Δ, and number of bits N_b) are most widely used for digitization of Rx input signals. Flash quantizers, which are fastest, require $2^{N_b} - 1$ comparators to achieve N_b-bit resolution. Thus, their complexity and power consumption grow exponentially as N_b increases. To avoid this growth, most modern quantizers are designed as composite structures that contain one or more internal quantizers, each with a relatively small number of bits n_b. Besides the quantization performed by the internal quantizers, composite quantizers carry out many other mixed-signal operations. Exploiting specific features of these operations and/or statistical properties of quantized signals enables improving composite quantizers' performance without increasing the burden imposed on their internal quantizers.

A huge variety of Rxs' input signals and their poorly known and unstable statistics usually prevent effective utilization of the quantized signals' properties for improving the performance of composite quantizers at the inputs of Rxs. Simultaneously, nothing prevents exploiting the features of mixed-signal operations in these quantizers, and this approach is discussed below. Four versions of this approach are probably most noticeable. The first of them, which can be called the generalized successive-approximation technique, is concisely described in Section 7.2.1.1. The second one, based on oversampling, is outlined in Section 7.2.1.2. The third version, the use of time-interleaved quantizers, is most effective with samplers based on the sampling theorem's hybrid or direct interpretation, as shown in Chapter 6. In that case, it allows not only increasing f_s of composite quantizers without changing their internal quantizers but also widening the analog bandwidths and improving the dynamic range and flexibility of digitization. The fourth version, based on joint processing of several signal samples [18, 19], is examined in Section 7.3.

Two figures of merit,

$$F_1 = 2^{N_{b.e}} f_s \text{ and } F_2 = \frac{2^{N_{b.e}} f_s}{P_c} \quad (7.1)$$

where $N_{b.e}$ is the ENOB and P_c is power consumption, are insufficient for characterizing performance of A/Ds because they do not reflect their analog bandwidths. However, since these bandwidths are determined by the A/Ds' sampling circuits, F_1 and F_2 adequately characterize quantization in A/Ds. When P_c is not of significant concern, F_1 can be used. When P_c is limited, F_2 is more adequate. Taking into account (7.1), it is implied below that any improvement of quantizers' sensitivity and resolution is achieved, in most cases, without reducing f_s and increasing P_c compared to alternative techniques.

7.2.1.1 Generalized Successive-Approximation Technique

Generalized successive-approximation technique decomposes quantization procedures into several steps for high-resolution digital representation of signals using quantizers with a relatively small number of bits. A conventional successive-approximation quantizer uses an internal 1-bit quantizer (comparator) whose digital outputs are stored in a successive-approximation register (SAR) over several cycles. At each cycle, the SAR output is converted to the analog domain by an internal D/A and compared to the sample analog value. The quantization is completed when the D/A output becomes equal to the sample value with required accuracy.

This technique can be generalized to include composite quantizers that contain multibit internal quantizers and structures where processing at each step is performed by a separate stage. Thus, the generalized technique can be realized using

7.2 Conventional Quantization

multi-iteration or multistage structures. Multi-iteration realization of this technique is illustrated by the simplified block diagram in Figure 7.1. Here, an analog signal $u(t)$ is sampled, and samples are quantized by a flash n_b-bit quantizer. Prior to entering this quantizer, each sample passes through a subtracting circuit and a scaling BA with controlled gain. During the first cycle (coarse conversion), nothing is subtracted from the sample, and the BA gain $g = 1$. The conversion result is saved in the output register of the correction and control logic as MSBs of the digital word corresponding to the sample. These bits are converted to the analog domain by an accurate D/A to be subtracted from the same sample at the second cycle.

The difference between the sample and the coarse conversion result is amplified by the BA with $g = 2^{n_b}$ at that cycle. The amplified difference is converted to the digital domain, and the result is saved in the output register as the next n_b bits. Every subsequent cycle, g is increased by factor 2^{n_b}. This quantizer would ideally provide the total resolution of $N_b = mn_b$ bits after m cycles. In practice, however, certain redundancy should be introduced for error correction. Therefore, g should be increased by a factor less than 2^{n_b} at each cycle, and the total resolution $N_b < mn_b$ bits. In this and other multi-iteration converters, N_b is increased at the expense of reducing f_s. Multistage converters enable much higher f_s for a given N_b.

While the connection of conventional successive-approximation quantizers to multi-iteration quantizers is obvious, their connection to multistage quantizers that also successively approximate $u(nT_s)$ is more obscure and usually not emphasized. Therefore, multistage realizations of generalized successive-approximation technique are known under different names. For instance, subranging quantizers are embodiments of this technique.

Pipelined versions of subranging quantizers are most important for digital radios because they provide high f_s and N_b due to simultaneous processing of several samples. In these versions (see Figure 7.2) [2, 5–10, 14–16], m consecutive stages perform quantization. Each of the first $m - 1$ stages contains a THA, a flash internal quantizer, a D/A, and a scaling BA. The quantizer and D/A of any stage have the same resolution, but the D/A accuracy corresponds to the composite quantizer's final resolution. The first-stage digital output is sent simultaneously to the correction and control logic to form the MSBs of the quantizer's output word and to its D/A, whose output is subtracted from the analog input sample.

The difference is amplified by the BA and fed into the second stage to be processed the same way as the input sample in the first stage. This procedure is repeated throughout the stages to achieve the desired resolution. Since the bits of each sample are generated by different stages at different times, the correction and control logic

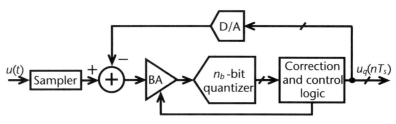

Figure 7.1 Multi-iteration composite quantizer.

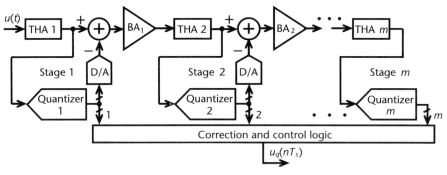

Figure 7.2 Pipelined composite quantizer.

aligns them in time, in addition to performing corrections and quantizer calibration. Each stage starts processing the next sample after it completes processing the previous one. Different stages of pipelined quantizers can have different resolutions, and these resolutions are slightly redundant to accommodate the D/As' offset and gain errors. The latency produced by pipelined processing is acceptable in most applications.

Note that subranging quantizers are not the only multistage quantizers using generalized successive-approximation technique, and not only subranging quantizers can be pipelined.

7.2.1.2 Oversampling Techniques

Sigma-delta quantizers [2–7, 12–17] are the most important type of quantizers based on oversampling. Their origin can be traced to the 1940s and 1950s when delta modulation and differential PCM (DPCM) were invented to increase the throughput of communications by transmitting the changes between consecutive samples rather than the samples themselves. The idea of oversampling and noise shaping for increasing resolution was introduced by C. Cutler in 1954 and improved by other researchers within the scope of direct transmission of oversampled digitized signals. In 1969, D. Goodman suggested sigma-delta modulation for universal A/Ds by adding digital filtering with downsampling. Sigma-delta quantizers are fully compatible with the IC technology and allow low-cost complementary metal-oxide semiconductor (CMOS) implementation. Initially, they were used mostly for high-resolution baseband applications. The invention of bandpass sigma-delta quantizers in 1988 made their implementation in digital radios more attractive.

Before discussing sigma-delta quantization, note that oversampling improves the resolution even for PCM quantizers. Indeed, when N_b in such a quantizer is sufficiently large and the input samples' rms values $\sigma_u \gg \Delta$, the quantization errors $\varepsilon(nT_s)$ are uniformly distributed within Δ and uncorrelated even if the corresponding samples $u(nT_s)$ are correlated. In this case, the sequence of $\varepsilon(nT_s)$ can be considered a realization of stationary discrete-time quantization noise $E(nT_s)$ with mean and rms values, respectively,

$$m_\varepsilon = 0 \text{ and } \sigma_\varepsilon = \left(\frac{1}{12}\right)^{0.5} \Delta \approx 0.2887\Delta \tag{7.2}$$

7.2 Conventional Quantization

if the quantizer's output data are rounded. The PSD of this noise is

$$N_q(f) \approx \frac{\Delta^2}{6f_s} \tag{7.3}$$

and, therefore, its power within the signal bandwidth B is

$$P_q \approx \Delta^2 \frac{B}{6f_s} \tag{7.4}$$

If the maximum and minimum acceptable values of $u(nT_s)$ are, respectively, $+U_m$ and $-U_m$,

$$\Delta = \frac{2U_m}{2^{N_b} - 1} \approx \frac{2U_m}{2^{N_b}} \tag{7.5}$$

Equation (7.5) allows rewriting (7.2)–(7.4) as follows:

$$m_\varepsilon = 0 \text{ and } \sigma_\varepsilon = \frac{U_m}{3^{0.5} \cdot 2^{N_b}} \tag{7.6}$$

$$N_q(f) \approx \frac{2U_m^2}{3 \cdot 2^{2N_b} f_s} \tag{7.7}$$

$$P_q \approx \left(\frac{2B}{f_s}\right) \frac{U_m^2}{3 \cdot 2^{2N_b}} \tag{7.8}$$

From (7.8), the ratio R of the rms value of a sinewave with amplitude U_m to $P_q^{0.5}$ is

$$R = \left(\frac{0.5 U_m^2}{P_q}\right)^{0.5} = 1.5^{0.5} \cdot 2^{N_b} \cdot \left(\frac{f_s}{2B}\right)^{0.5} \tag{7.9}$$

and

$$R_{dB} = 1.76 + 6.02 N_b + 10 \log_{10}\left(\frac{f_s}{2B}\right) \tag{7.10}$$

As follows from (7.9) and (7.10), an increase of f_s for a given B improves the PCM quantizer sensitivity and resolution due to P_q reduction, but this oversampling is inefficient because it requires quadrupling f_s for doubling R (i.e., for a 1-bit increase of N_b). Sigma-delta quantizers use oversampling more efficiently.

A block diagram of a first-order sigma-delta quantizer is shown in Figure 7.3. Here, the D/A output signal is subtracted from the input analog signal $u(t)$, and the difference is integrated. The integrator output is quantized by a low-resolution internal quantizer (1-bit quantizers were initially used). The digital words (each containing one or a few bits) from this quantizer output enter the D/A and the digital filter-decimator. As mentioned above, the D/A output signal is subtracted from $u(t)$, whereas the digital filter-decimator processes the internal quantizer output words, increasing their resolution and reducing the sampling rate from the initial $f_{s1} = 1/T_{s1}$ to $f_{s2} = 1/T_{s2}$. The sigma-delta quantizer's feedback loop forces the loop input and output signals to be almost equal within the integrator bandwidth, pushing the quantization noise out of the band. This quantization noise shaping makes the exchange of f_s for N_b in sigma-delta quantizers much more efficient than in PCM quantizers.

It has been proven that every doubling of $f_s/(2B)$ in a first-order sigma-delta quantizer improves R_{dB} by approximately 9 dB (i.e., increases N_b by 1.5 bits). Sigma-delta quantizers of higher orders exchange f_s for N_b even more efficiently due to better shaping of quantization noise. In an ideally realized Lth-order sigma-delta quantizer, every doubling of $f_s/(2B)$ improves R_{dB} by approximately $(6L + 3)$ dB, and, consequently, increases N_b by $(L + 0.5)$ bits. However, nonideal realization lowers the improvement in high-order sigma-delta quantizers. Employment of multibit internal quantizers and development of multistage sigma-delta quantizers further improve sensitivity and resolution of quantization for a given ratio $f_s/(2B)$.

Enhanced efficiency of exchanging f_s for N_b not only increases the sensitivity and resolution of sigma-delta quantizers but also allows their use for signals with wider B. This motivated the development of their parallel structures, such as multiband, time-interleaved, and Hadamard-modulation-based structures. Sigma-delta quantizers are effective in superconductor Rxs where superconductivity enables high f_s by N_b products. Implementation of sigma-delta quantizers in Rxs has its specifics. For instance, low requirements for antialiasing filtering are usually considered their advantage. This advantage cannot be used to its full extent in overcrowded frequency bands because widening the Rx bandwidth in such a band requires increasing the Rx dynamic range (see Section 4.3.3).

7.2.2 Quantization of Tx Input Signals

As mentioned at the beginning of this section, poorly known and unstable statistical properties of Rxs' input signals usually prevent utilization of these properties for improving and/or simplifying the quantization. In contrast with Rxs, statistical properties of Txs' input signals are usually well known and, therefore, are often used for improving and/or simplifying their quantization by combining it with the source

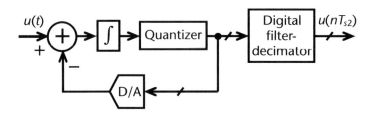

Figure 7.3 First-order sigma-delta quantizer.

encoding. Although digital source encoding is more accurate and effective than mixed-signal encoding, it cannot improve and/or simplify the signals' quantization.

Thus, the main motivation for performing at least part of the Tx analog signals' source encoding in the mixed-signal domain is the possibility of improving and/or simplifying their digitization. Two techniques, namely nonuniform quantization and various forms of predictive quantization, are most widely used for reducing the required N_b of quantizers without worsening the digitization quality when statistical properties of signals are known.

7.2.2.1 Nonuniform Quantization

The key idea of nonuniform quantization is to make quantization steps dependent on signal levels (small for low levels and large for high levels), providing almost the same signal-to-quantization-noise ratio for weak and strong signals. This approach leads to an almost logarithmic quantization scale. Logarithmic quantization also makes the distribution of code words corresponding to different signal levels closer to uniform, reducing the signal redundancy. This quantization is effective when interference and noise are relatively low within the digitized signal bandwidth. Otherwise, it can reduce the signal-to-noise-and-interference ratio.

Logarithmic quantization reduces N_b by up to 1.5 times for telephone signals. The compressed signals are represented by a smaller number of bits throughout their transmission and reception. After the reception, they are expanded in Rxs for their correct perception by end users. Such processing, called companding, is also used for transmitting images and some other types of signals. Modern quantization and source coding techniques enable more sophisticated and effective companding.

7.2.2.2 Predictive Quantization

When samples of Txs' input signals are dependent, predictive quantization reduces the number of bits needed for their representation and transmission. Delta modulation and DPCM are the simplest and earliest versions of predictive quantization. In DPCM, the next sample's predicted value is equal to the current one. Thus, only the differences between neighboring samples must be quantized. When the samples are highly correlated, N_b can be reduced because the differences are much smaller than the samples' values. In the initial versions of delta modulation invented in the 1940s (see Section 7.2.1.2), only the signs of the differences between neighboring samples were taken into account. Thus, 1 bit of information was transmitted by each output sample of a delta quantizer. Later, more complex delta-modulation systems became very similar to DPCM ones.

Linear or nonlinear prediction can be used for quantization. Knowledge of signal spectral density or correlation function is sufficient for linear prediction. Nonlinear prediction requires more detailed information about quantized signals. For many reasons, linear prediction is used more often than nonlinear one. The quality of prediction is determined by the types and parameters of the dependences among samples, dimensionality of predicted signals, optimality of employed prediction algorithms, and by the number and positions of samples used for prediction.

Predictive quantization can be effectively combined with other techniques, as shown in Section 7.4.

7.2.2.3 Dithering

Dithering is an additional technique commonly used for quantizing (and processing) Txs' input signals (as well as signals in other applications). It means applying random or pseudorandom signals together with desired signals to the quantizer input to improve quantization quality by reducing the impact of small nonlinearities. Dithering can be subtractive or nonsubtractive. In digital Rxs with high-resolution quantization of input signals, the presence of noise usually makes it unnecessary. However, it can be useful in Rxs with low-resolution quantization.

7.3 Joint Quantization of Samples

7.3.1 Principles of Joint Quantization

Properly used methods of multiple-symbol demodulation in communications and joint processing of signals in radar improve signal reception quality in these applications. In quantization, joint processing of samples is also used. In sigma-delta quantizers, for instance, it (combined with oversampling) effectively reshapes the quantization noise spectrum, improving sensitivity and resolution. A different approach to this improvement is used in the joint-processing quantizers (JPQs) described below [18, 19], which can be considered a special case of vector quantizers. The approach does not utilize any dependences among the samples; rather, it exploits the fact that their sums can be measured with higher relative accuracy than the separate samples.

The employment of Walsh spectral coefficients as such sums allows distributing the achieved resolution increase among the M samples involved in the joint processing and simplifies the JPQs' realization. This realization is further simplified by the use of only the samples' LSBs for the joint processing. Although the dependences among the samples are not utilized, and the samples should not necessarily be successive, the use of successive samples is convenient. The sets of M successive samples involved in the processing can be formed using a hopping or a sliding window, as illustrated in Figure 7.4 where dots correspond to samples following at the rate $f_s = 1/T_s$.

When a hopping window is used (see Figure 7.4(a)), M previously quantized, preprocessed, and stored successive samples undergo joint processing during M consecutive intervals T_s. Simultaneously, M new successive samples are quantized, preprocessed, and stored one by one. When the first M samples with improved resolution leave the window, the next set of quantized, preprocessed, and stored successive samples replaces them. The samples of that set are jointly processed during the next M intervals T_s, and so on.

When a sliding window is used (see Figure 7.4(b)), a new sample enters the window, and the oldest sample with increased resolution leaves it every T_s. Prior to entering the window, each sample is quantized, preprocessed, and stored individually. In the window, it is processed jointly with other samples throughout M

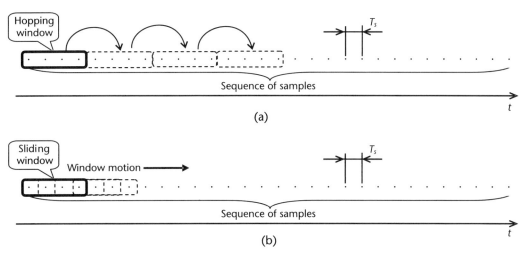

Figure 7.4 Motion of a window with length $M = 4$: (a) hopping window and (b) sliding window.

consecutive intervals T_s. The joint processing is the same for all the intervals, but the samples within the window differ for each sampling interval.

7.3.1.1 Hopping-Window JPQs

The block diagram of a hopping-window JPQ is shown in Figure 7.5. During the first M sampling intervals, each sample u_i from the sampler output is stored in analog memory AM_0 and quantized by main quantizer MQr with N_{b1}-bit resolution. The quantized value u_{qi} of u_i is stored in the DSP memory. It is clear that

$$u_i = u_{qi} + \varepsilon_{1i} \tag{7.11}$$

where ε_{1i} is the MQr quantization error. The sequence of ε_{1i} can be considered a realization of stationary discrete-time quantization noise E_1. Since N_{b1} is large enough, samples ε_{1i} of E_1 are uniformly distributed within the MQr quantization step Δ_1 and uncorrelated even if the corresponding u_i are correlated. For quantization with rounding, the mean and rms values of all ε_{1i} are, respectively,

$$m_{\varepsilon 1} = 0 \quad \text{and} \quad \sigma_{\varepsilon 1} = \left(\frac{1}{12}\right)^{0.5} \Delta_1 \approx 0.2887\Delta_1 \tag{7.12}$$

To simplify the explanation of the JPQ principles, the delays in its blocks (reflected in Figure 7.5) are neglected in description below. The analog value x_i corresponding to the LSB x_{qi} of $u_{qi} = u_{qsi} + x_{qi}$ is generated by reconstructing the analog value u_{si}, corresponding to u_{qsi} represented by $(N_{b1} - 1)$ MSBs of u_{qi} and zero-valued LSB, and subtracting u_{si} from u_i saved in AM_0. If the reconstructing D/A is accurate,

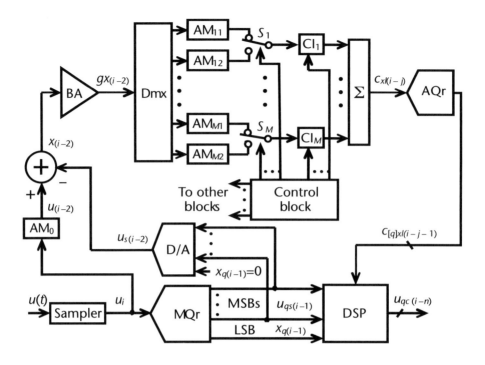

Figure 7.5 Hopping-window JPQ.

$$x_i = u_i - u_{si} = x_{qi} + \varepsilon_{1i} \tag{7.13}$$

as follows from (7.11). In principle, x_i can correspond to $n_0 \neq 1$ LSBs. After passing through the scaling BA with gain g, gx_i is sent to an appropriate analog memory cell AM_{m1} where $m = 1, ..., M$. Each AM_{m1} belongs to the first group of analog memory cells intended for storing all odd sets of M successive gx_i (each gx_i in a separate AM_{m1}). Once the first M gx_i are stored, their MT_s-long joint processing starts. During that time, M new successive samples entering the JPQ are quantized, preprocessed, and stored similarly to the previous M samples. Their gx_i are stored in cells AM_{m2} intended for all even sets of M successive gx_i. Analog memory cells AM_0, AM_{m1}, and AM_{m2} can be, for instance, SHAs.

When the joint processing of the first set of M successive samples is completed, their digital values $u_{qc(i-n)}$ (where n reflects the delay in the JPQ) with improved resolution leave the DSP. Then all AM_{m1} are cleared, and the joint processing of the second set of M samples starts. During that processing, next M samples are quantized, preprocessed, and stored the same way as the first set. The joint processing of M samples within a window is the same for both groups $\{AM_{m1}\}$ and $\{AM_{m2}\}$ and comprises the following six steps.

1. The Walsh spectrum of a current set $\{gx_i\}$ is determined in the analog domain. Switches S_m where $m = 1, ..., M$ connect the appropriate group of

memory cells ($\{AM_{m1}\}$ or $\{AM_{m2}\}$) to the controlled inverters CI_m. Expressing $i = (\eta - 1)M + m$ (here, η is the current sample set number and m is the sample number within the set) allows representing the Walsh coefficients as

$$c_{x\eta l} = \sum_{m=1}^{M} h_{lm} \cdot gx_{\eta m} = g \cdot \sum_{m=1}^{M} h_{lm} \cdot x_{\eta m} \quad (7.14)$$

where $l = 1, \ldots, M$ is the Walsh coefficient index and h_{lm} is an element of the Hadamard matrix H_M of order M. Since $h_{lm} = +1$ or -1, the multiplications in (7.14) are reduced to inverting the signs of $gx_{\eta m}$ performed by CI_m.

2. The Walsh spectrum of the corresponding set $\{gx_{q\eta m}\}$ is determined in the DSP:

$$c_{qx\eta l} = g \cdot \sum_{m=1}^{M} h_{lm} \cdot x_{q\eta m} \quad (7.15)$$

3. Coefficients $c_{x\eta l}$ are quantized by auxiliary quantizer AQr with resolution N_{b2} bits and quantization step Δ_2. The quantized coefficients are

$$c_{[q]x\eta l} = c_{x\eta l} + \varepsilon_{2\eta l} \quad (7.16)$$

where $\varepsilon_{2\eta l}$ are the AQr quantization errors. Their sequence can be considered a realization of stationary discrete-time quantization noise E_2. Although $N_{b2} < N_{b1}$, it is sufficient to make samples $\varepsilon_{2\eta l}$ of E_2 uncorrelated and uniformly distributed within Δ_2 with the mean and rms values

$$m_{\varepsilon 2} = 0 \text{ and } \sigma_{\varepsilon 2} = \left(\frac{1}{12}\right)^{0.5} \Delta_2 \approx 0.2887\Delta_2 \quad (7.17)$$

Since 1 bit is required for $h_{lm} = +1$ or -1 and x_i may correspond to n_0 LSBs,

$$N_{b2} = 1 + n_0 + \text{ceil}\left[\log_2\left(\frac{Mg\Delta_1}{\Delta_2}\right)\right] \quad (7.18)$$

Usually, $N_{b2} = 6 \ldots 8$ is optimum. As follows from (7.18), increasing M allows reducing n_0, and when $M > 32$, $n_0 = 0$ can be selected. In that case, $x_{qi} = x_{q\eta m} = 0$, $x_i = x_{\eta m} = \varepsilon_{1i} = \varepsilon_{1\eta m}$, and calculation of the Walsh coefficients in the DSP is unnecessary.

4. In the general case, however, $c_{qx\eta l}$ are subtracted from $c_{[q]x\eta l}$. As follows from (7.13)–(7.15), the obtained system of equations would allow determining $\varepsilon_{1i} = \varepsilon_{1\eta m}$ precisely if $c_{[q]x\eta l}$ were equal to $c_{x\eta l}$. Indeed,

$$c_{x\eta l} - c_{qx\eta l} = g \cdot \sum_{m=1}^{M} h_{lm} \cdot \varepsilon_{1\eta m} \quad (7.19)$$

Since $c_{x\eta l}$ are actually quantized with errors $\varepsilon_{2\eta l}$, this subtraction gives

$$c_{[q]x\eta l} - c_{qx\eta l} = g \cdot \sum_{m=1}^{M} h_{lm} \cdot \varepsilon_{1\eta m} + \varepsilon_{2\eta l} \qquad (7.20)$$

instead of (7.19). Substituting $m_{\varepsilon 2} = 0$ for unknown $\varepsilon_{2\eta l}$ in (7.20) produces the system of M independent linear equations that allows calculating the estimates $\varepsilon_{1\eta me}$ of $\varepsilon_{1\eta m}$:

$$g\mathbf{H}_M \boldsymbol{\varepsilon}_{1\eta e} = \mathbf{c}_\eta \qquad (7.21)$$

where components of $\mathbf{c}_\eta$ are $c_{[q]x\eta l} - c_{qx\eta l}$.

5. The solution of (7.21) is

$$\boldsymbol{\varepsilon}_{1\eta e} = \frac{\mathbf{H}_M^{-1}\mathbf{c}_\eta}{g} = \frac{\mathbf{H}_M^T \mathbf{c}_\eta}{Mg} \qquad (7.22)$$

where $\mathbf{H}_M^{-1}$ and $\mathbf{H}_M^T$ are the inverse and transposed Hadamard matrices respectively.

6. Estimates $\varepsilon_{1\eta me} = \varepsilon_{1ie}$ are added to the corresponding u_{qi} previously stored in the DSP memory:

$$u_{qci} = u_{qi} + \varepsilon_{1ie} \qquad (7.23)$$

Samples u_{qci} have improved resolution and are the final results of the joint quantization. In Figure 7.5, they are shown as $u_{qc(i-n)}$ where n reflects the nT_s-delay in the JPQ. After that, the used $\{AM_{m1}\}$ or $\{AM_{m2}\}$ are cleared, and the corresponding u_{qi} are erased from the DSP memory. By that time, the next M samples u_{qi} are already stored in the DSP memory, and the corresponding M gx_i are stored in $\{AM_{m2}\}$ or $\{AM_{m1}\}$, respectively.

While the MQr accuracy is determined by ε_{1i}, the JPQ accuracy is determined by the errors of ε_{1i} calculation. The latter can be found as follows. The components of $\mathbf{c}_\eta$ are finite sums of $g\varepsilon_{1ie}$. Since all samples ε_{1i} of $\mathbf{E}_1$ are identically distributed and uncorrelated, the variance of each sum is $Mg^2\sigma_{\varepsilon 1e}^2$ where $\sigma_{\varepsilon 1ie}$ is the rms error of determining ε_{1ie} and, consequently, u_{qci}. At the same time, this variance is equal to $\Delta_2^2/12$. Therefore,

$$\sigma_{\varepsilon 1e} = \frac{\Delta_2}{g(12M)^{0.5}} \qquad (7.24)$$

Since $m_{\varepsilon 1e} = 0$ and $m_{\varepsilon 1} = 0$, the increase α in the MQr resolution provided by the joint processing is

$$\alpha = \frac{\sigma_{\varepsilon 1}}{\sigma_{\varepsilon 1e}} = gM^{0.5}\frac{\Delta_1}{\Delta_2} \qquad (7.25)$$

7.3 Joint Quantization of Samples

It was assumed above that only quantization in both MQr and AQr is inaccurate, while all other analog, digital, and mixed-signal operations within the JPQ are precise. Their nonideal realization reduces α (see Section 7.3.2).

7.3.1.2 Specifics of Sliding-Window JPQs

In the JPQ shown in Figure 7.5, the maximum delay $T_{d\max}$ between the instant t_{in} when u_i enters MQr and the instant t_{out} when the corresponding u_{qci} leaves the JPQ is

$$T_{d\max 1} = t_{out} - t_{in} = 2MT_s \qquad (7.26)$$

The desire to reduce $T_{d\max}$ motivates the development of sliding-window JPQs.

The block diagram of a JPQ with a short sliding window (i.e., small M) is shown in Figure 7.6. In this JPQ, like in the one in Figure 7.5, u_i is stored in AM_0 and quantized by MQr with N_{b1}-bit resolution. The analog gx_i are also generated the same way as in the JPQ in Figure 7.5. However, the JPQ in Figure 7.6 has only one group of M cells AM_m, and $n_0 > 1$ because M is small. Also in contrast with the JPQ in Figure 7.5, where the joint processing cycle starts after storing the first M gx_i and has length MT_s, the JPQ in Figure 7.6 begins the joint processing after storing gx_1 and gx_2, and executes the six-step joint processing cycle every T_s. Thus, full-length cycles are executed only for $i \geq M$. For $i > M$, every new gx_i, stored in

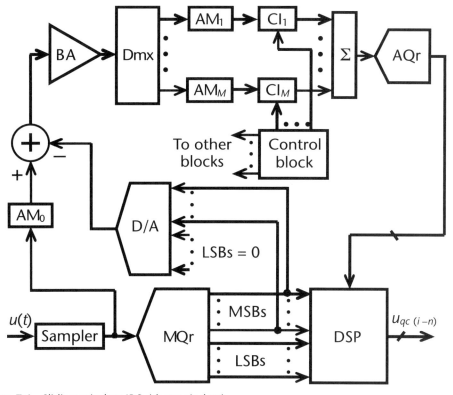

Figure 7.6 Sliding-window JPQ (short window).

AM_m, erases previously stored gx_{i-M}, and the CI_m, AQr, and DSP must operate M times faster than in a hopping-window JPQ with the same M.

Another difference is that ε_{1ie} in the JPQ in Figure 7.6 are initially added to the corresponding values x_{qi} (rather than to the corresponding values u_{qi} as in the JPQ in Figure 7.5), and the corrected values are used for the joint processing at the subsequent $(M - 1)$ intervals T_s. At the last of these intervals, x_{qi} are added to the corresponding values u_{qsi} to obtain u_{qci}. Despite M corrections of x_{qi} in the JPQ in Figure 7.6, both JPQs provide the same MQr resolution increase α, reflected by (7.25). Zero-padding is needed at the beginning of sliding-window joint processing, to compensate an insufficient number of samples within the window.

In the JPQ in Figure 7.6, $T_{d\max}$ is

$$T_{d\max 2} = t_{out} - t_{in} = (M + 1)T_s \qquad (7.27)$$

Thus, the JPQ in Figure 7.6 has approximately two times shorter $T_{d\max}$ and two times smaller number of the analog memory cells AM_m compared to the JPQ in Figure 7.5. These advantages have been obtained at the cost of the M-fold speed increase of CI_m, AQr, and DSP. When the window is short, such an increase is tolerable because $N_{b2} < N_{b1}$. However, a long window makes AQr realization difficult or impossible.

The problem can be solved by employing several AQrs, as in the JPQ in Figure 7.7 where K groups of CI_m and K AQrs are used for the joint processing of M gx_i stored in AM_m. Each group of M CI_m with a separate analog adder calculates M/K Walsh coefficients that are quantized by a separate AQr. In this JPQ, the required speed of the CI_m, analog adders, and AQrs is K times slower than in the JPQ in Figure 7.6 for the same M, but their number is K times larger. The DSP speed is not reduced compared to that in the JPQ in Figure 7.6, but this is not an issue because only LSBs of u_{qi} are processed in the DSP, and n_0 can be equal to one or even zero.

7.3.2 Design Considerations

Although the JPQs in Figures 7.5 through 7.7 have different delays and complexities of realization, they provide the same resolution increase α reflected by (7.25) if the finite quantization steps of the MQr and AQrs are the only sources of the errors within the JPQs. The increase in resolution expressed in the number of extra bits ΔN is

$$\Delta N = \log_2 \alpha = 0.5 \log_2 M + \log_2 \left(\frac{g\Delta_1}{\Delta_2} \right) \qquad (7.28)$$

As follows from (7.28), a 4-sample-long window ($M = 4$) increases the JPQ resolution by 1 bit compared to N_{b1} when $g\Delta_1/\Delta_2 = 1$. Each quadrupling of M and each doubling of $g\Delta_1/\Delta_2$ add an extra bit to ΔN. The potential ΔN for various M and $g\Delta_1/\Delta_2$ are shown in Table 7.1.

Since practical realization of the proposed method is not ideal, ΔN reflected by (7.28) and Table 7.1 cannot always be achieved. The D/A inaccuracy is the major factor limiting ΔN. Although D/As have better accuracy and resolution than quantizers

7.3 Joint Quantization of Samples

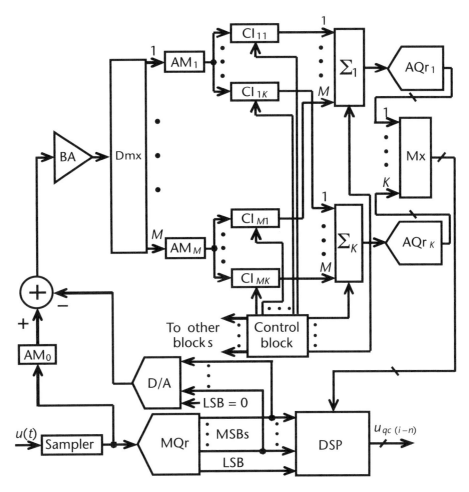

Figure 7.7 Sliding-window JPQ (long window).

for the same f_s, N_b, and technology, development of D/As with, for instance, four bits better resolution than that of quantizers requires significant efforts. Other factors limiting ΔN are the realizable ratios Δ_1/Δ_2, acceptable g, and influence of the mismatch among AM_m and CI_m.

It is easy to make $\Delta_2 \leq \Delta_1$ in hopping-window JPQs where the MQrs and AQrs have the same f_s and $N_{b2} < N_{b1}$. In sliding-window JPQs, providing $\Delta_2 \leq \Delta_1$ is more complex. However, this complexity is moderate when $M \leq 4$ in the JPQ in Figure 7.6 and when $M/K \leq 4$ in the JPQ in Figure 7.7. In principle, g can compensate insufficient Δ_1/Δ_2. The value of $g\Delta_1/\Delta_2$ is limited by mismatch among different AM_m and CI_m pairs (in the JPQ in Figure 7.7, the mismatch among the AQrs should also be taken into account). Placing all AM_m and CI_m on the same die, other technological measures, and employment of adaptive mismatch compensation allow reducing this mismatch to 1% or less. Such mismatch is acceptable if $\Delta N \leq 4$.

When $\Delta N \leq 4$, realization of JPQs has moderate complexity. Achieving $\Delta N > 4$ requires technological improvements directed mostly towards increasing the D/A accuracy. When the required $\Delta N = 4$, it is reasonable to provide 2-bit or 3-bit increase

Table 7.1 Increase in JPQ Resolution

$g\Delta_1/\Delta_2$	$M \to$	4	16	64	256
1	ΔN	1	2	3	4
2	ΔN	2	3	4	5
4	ΔN	3	4	5	6
8	ΔN	4	5	6	7

by selecting $M = 16$ or $M = 64$ and additional 2-bit or 1-bit increase by selecting $g = 4$ or $g = 2$ with $\Delta_1/\Delta_2 = 1$.

The resolution increase (7.25) in JPQs was confirmed by simulations for $g\Delta_1/\Delta_2 = 1$ and $M = 4$, 16, and 64. In all situations where $T_{d\max} = 2MT_s$ is acceptable, hopping-window JPQs are preferable because their practical realization is simpler and less expensive than that of sliding-window JPQs.

7.4 Compressive Quantization of Images

7.4.1 Basic Principles

Compressive quantization [20–23] was initially suggested for very large optical sensors that may contain many millions (or even billions) of pixels and require quantization of many frames per second with effective resolution $N_{b.e} \geq 16$ bits. The use of conventional PCM quantizers (or their systems) with this resolution and f_s of many gigasamples per second is a very expensive and power-consuming solution. As shown below, combining quantization and mixed-signal image compression enables more efficient solutions.

Large optical sensors have many scientific, medical, industrial, military, and law-enforcement applications. Although these systems are mostly located on platforms with sufficient energy supplies, increasing their energy efficiency lowers their power dissipation and, consequently, the required size of their compartments. The information obtained by the sensors, after its initial processing, is usually transmitted to the control station and typically undergoes lossless compression to reduce the time and energy needed for the transmission. Being partially performed in the mixed-signal domain, the compression also allows simplifying image quantization.

Conventional mixed-signal techniques (see Section 7.2.2), however, are unacceptable in this case. Nonuniform quantization is unsuitable because its resolution depends on the brightness of different objects in view. Predictive quantization, which reduces the required $N_{b.e}$ by utilizing the statistical dependences among pixels within frames and corresponding pixels of subsequent frames, introduces slope-overload distortions that cause loss of information, especially at discontinuities. Meanwhile, the discontinuities carry disproportionately large amount of information for the abovementioned applications because they correspond to the edges of sensed objects or are caused by sharp changes in materials, orientation of surfaces, color, depth, and/or illumination.

7.4.1.1 Compressive Quantizer for Large Sensors

Compressive quantizers solve the aforementioned problems by adaptively combining predictive quantization with instantaneous adjustment of their resulting N_b and f_s to utilize the input signals' statistical properties. A simplified block diagram of such a quantizer is shown in Figure 7.8. It includes two internal quantizers: a fast quantizer FQr with a very high sampling rate f_{s1} and a relatively small number of bits N_{b1} (e.g., $N_{b1} = 4$) and a multibit quantizer MBQr with sampling rate $f_{s2} \ll f_{s1}$ but a significantly larger number of bits N_{b2} (e.g., $N_{b2} = 16$).

Signals from the image sensor pixels are amplified by buffer amplifiers BA_1, BA_2, ..., BA_K where K is the total number of pixels in the frame. Multiplexer Mx sequentially connects the outputs of BAs to a subtractor that determines analog prediction errors $e_{a.p}(k)$, that is, the differences between the actual pixel signals $u(k)$ and their analog predicted values $u_{a.p}(k)$ from the D/A output. Errors $e_{a.p}(k)$ enter MBQr directly and FQr through a scaling buffer amplifier BA_s that equalizes the quantization steps of FQr and MBQr. By default, FQr is in the active mode and MBQr is in the idle mode. Due to the statistical dependences among pixel signals, most $|e_{a.p}(k)|$ are small and quantized by FQr. At discontinuities, $|e_{a.p}(k)|$ become large, and FQr sends overflow signals to the control block that switches FQr into the idle mode and MBQr into the active one. Upon completing the $e_{a.p}(k)$ quantization, FQr returns to the active mode, and MBQr to the idle one.

Thus, the resulting sampling rate f_{sr} of the compressive quantizer is variable. Since discontinuities are sparse, most $e_{a.p}(k)$ are quantized by FQr and the mean resulting sampling rate $f_{s2} \ll f_{sr.m} < f_{s1}$, whereas the resulting resolution $N_{br} = N_{b2}$. Digital multipole switch DMS connects the currently active internal quantizer (FQr or MBQr) to the output. To accommodate variable f_{sr}, the control block synchronizes Mx, FQr, MBQr, DMS, and other circuits. The previous pixel's quantized

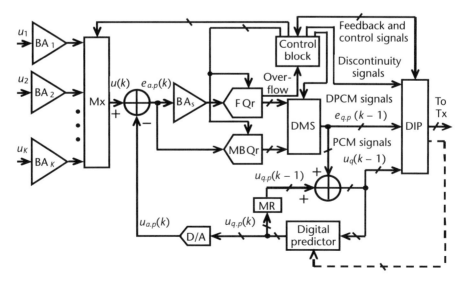

Figure 7.8 Compressive quantizer with predictive quantization.

prediction error $e_{q.p}(k-1)$ at the DMS output is added to the quantized predicted value $u_{q.p}(k-1)$ of that pixel to obtain the actual quantized pixel value

$$u_q(k-1) = u_{q.p}(k-1) + e_{q.p}(k-1) \qquad (7.29)$$

The PCM value $u_q(k-1)$ is sent to the digital predictor that forms $u_{q.p}(k)$ based on several previously quantized pixel signals. Simultaneously, $u_q(k-1)$ is sent to digital image processor DIP, along with $e_{q.p}(k-1)$ and discontinuity signals. Although the term DPCM can be strictly applied only to differential quantization errors, here, it is also applied to predictive quantization errors $e_{q.p}(k-1)$ for conciseness. Image processing in DIP is application-specific. Digital bus "Feedback and control signals" supports communication between the control block and DIP, including the feedback adaptation instructions from DIP to the control block.

In particular, DIP can change N_{b1} if FQr is adjustable. Indeed, analysis of images [22, 23] has shown the existence of optimal N_{b1} that depends on the image type. When N_{b1} is lower than optimal, the number of supposed discontinuities increases and MBQr is used more often than necessary, reducing $f_{sr.m}$ and increasing the total power consumption. Selection of N_{b1} higher than optimal may reduce the attainable f_{s1} and increase the FQr power consumption, producing the same result as an insufficient N_{b1}. Adjustable FQrs with adaptively set N_{b1} solve the problem. A reasonable initial setting is $N_{b1} = 4$ bits.

The quantizer in Figure 7.8 is intended for large image sensors. Its primary design objective is achieving high N_{br} and $f_{sr.m}$ with minimum energy consumption. Its contribution to the subsequent processing in DIP is a byproduct in this case. Sufficient energy resources of a large sensor platform allow performing many complex operations (e.g., filtering, detection, tracking, recognition, and registration of images) in DIP prior to compressing the information intended for transmission. DIP also generates many signals needed for the compressive quantizer adaptation. The quantizer sends discontinuity, DPCM, and PCM signals to DIP. The discontinuity signals allow fast initial feature detection and extraction, while the possibility to select the most convenient signal representation (PCM or DPCM) simplifies the digital processing.

7.4.1.2 Compressive Quantizer for Small Sensors

Compressive quantizers are also beneficial for miniature optical cameras located on small energy-constrained platforms (such as micro-UAVs). Digitization is not a problem in such cameras due to a relatively small number of pixels there, but compressive quantizers allow minimizing the overall energy required for sensing, processing, and transmitting images to the control center. When such a quantizer operates jointly with a colocated DIP, the limited energy requires extreme simplification of the architecture and algorithms of both quantizer and DIP. For example, processing in DIP should be restricted to completing the image compression, and predictive quantization should be reduced to differential one.

A simplified block diagram of a compressive quantizer for small optical sensors is shown in Figure 7.9. Here, predictive quantization is replaced with a differential

7.4 Compressive Quantization of Images

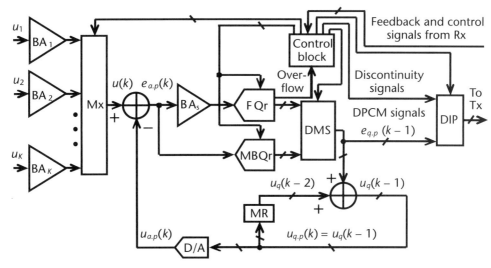

Figure 7.9 Compressive quantizer with differential quantization.

one. Although only discontinuity and DPCM signals are sent to DIP, PCM signals can also be sent there. Many feedback and control signals needed for the adaptation are sent to the platform from the control center. Minimizing energy consumption on the platform requires two additional measures. First, channel encoding and modulation techniques in the platform's Tx should be relatively simple because their weaknesses can be at least partially compensated by effective signal reception methods in the control center Rx where energy is less restricted. Second, the signals sent from the control center should be sufficiently strong and allow simple reception techniques on the platform due to the energy and weight limitations.

7.4.2 Design Considerations

The most challenging problem of the compressive quantizer design is reconstruction of $u_{a.p}(k)$ from $u_{q.p}(k)$ that are represented with resolution N_{b2} bits and can enter the D/A at rate f_{s1}. There are several solutions to this problem, but the simplified block diagrams in Figures 7.8 and 7.9 do not reveal them.

7.4.2.1 Reconstruction of Predicted Signals

The block diagram in Figure 7.10, corresponding to that in Figure 7.9, illustrates a solution for compressive quantizers with differential quantization. In this case, two predicted values $u_{a.p1}(k)$ and $u_{a.p2}(k)$, formed in the mixed-signal domain, can be utilized using one of two methods.

The first method is as follows. When FQr quantizes $e_{a.p}(k)$, $u_{a.p1}(k)$, formed by summing the outputs of fast D/A and SHA, is sent as $u_{a.p}(k)$ through switch S to the subtractor and to the SHA input. The N_{b1}-bit (or slightly higher) resolution of fast D/A and its sampling rate f_{s1} (or slightly higher) allow supporting the FQr operation. When MBQr quantizes $e_{a.p}(k)$, $u_{a.p2}(k)$, formed at the output of MB D/A, is sent as

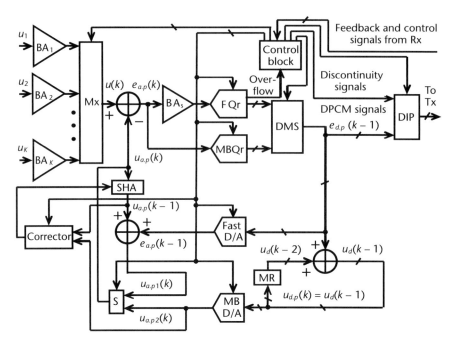

Figure 7.10 Compressive quantizer with differential quantization (reconstruction circuit shown in detail).

$u_{a.p}(k)$ through switch S to the subtractor and to the SHA input. The N_{b2}-bit (or slightly higher) resolution of MB D/A and its sampling rate f_{s2} (or slightly higher) allow supporting the MBQr operation. Using $u_{a.p2}(k)$ as $u_{a.p}(k)$ and storing $u_{a.p2}(k)$ in SHA prevent significant divergence of $u_{a.p1}(k)$ and $u_{a.p2}(k)$. The method's drawback is that the divergence can still become substantial when discontinuities are very sparse.

To avoid this, $u_{a.p2}(k)$ is not sent to the subtractor in the second method. Instead, $u_{a.p2}(k)$ is used for recurring corrections of $u_{a.p1}(k)$, which is sent to the subtractor. To this end, the control block periodically instructs the compressive quantizer to use MBQr and MB D/A even in the absence of discontinuities when the time between neighboring discontinuities exceeds a certain threshold. As a result, the correction frequency is sufficient for keeping $u_{a.p}(k)$ accurately reconstructed.

These solutions, being appropriate for compressive quantizers with differential quantization, cannot be used when prediction is based on several previous pixels. A possible solution for that case is illustrated by the block diagram in Figure 7.11 where two predictors, digital and digitally controlled mixed-signal, are used. Both predictors execute the same prediction algorithm that can be adaptively changed according to the control block instructions.

The number of pixels used for prediction is variable in both predictors: only one previous pixel is used immediately after any discontinuity, two preceding pixels are used at the second step, and the maximum intended number i_{max} of pixels is used only after $(i_{max} - 1)$ steps. Analog predicted values $u_{a.p1}(k)$ of pixel signals generated by the mixed-signal predictor are sent to the subtractor, while analog predicted values $u_{a.p2}(k)$ generated by the digital predictor are used for periodical corrections

7.4 Compressive Quantization of Images

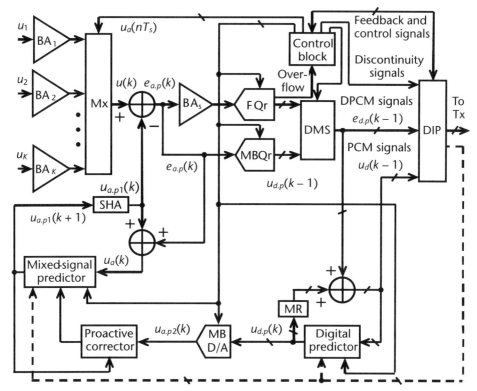

Figure 7.11 Compressive quantizer with predictive quantization (reconstruction circuit shown in detail).

of the mixed-signal predictor. The corrector should be proactive because $u_{a.p1}(k)$ is generated earlier than $u_{a.p2}(k)$.

Although the block diagrams in Figures 7.10 and 7.11 are more detailed than those in Figures 7.8 and 7.9, they cannot reflect all circuits of compressive quantizers due to their application-specific nature. For example, filtering of the pixel signals can be carried out between Mx and subtractor in noisy environment. This filtering may require a certain trade-off between the noise immunity and spatial resolution of the image digitization. Also, FQr and MBQr can, in principle, be replaced by one reconfigurable structure that instantaneously adjusts its N_b and f_s, depending on the relations between the neighboring $u(k)$.

7.4.2.2 Adaptation in Compressive Quantizers

The capabilities and parameters of compressive quantizers are largely determined by their adaptation algorithms. Since these algorithms depend on the quantizers' applications, only the frame quantization time adjustment algorithm is described below because it is more universal.

The frame quantization time determines the maximum permitted frame rate. A reasonable approach to its adjustment is as follows. An initial quantization time

T_{fq0} is allocated for a frame represented by PCM or DPCM signals based on prior experience. It should meet condition

$$KT_{q1} < T_{fq0} \ll KT_{q2} \tag{7.30}$$

where $T_{q1} = 1/f_{s1}$ and $T_{q2} = 1/f_{s2}$ are, respectively, the pixel quantization times of FQr and MBQr.

Throughout the frame quantization, a counter in the control block monitors the time t_{sk} spent on quantizing the first k pixels and, consequently, the time $t_{r(K-k)}$ left for quantizing the remaining $K - k$ pixels of the frame:

$$t_{r(K-k)} = T_{fq0} - t_{sk} \tag{7.31}$$

While $t_{r(K-k)} > (K - k)T_{q1}$, quantization is performed as described in Section 7.4.1. Then, three situations are possible.

In the first one, the actual frame quantization time $T_{fqa} \leq T_{fq0}$. In this case, the quantization time predicted for the next frame is

$$T_{fq1} = T_{fqa} + T_{fb} \tag{7.32}$$

where T_{fb} is relatively small backup time ($T_{fb} \ll T_{fqa}$).

In the second situation, the time required for the frame quantization exceeds T_{fq0}, and correcting T_{fq0} is not allowed for the current frame. In that case, after quantizing the first k_0 pixels where k_0 corresponds to

$$t_{r(K-k_0)} = (K - k_0)T_{q1} \tag{7.33}$$

the remaining pixels should be quantized only by FQr and

$$T_{fqa} = T_{fq0} = t_{sk_0} + (K - k_0)T_{q1} \tag{7.34}$$

Since the quantization only by FQr may introduce distortions, the pixels quantized this way should be marked as less accurate, and the next frame quantization time T_{fq2} should be

$$T_{fq2} = \frac{K}{k_0}t_{sk_0} + T_{fb} \tag{7.35}$$

In the third situation, the time required for the frame quantization also exceeds T_{fq0}, but correcting T_{fq0} for the current frame is allowed. Then T_{fq0} should be corrected according to (7.35) after quantizing the first k_0 pixels.

7.4.3 Assessment of Benefits

Compressive quantization utilizes statistical dependences among pixel signals, and the simplest measure of their intra-frame dependences is ratio K_d/K of the number

K_d of discontinuities in a frame to the number K of pixels in it (larger K_d/K indicates shorter correlation intervals). Two sites with different K_d/K (for 4-bit overflow level), shown in Figures 7.10 and 7.11, are represented by grayscale (GS) images and binary images displaying only discontinuities. They illustrate the fact that positions of discontinuities, which are so important, require much smaller number of bits than that needed for GS images. In most practical cases, $0 < K_d/K < 0.25$.

Compressive quantization supports many ways of utilizing statistical dependences among pixels of the same and/or consecutive frames. Some methods of utilizing intra-frame dependences among pixels have been described in Section 7.4.1. As to various methods of utilizing the dependences among the pixels of consecutive frames, the simplest one is outlined below. It also illustrates the compression algorithms with controlled losses, which eliminate only the data that are irrelevant or unimportant for a current application. Such algorithms are used when lossy compression is unacceptable, but lossless compression is insufficient.

This method subjects just one out of M consecutive frames to full-scale compressive quantization and represents it by DPCM signals, whereas the images in the remaining $(M - 1)$ frames undergo only discontinuity detection and, therefore, are represented by only 1 bit per pixel. If necessary, the images in the frames represented by discontinuity signals can be accurately reconstructed on the Rx side by interpolating the images from the frames represented by DPCM signals and using the discontinuity signals as reference points to track the image evolution.

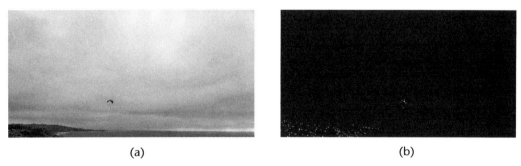

Figure 7.12 Coast, $K_d/K = 0.0035$: (a) GS representation and (b) representation by discontinuities.

Figure 7.13 Street, $K_d/K = 0.0996$: (a) GS representation and (b) representation by discontinuities.

7.4.3.1 Quantization Time Reduction

The frame quantization time reduction reflects the increase in speed provided by compressive quantization. For a conventional quantizer with N_{b2}-bit resolution, frame quantization time is

$$\tilde{T}_{fq} = KT_{q2} \tag{7.36}$$

and for a compressive quantizer that utilizes only intra-frame statistical dependences among pixels, this time is

$$\hat{T}_{fq} = KT_{q1} + K_d\left(T_{q2} + 2T_{sw}\right) \tag{7.37}$$

where T_{q1} and T_{q2} are, respectively, the pixel quantization times of FQr and MBQr, K_d is the number of discontinuities per frame, and T_{sw} is the switching time needed for changing the modes of FQr and MBQr. As follows from (7.36) and (7.37), the frame quantization time reduction is

$$R_{T1} = \frac{\tilde{T}_{fq}}{\hat{T}_{fq}} = \frac{\left(\dfrac{T_{q2}}{T_{q1}}\right)}{1 + \left(\dfrac{K_d}{K}\right)\left[\left(\dfrac{T_{q2}}{T_{q1}}\right) + 2\left(\dfrac{T_{sw}}{T_{q1}}\right)\right]} \tag{7.38}$$

In practice, $T_{q2}/T_{q1} \approx 20 \ldots 60$ and $T_{sw}/T_{q1} \leq 0.25$. The $R_{T1}(K/K_d)$ curves in Figure 7.14 are calculated according to (7.38) for several values of T_{q2}/T_{q1} and $T_{sw}/T_{q1} = 0.25$. Knowledge of the K_d/K probability distribution allows determining the average

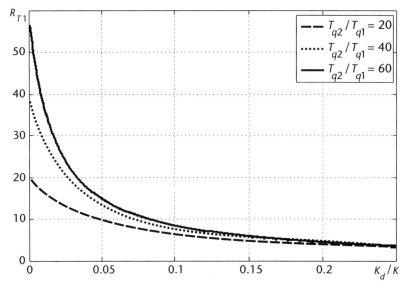

Figure 7.14 Quantization time reduction factor R_{T1} provided by utilizing intra-frame statistical dependences.

7.4.3.2 Compression Factor

The compression factor R_D characterizes the reduction of the number of bits per frame. Consequently, it influences the reduction of the time, bandwidth, and/or energy required for the data transmission. When a compressive quantizer utilizes only intra-frame statistical dependences among pixels,

$$R_{D1} = \frac{KN_{b2}}{K_d N_{b2} + (K - K_d) N_{b1}} = \frac{\left(\frac{N_{b2}}{N_{b1}}\right)}{\left(\frac{K_d}{K}\right)\left(\frac{N_{b2}}{N_{b1}}\right) + \left[1 - \left(\frac{K_d}{K}\right)\right]} \tag{7.39}$$

The $R_{D1}(K_d/K)$ curves in Figure 7.15 are calculated according to (7.39) for several values of N_{b2}/N_{b1}.

When statistical dependences among pixels of consecutive frames are utilized (in addition to their intra-frame dependences) by representing the images in one frame out of M by DPCM signals and the images in the remaining $(M - 1)$ frames by discontinuity signals, the average number of bits per frame is

$$\bar{N}_{b.f} = \frac{K_d N_{b2} + (K - K_d) N_{b1} + (M - 1)K}{M} \tag{7.40}$$

Therefore, the compression factor is

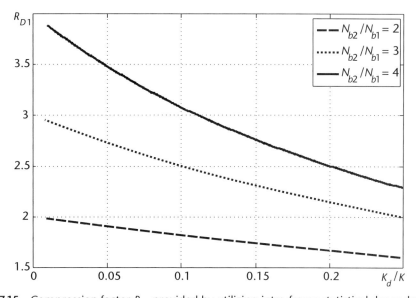

Figure 7.15 Compression factor R_{D1} provided by utilizing intra-frame statistical dependences.

$$R_{D2} = \frac{KN_{b2}}{\bar{N}_{b.f}} = \frac{1}{\frac{1}{MR_{D1}} + \frac{M-1}{MN_{b2}}} \qquad (7.41)$$

The $R_{D2}(K_d/K)$ curves in Figure 7.16 are calculated according to (7.41) for several values of N_{b2}/N_{b1} and M. They show, in particular, that increasing M reduces the influence of K_d/K.

Note that R_{D1} and R_{D2} reflect only mixed-signal-domain compression of images. Their overall compression in the mixed-signal and digital domains can be much more significant.

7.4.3.3 Energy Consumption Reduction

As shown above, compressive quantization influences not only the digitization of pixel signals but also the subsequent digital processing and compression of images. These factors can be taken into account only on a case-by-case basis and require knowledge of the equipment and algorithms used for sensing, processing, and transmitting images. However, it has been shown, based on several reasonable assumptions, that reduction R_{E1} of the energy required for quantization is $R_{E1} \approx 0.5 R_{T1}$.

7.4.3.4 Other Applications

Poorly known and unstable statistical properties of Rxs' input signals complicate their utilization, especially in the mixed-signal domain. Still, compressive quantization can be effective in some Rxs where sharp changes in signal level carry

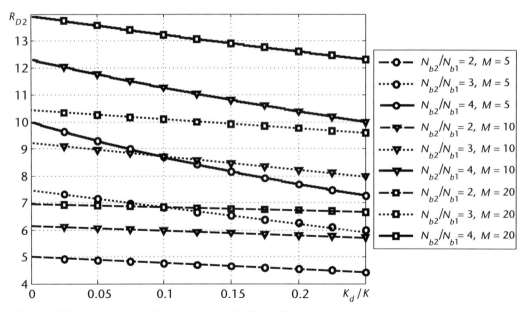

Figure 7.16 Compression factor R_{D2} provided by utilizing intra-frame and inter-frame statistical dependences.

important information. Compressive quantization in Rxs requires buffering in the mixed-signal and digital domains [21].

7.4.3.5 Comparison with Compressive Sampling

Compressive sampling (or sensing) is an approach to sampling proposed by E. Candès, J. Romberg, T. Tao, and D. Donoho in the early 2000s and widely discussed in mathematical and engineering publications. It utilizes sparsity of signals and images for reducing the number of samples (or projections) required for their recovery.

Compressive quantization described above differs from compressive sampling in many respects. First, from the mathematical standpoint, compressive sampling is a new approach to sampling, whereas compressive quantization combines older approaches to solve quantization problems. Second, compressive sampling reduces the required average number of samples (or projections), whereas compressive quantization reduces the required number of bits (often without reducing the number of samples). Third, compressive sampling utilizes only sparsity of images and signals and neglects other statistical properties, whereas compressive quantization opportunistically utilizes all their statistical properties including sparsity. Fourth, compressive sampling performs lossy compression of signals and images, whereas compressive quantization uses only lossless algorithms or algorithms with controlled losses.

These and some other differences do not allow general comparison of compressive sampling and compressive quantization. Still, the effects of their use can be compared for the scenarios where the final result of compression is the total reduction in the number of bits. Simulations have shown that when compressive sampling and compressive quantization were applied to the same images, the latter provided at least two times better compression. Compressive sampling and compressive quantization are, in principle, compatible because compressive quantization can exploit statistical properties of signals and images that are not utilized by compressive sampling.

7.5 Summary

To avoid the exponential growth of complexity and power consumption with an increase of N_b, most modern quantizers are composite structures that contain one or more internal quantizers (each with relatively small n_b) and perform mixed-signal processing intended to efficiently utilize the resolution of internal quantizers.

Exploiting specific features of this processing and/or statistical properties of quantized signals enables improving performance of composite quantizers without increasing the burden imposed on their internal ones.

A huge variety of Rxs' input signals and their poorly known and unstable statistics usually prevent effective utilization of the quantized signals' properties for improving the performance of composite quantizers at the inputs of Rxs. However, nothing prevents exploiting the features of mixed-signal operations there.

Among the currently used versions of this approach, pipelined subranging quantizers and sigma-delta quantizers are most effective. Time-interleaved quantizers can realize their full potential only in combination with samplers based on the sampling theorem's hybrid or direct interpretation.

In contrast with Rxs, statistical properties of Txs' input signals are usually well known. The source encoding of these signals in the mixed-signal domain allows improving and/or simplifying their quantization. Logarithmic and various versions of predictive quantization are most widely used for this purpose.

This chapter shows that despite the outstanding progress of quantization techniques throughout the last three decades, new quantization concepts can still be suggested.

JPQs exploit the fact that sums of several samples can be measured with higher relative accuracy than the separate samples. Employment of Walsh spectral coefficients as such sums allows distributing the achieved resolution increase among the samples and simplifies the JPQ realization. Using only the samples' LSBs for the joint processing further simplifies it.

Although dependences among the samples are not utilized in JPQs, and the samples should not necessarily be successive, the use of successive samples is convenient. The sets of samples involved in the joint processing can be formed using a hopping or sliding window. Both windows provide the same increase in resolution. However, sliding-window JPQs have twice shorter maximum delay at the cost of more complex realization.

Digitization of pixel signals in large optical sensors prior to their transmission often requires quantizers with extremely high $N_{b.e}$ and f_s. Utilization of intra-frame and inter-frame statistical dependences among pixel signals in the mixed-signal domain, in principle, allows solving this problem. However, conventional methods of their utilization cannot be used because they may cause loss of critically important information.

Therefore, compressive quantization that adaptively combines predictive quantization with instantaneous adjustment of quantizer speed and resolution was suggested. Later it was found that this technique is also beneficial for miniature optical cameras located on small energy-constrained platforms (such as micro-UAVs). Realization of compressive quantization and its advantages are described.

References

[1] Razavi, B., *Principles of Data Conversion System Design*, New York: Wiley-IEEE Press, 1995.

[2] Walden, R. H. "Analog-to-Digital Converter Survey and Analysis," *IEEE J. Select. Areas Comm.*, Vol. 17, No. 4, 1999, pp. 539–550.

[3] Medeiro, F., B. Perez-Verdu, and A. Rodriguez-Vazquez, *Top-Down Design of High-Performance Sigma-Delta Modulators*, Boston, MA: Kluwer, 1999.

[4] De la Rosa, J. M., B. Perez-Verdu, and A. Rodriguez-Vazquez, *Systematic Design of CMOS Switched-Current Bandpass Sigma-Delta Modulators for Digital Communication Chips*, Boston, MA: Kluwer, 2002.

[5] Merkel, K. G., and A. L. Wilson, "A Survey of High Performance Analog-to-Digital Converters for Defense Space Applications," *Proc. IEEE Aerosp. Conf.*, Big Sky, MT, March 8–15, 2003, pp. 1–13.

References

[6] Van de Plassche, R., *CMOS Integrated Analog-to-Digital and Digital-to-Analog Converters*, 2nd ed., Norwell, MA: Kluwer Academic Publishers, 2003.

[7] Kester, W. (ed.), *The Data Conversion Handbook*, Norwood, MA: Analog Devices and Newnes, 2005.

[8] Cao, Z., and S. Yan, *Low-Power High-Speed ADCs for Nanometer CMOS Integration*, New York: Springer, 2008.

[9] Ahmed, I., *Pipelined ADC Design and Enhancement Techniques*, New York: Springer, 2010.

[10] Zjajo, A., and J. de Gyvez, *Low-Power High-Resolution Analog to Digital Converters*, New York: Springer, 2011.

[11] El-Chammas, M., and B. Murmann, *Background Calibration of Time-Interleaved Data Converters*, New York: Springer, 2012.

[12] Pandita, B., *Oversampling A/D Converters with Improved Signal Transfer Functions*, New York: Springer, 2013.

[13] Ohnhäuser, F., *Analog-Digital Converters for Industrial Applications Including an Introduction to Digital-Analog Converters*, Berlin, Germany: Springer-Verlag, 2015.

[14] Harpe, P., A. Baschirotto, and A. Makinwa (eds.), *High-Performance AD and DA Converters, IC Design in Scaled Technologies, and Time-Domain Signal Processing*, Cham, Switzerland: Springer, 2015.

[15] Ali, A., *High Speed Data Converters*, London, U.K.: IET, 2016.

[16] Pelgrom, M., *Analog-to-Digital Conversion*, 3rd ed., Cham, Switzerland: Springer, 2017.

[17] Pavan, S., R. Schreier, and G. Temes, *Understanding Delta-Sigma Data Converters*, 2nd ed., New York: Wiley-IEEE Press, 2017.

[18] Poberezhskiy, Y. S., "Multiple-Sample Processing for Increasing the A/D Resolution," *Proc. IEEE Milcom*, San Diego, CA, November 17–29, 2008, pp. 1–7.

[19] Poberezhskiy, Y. S., "A New Approach to Increasing Sensitivity and Resolution of A/Ds," *Proc. IEEE Aerosp. Conf.*, Big Sky, MT, March 7–14, 2009, pp. 1–15.

[20] Poberezhskiy, Y. S., "Adaptive High-Speed High-Resolution Quantization for Image Sensors," *Proc. SPIE Conf.*, San Diego, CA, August 2–6, 2009, pp. 1–12.

[21] Poberezhskiy, Y. S., and G. Y. Poberezhskiy, "Compressive Quantization in Software Defined Receivers," *Proc. IEEE Aerosp. Conf.*, Big Sky, MT, March 6–13, 2010, pp. 1–16.

[22] Poberezhskiy, Y. S., "Compressive Quantization of Images," *Proc. IEEE Aerosp. Conf.*, Big Sky, MT, March 5–12, 2011, pp. 1–17.

[23] Poberezhskiy, Y. S., "Compressive Quantization Versus Compressive Sampling in Image Digitization," *Proc. IEEE Aerosp. Conf.*, Big Sky, MT, March 3–10, 2012, pp. 1–20.

APPENDIX A
Functions Used in the Book

This appendix contains concise information on the functions most often used in the book: the rectangular and some related functions in Section A.1, the delta function in Section A.2, and B-splines in Section A.3.

A.1 Rectangular and Related Functions

Rectangular functions of time are widely used due to the ease of generating rectangular signals and multiplying other signals by them. Rectangular functions of frequency often reflect desirable (but unfeasible) shapes of signals' amplitude spectra or band-limiting circuits' AFRs. Rectangular and several related functions are considered here.

The sign (or signum) function shown in Figure A.1(a) is defined as

$$\text{sgn}(t) = \begin{cases} t/|t| & \text{for } t \neq 0 \\ 0 & \text{for } t = 0 \end{cases} = \begin{cases} -1 & \text{for } t < 0 \\ 0 & \text{for } t = 0 \\ 1 & \text{for } t > 0 \end{cases} \quad (A.1)$$

The unit step function (or Heaviside step function), shown in Figure A.1(b), can be defined through the sign function:

$$H(t) = 0.5[1 + \text{sgn}(t)] = \begin{cases} 0 & \text{for } t < 0 \\ 0.5 & \text{for } t = 0 \\ 1 & \text{for } t > 0 \end{cases} \quad (A.2)$$

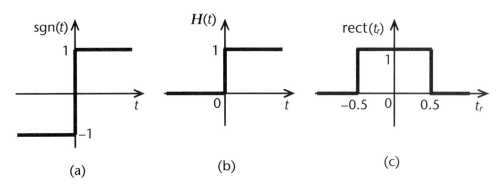

Figure A.1 Functions related to rectangular signals: (a) sign function, (b) unit step function, and (c) gating function.

Thus, $H(t)$ is a halved sum of the sign function and unit dc component.

The gating (or normalized rectangular) function, shown in Figure A.1(c), can be defined from (A.2), (1.27), and (1.28):

$$\text{rect}(t_r) = H(t_r + 0.5) - H(t_r - 0.5) = \begin{cases} 1 & \text{for } |t_r| < 0.5 \\ 0.5 & \text{for } |t_r| = 0.5 \\ 0 & \text{for } |t_r| > 0.5 \end{cases} \quad (A.3)$$

The relative time $t_r = t/\tau$ in (A.3) simplifies scaling of rectangular functions with arbitrary length τ. Using (A.3), (1.27), and (1.28), any rectangular signal $u_1(t)$ with amplitude U and duration τ, centered at $t = t_0$ (see Figure A.2(a)) can be expressed through the gating function:

$$u_1(t) = U\text{rect}\left(\frac{t - t_0}{\tau}\right) \quad (A.4)$$

The normalized triangular function, shown in Figure A.2(b), can be defined as

$$\text{tri}(t_r) = \begin{cases} 1 - |t_r| & \text{for } |t_r| \leq 1 \\ 0 & \text{for } |t_r| > 1 \end{cases} \quad (A.5)$$

This function is the convolution of two gating functions:

$$\text{tri}(t_r) = \text{rect}(t_r) * \text{rect}(t_r) \quad (A.6)$$

As follows from (A.5), (1.27), and (1.28), any triangular signal $u_2(t)$ with amplitude U and duration 2τ, centered at $t = t_0$ (see Figure A.2(c)) can be expressed through the normalized triangular function:

$$u_2(t) = U\text{tri}\left(\frac{t - t_0}{\tau}\right) \quad (A.7)$$

As follows from (A.3) and (1.47), the spectral density of $\text{rect}(t_t)$ is

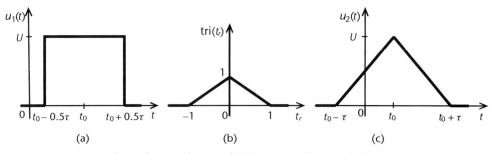

Figure A.2 Rectangular and triangular signals: (a) rectangular signal, (b) normalized triangular function, and (c) triangular signal.

$$S_{\text{rect}}(f) = \int_{-\infty}^{\infty} \text{rect}(t_r)\exp(-j2\pi f t_r)\,dt_r = \frac{\sin(\pi f)}{\pi f} = \text{sinc}(\pi f) \qquad (A.8)$$

Equation (A.8) follows from (1.50) when $U = 1$ and $\tau = 1$. Hence, Figure 1.15(b) after scaling can illustrate the gating function spectral density. In (1.50) and (A.8), sinc is a function of frequency.

It also often emerges as a function of time. The fact of primary importance for this book is that the sinc functions form the ideal orthogonal basis $\{\varphi_{nBB}(t)\}$ for uniform sampling of bandlimited baseband signals:

$$\varphi_{nBB}(t) = \text{sinc}\left[2\pi B(t - nT_s)\right] \qquad (A.9)$$

where B is the one-sided signal bandwidth, $T_s = 1/(2B)$ is the sampling period, and n is an integer. As follows from the time-frequency duality of the Fourier transform (see Section 1.3.3), $\varphi_{nBB}(t)$ has rectangular amplitude spectrum $|S_{\varphi nBB}(f)|$ and linear phase spectrum $\exp(-j2\pi f n T_s)$. According to the time convolution property of the Fourier transform (also see Section 1.3.3), the convolution of a signal $u(t)$ and $\varphi_{nBB}(t)$ is equivalent to multiplying their spectra $S(f)$ and $S_{\varphi nBB}(f)$. Due to the rectangular amplitude and linear phase spectra of $\varphi_{nBB}(t)$, such multiplications do not distort the $u(t)$ spectral components within the band $[-B, B]$ but reject them outside the band. This property makes sinc ideal for bandlimited sampling. Because it is physically unrealizable, in practice it is approximated by physically realizable functions.

A.2 Delta Function

The delta function $\delta(t)$ (also called δ-function, Dirac delta function, or unit impulse) is a generalized function that represents a pulse with infinitely large amplitude, infinitesimal duration, and unit area. It is defined as

$$\delta(t) = \begin{cases} \infty & \text{for } t = t_0 \\ 0 & \text{for } t \neq t_0 \end{cases} \quad \text{and} \quad \int_{-\infty}^{\infty} \delta(t)\,dt = 1 \qquad (A.10)$$

A rigorous treatment of $\delta(t)$ is based on the measure or distributions theories. Considering it a limit, to which a unit-area pulse $s_1(t)$ tends when its duration approaches zero while the pulse area remains constant, clarifies its nature. The initial pulse shape is unimportant, but it is usually convenient to consider it rectangular, triangular, sinc, or Gaussian. As any of these unit-area initial pulses becomes shorter, its spectrum widens and flattens with the same spectral density $S_1(0) = 1$ at zero frequency. As $s_1(t)$ tends to $\delta(t)$, its spectral density $S_1(f) \to S_\delta(f) = 1$, which is the $\delta(t)$ spectral density. Time delay t_0 requires rewriting (A.10) as

$$\delta(t - t_0) = \begin{cases} \infty & \text{for } t = t_0 \\ 0 & \text{for } t \neq t_0 \end{cases} \quad \text{and} \quad \int_{-\infty}^{\infty} \delta(t - t_0)\,dt = 1 \qquad (A.11)$$

Since $\delta(t - t_0)$ is nonzero only at $t = t_0$,

$$u(t)\delta(t-t_0) = u(t_0)\delta(t-t_0) \tag{A.12}$$

Integrating both sides of (A.12) yields

$$\int_{-\infty}^{\infty} u(t)\delta(t-t_0)dt = u(t_0)\int_{-\infty}^{\infty} \delta(t-t_0)dt \tag{A.13}$$

Taking into account (A.11) transforms (A.13) into

$$\int_{-\infty}^{\infty} u(t)\delta(t-t_0)dt = u(t_0) \tag{A.14}$$

Equation (A.14) reflects the sifting (or sampling) property of $\delta(t)$, which allows using $\delta(t-t_0)$ for determining the value of a signal $u(t)$ at the instant $t=t_0$. This property is most important for all applications, including S&I.

It was heuristically shown above that $S_\delta(f) = 1$. The sifting property allows the formal proof of this result:

$$S_\delta(f) = \int_{-\infty}^{\infty} \delta(t)\exp(-j2\pi ft)dt = \exp(-j2\pi f 0) = 1 \tag{A.15}$$

Besides single delta functions, the trains of uniformly spaced delta functions $\delta(t-nT)$ often emerge in the sampling theory, spectral analysis, and other applications:

$$\delta_T(t) = \sum_{n=-\infty}^{\infty} \delta(t-nT) \tag{A.16}$$

Diagrammatically, $\delta(t-t_0)$ is represented by a vertical arrow at the instant t_0 as shown in Figure A.3. In this book, delta functions are used as generalized functions of time or frequency. In the latter role, they reflect the spectral densities of dc and periodic signals. This is important for aperiodic signals containing dc and periodic components because only spectral densities can adequately characterize their frequency distributions. For this reason, the spectral densities of the dc and periodic components are determined below.

Since the inverse Fourier transforms of $\delta(f)$ and $\delta(\omega)$ are, respectively,

$$\begin{aligned} F^{-1}[\delta(f)] &= \int_{-\infty}^{\infty} \delta(f)\exp(j2\pi ft)df = \exp(j2\pi 0 t) = 1 \text{ and} \\ F^{-1}[\delta(\omega)] &= \frac{1}{2\pi}\int_{-\infty}^{\infty} \delta(\omega)\exp(j\omega t)d\omega = \frac{1}{2\pi}, \end{aligned} \tag{A.17}$$

the spectral densities of the unit dc signal $u_{dc}(t) = 1$ expressed as functions of f and ω are, respectively,

A.2 Delta Function

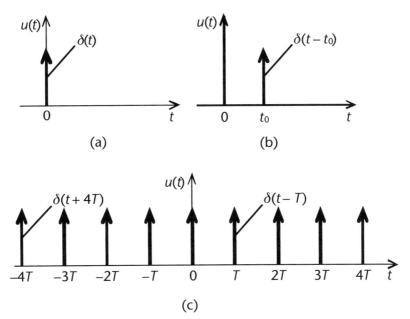

Figure A.3 Delta functions and train of delta functions: (a) $\delta(t)$, (b) $\delta(t - t_0)$, and (c) train $\delta_T(t)$ of $\delta(t - nT)$.

$$S_{\text{dc}}(f) = \delta(f) \quad \text{and} \quad S_{\text{dc}}(\omega) = 2\pi\delta(\omega) \tag{A.18}$$

Figure A.4(a) illustrates $S_{\text{dc}}(f)$.

The same approach allows determining the spectral density of a complex exponential $u_{\text{exp}}(t) = \exp(j2\pi f_0 t)$:

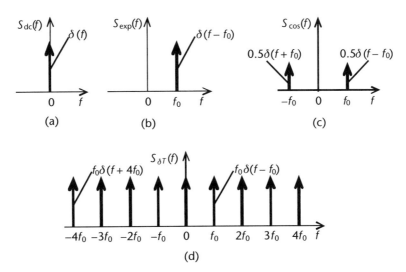

Figure A.4 Spectral densities of dc, complex exponential, cosine, and train of delta functions: (a) $S_{\text{dc}}(f)$, (b) $S_{\text{exp}}(f)$, (c) $S_{\cos}(f)$, and (d) $S_{\delta T}(f)$.

$$F^{-1}\left[\delta(f-f_0)\right] = \int_{-\infty}^{\infty} \delta(f-f_0)\exp(j2\pi ft)df = \exp(j2\pi f_0 t) \quad \text{and}$$

$$F^{-1}\left[\delta(\omega-\omega_0)\right] = \frac{1}{2\pi}\int_{-\infty}^{\infty} \delta(\omega-\omega_0)\exp(j\omega t)d\omega = \frac{1}{2\pi}\exp(j\omega_0 t) \tag{A.19}$$

Consequently, the spectral densities of $u_{\exp}(t)$ expressed as functions of f and ω are, respectively,

$$S_{\exp}(f) = \delta(f-f_0) \quad \text{and} \quad S_{\exp}(\omega) = 2\pi\delta(\omega-\omega_0) \tag{A.20}$$

Figure A.4(b) illustrates $S_{\exp}(f)$.

The spectral density $S_{\cos}(f)$ of a cosine signal $u_{\cos}(t) = \cos(2\pi f_0 t)$ follows from (A.20) and (1.40):

$$S_{\cos}(f) = 0.5\left[\delta(f+f_0) + \delta(f-f_0)\right] \quad \text{and}$$
$$S_{\cos}(\omega) = \pi\left[\delta(\omega+\omega_0) + \delta(\omega-\omega_0)\right] \tag{A.21}$$

Figure A.4(c) illustrates $S_{\cos}(f)$.

These results allow determining the spectral density of any periodic signal $u(t)$ (with period $T = T_0 = 1/f_0$) represented by its Fourier series. If $u(t)$ is represented by its complex exponential Fourier series (1.44), its spectral densities are

$$S(f) = \sum_{n=-\infty}^{\infty} D_n \delta(f-nf_0) \quad \text{and} \quad S(\omega) = 2\pi \sum_{n=-\infty}^{\infty} D_n \delta(\omega-n\omega_0) \tag{A.22}$$

To find the spectral density $S_{\delta T}(f)$ of the uniform train $\delta_T(t)$ (A.16), let us first represent $\delta_T(t)$ by the complex exponential Fourier series (1.44):

$$\delta_T(t) = \sum_{n=-\infty}^{\infty} D_n \exp(jn2\pi f_0 t) = \frac{1}{T}\sum_{n=-\infty}^{\infty} \exp(jn2\pi f_0 t) \tag{A.23}$$

where $D_n = 1/T$ and $f_0 = 1/T$. Taking into account (A.18) and (A.20), we obtain

$$S_{\delta T}(f) = \frac{1}{T}\sum_{n=-\infty}^{\infty} \delta(f-nf_0) = f_0 \sum_{n=-\infty}^{\infty} \delta(f-nf_0) \quad \text{and}$$
$$S_{\delta T}(\omega) = \frac{2\pi}{T}\sum_{n=-\infty}^{\infty} \delta(\omega-n\omega_0) = \omega_0 \sum_{n=-\infty}^{\infty} \delta(\omega-n\omega_0) \tag{A.24}$$

Thus, the spectral density of a uniform train of delta functions in the time domain (see Figure A.3(c)) is a uniform train of delta functions in the frequency domain (see Figure A.4(d)).

A.3 B-Splines

Splines are functions defined piecewise by polynomials and intended to provide a high degree of smoothness at the connection points (knots) of the polynomial pieces. They are widely used in computer graphics and computer-aided design due to their capability to accurately approximate complex shapes. B-splines (short for basis splines), which have minimal support for given degree and smoothness, are most attractive for S&I. Therefore, they are outlined below. A B-spline $\beta^k(t_r)$ of degree k (where $k = 0, 1, 2, \ldots$) and order $k + 1$ is a convolution of $k + 1$ gating functions. Thus, the gating function (A.3) is a B-spline of degree zero and order one, while the normalized triangular function (A.6) is a B-spline of degree one and order two, that is,

$$\text{rect}(t_r) = \beta^0(t_r) \quad \text{and} \quad \text{tri}(t_r) = \beta^1(t_r) \tag{A.25}$$

A B-spline of degree k can be expressed as:

$$\beta^k(t_r) = \beta^{k-1}(t_r) * \beta^0(t_r) \tag{A.26}$$

B-splines are non-negative, and the area under any B-spline (A.26) is equal to one. B-splines can be scaled for the absolute time $t = t_r\tau$ if τ is the length of the original rectangle. All $\beta^k(t_r)$ and $\beta^k(t)$ do not have discontinuities when $k > 1$.

B-splines are attractive as basis (or frame) functions for nonideal S&I of band-limited baseband signals for the following reasons: (1) relative simplicity of B-spline generation, (2) possibility to perform B-spline-weighted integration without multipliers, (3) proper locations of the $\beta^k(t_r)$ and $\beta^k(t)$ spectral nulls (at the stopbands' midpoints) when $\beta^0(t)$ length τ is equal to $T_s = 1/f_s$, and (4) improvement of B-splines' filtering properties with an increase in their degree k. To explain the statements 3 and 4, recall that the spectral density of $\beta^0(t_r)$ is $S_{\beta r0}(f) = \text{sinc}(\pi f)$ and the spectral density of $\beta^0(t)$ is $S_{\beta 0}(f) = \tau\text{sinc}(\pi f\tau) = T_s \text{sinc}(\pi f T_s)$ according to (A.8) and (A.25). Taking into account the time convolution property of the Fourier transform (see Section 1.3.3), the spectral densities of $\beta^k(t_r)$ and $\beta^k(t)$ are, respectively:

$$S_{\beta rk}(f) = [\text{sinc}(\pi f)]^{k+1} \quad \text{and} \quad S_{\beta k}(f) = \left[T_s \text{sinc}(\pi f T_s)\right]^{k+1} \tag{A.27}$$

It follows from (A.27) that the first k derivatives of $S_{\beta rk}(f)$ and $S_{\beta k}(f)$ have nulls at the midpoints of the frequency intervals that should be suppressed during S&I. This means increasing B-spline degree improves its filtering properties, and a trade-off between the complexity of realization and quality of filtering determines k.

APPENDIX B
Sampling Rate Conversion in Digital Radios

This appendix provides concise information on downsampling and upsampling of signals in digital radios. Sections B.1 to B.3 describe the principles of sampling rate conversion for baseband real-valued signals. Optimal realization of these principles is considered in Section B.4. Section B.5 extends the obtained results to baseband complex-valued and bandpass real-valued signals.

B.1 Downsampling by an Integer Factor

Downsampling (i.e., sampling rate reduction or decimation) of a baseband digital signal by an integer factor $L = f_{s1}/f_{s2}$ requires two steps (see Figure B.1(a)). First, decimating filtering by an LPF cleans up the frequency intervals where the signal spectrum replicas from higher Nyquist zones will appear after downsampling (see Figures B.1(b, c)). Then, $L - 1$ out of every L sequential samples are discarded. In a FIR LPF, these steps can easily be combined, and there is no need to calculate the samples that must be discarded. As shown in Figure B.1, the input digital signal $u_{q1}(nT_{s1})$ is an additive mixture of a desired signal and interference. Ideally, the decimating LPF must reject interference within intervals $[(k + r_L/L)f_{s1} - B, (k + r_L/L)f_{s1} + B]$ without distorting the signal spectrum within intervals $[kf_{s1} - B, kf_{s1} + B]$ where B is a one-sided signal bandwidth, k is any integer, and $r_L = 1, 2, ..., L - 1$. When a desired signal is the strongest part of $u_{q1}(nT_{s1})$, the decimating LPF should have unit gain to preserve the signal magnitude. Otherwise, either the decimating LPF gain or the signal should be scaled. In digital Rxs, this scaling is often performed automatically by an AGC.

Although both FIR and IIR filters can be used in sampling rate converters, only FIR filters are discussed below because they simplify digital radio design due to the ease of combining the steps of down- or upsampling and achieving perfectly linear PFRs, as well as the absence of round-off error accumulation.

B.2 Upsampling by an Integer Factor

Upsampling (i.e., sampling rate increase) of a baseband digital signal by an integer factor $M = f_{s2}/f_{s1}$ also requires two steps (see Figure B.2(a)). First, a sequence of the original samples separated by $M - 1$ zeros is formed. Then this sequence undergoes

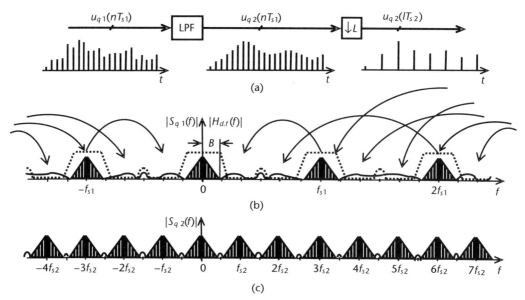

Figure B.1 Downsampling by factor $L = f_{s1}/f_{s2} = 3$ with decimating filtering: (a) block and timing diagrams, (b) amplitude spectrum $|S_{q1}(f)|$ of $u_{q1}(nT_{s1})$ and AFR $|H_{d.f}(f)|$ of decimating LPF (dotted line), and (c) amplitude spectrum $|S_{q2}(f)|$ of $u_{q2}(lT_{s2})$

interpolating filtering by an LPF that calculates the samples at the positions of zeros (note that $u_{q1}(nT_{s1})$ does not contain interference). In the frequency domain, upsampling with interpolating filtering rejects the signal spectral replicas within intervals $[(k + r_M/M)f_{s2} - B, (k + r_M/M)f_{s2} + B]$ without distorting them within intervals $[kf_{s2} - B, kf_{s2} + B]$ (see Figures B.2(b, c)) where B is one-sided signal bandwidth, k is any integer, and $r_M = 1, 2, \ldots, M - 1$. Since the interpolating LPF suppresses $M - 1$ of M spectral replicas, its gain or the signal should be scaled by factor M to preserve the signal magnitude. Thus, in contrast with downsampling where scaling may be needed only in the presence of interference, upsampling always requires it. In properly designed interpolating FIR LPFs, inserting zeros between the original samples does not increase the amount of calculations, and both upsampling steps can easily be combined. This advantage has the same nature as avoiding the calculation of discarded samples in decimating FIR LPFs.

B.3 Sampling Rate Conversion by a Noninteger Factor

The most straightforward way of changing the sampling rate of a baseband digital signal by a rational noninteger factor M/L where M and L are mutually prime integers is shown in Figure B.3(a). Here the sampling rate $f_{s1} = 1/T_{s1}$ of the input signal $u_{q1}(nT_{s1})$ is first increased by factor M (i.e., $f_{s2} = Mf_{s1}$), and then the sampling rate $f_{s2} = 1/T_{s2}$ of the obtained signal $u_{q2}(mT_{s2})$ is decreased by factor L (i.e., $f_{s3} = f_{s2}/L$). Thus, the sampling rate of the output signal $u_{q3}(lT_{s3})$ is $f_{s3} = (M/L)f_{s1}$.

It is beneficial to combine the interpolating and decimating LPFs as shown in Figure B.3(b). The impulse response of the combined LPF is a convolution of the impulse responses of both LPFs. Ideally, the combined LPF rejects the signal

B.3 Sampling Rate Conversion by a Noninteger Factor

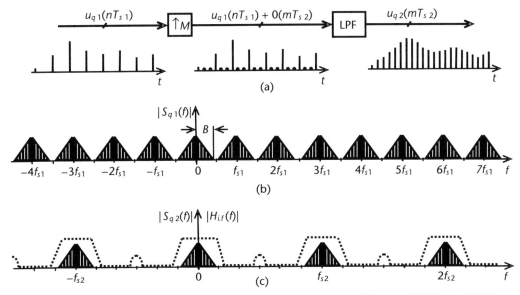

Figure B.2 Upsampling by factor $M = f_{s2}/f_{s1} = 3$ with interpolating filtering: (a) block and timing diagrams, (b) amplitude spectrum $|S_{q1}(f)|$ of $u_{q1}(nT_{s1})$ and AFR $|H_{i,f}(f)|$ of interpolating LPF (dotted line), and (c) amplitude spectrum $|S_{q2}(f)|$ of $u_{q2}(mT_{s2})$.

spectral components within intervals $[(k + r_M/M)f_{s2} - B, (k + r_M/M)f_{s2} + B]$ as well as interference within intervals $[(k + r_L/L)f_{s2} - B, (k + r_L/L)f_{s2} + B]$ without distorting the signal spectrum within intervals $[kf_{s2} - B, kf_{s2} + B]$ where B is one-sided signal bandwidth, k is any integer, $r_M = 1, 2, ..., M - 1$ and $r_L = 1, 2, ..., L - 1$. For the reasons mentioned in Sections B.1 and B.2, the combined filter gain or the signal must be properly scaled to preserve the signal magnitude.

The described approach is optimal when Mf_{s1} is not excessively large. Otherwise, nonexact rate conversion is employed. Note that the accuracy of the "exact" rate conversion described above depends on the quality of decimating and interpolating filtering that cannot be ideal. Conversely, the nonexact methods outlined

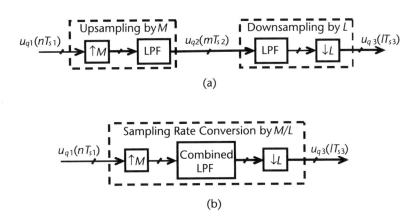

Figure B.3 Block diagrams of sampling rate conversion by a rational noninteger factor M/L: (a) conceptual structure and (b) practical structure.

below can, in principle, be implemented with any required accuracy. Thus, attainable accuracy of sampling rate conversion is determined by acceptable complexity of converters in both cases.

Nonexact rate conversion requires calculating the values of new samples located between the existing ones for both downsampling and upsampling. This problem, known as fractional-delay approximation, is solved using an interpolating filter with impulse response that should approximate the ideal one with accuracy sufficient to meet the signal distortion and interference suppression requirements. In most cases, low-order Lagrange polynomial approximation is utilized. Zero-order approximation assigns the approximated sample the value of the closest preceding original sample. First-order approximation sums two original samples surrounding the approximated one with the weights $\mu = T_D/T_{s1}$ and $(1 - \mu)$, respectively, where T_D is the delay of the approximated sample relative to the closest original sample, and μ is called fractional interval. Second-order approximation uses three surrounding samples for parabolic approximation, and third-order approximation uses four surrounding samples for cubic approximation. Higher-order approximations are used less often. Every μ requires its own set of the filter coefficients. An efficient filter implementation, called Farrow structure, has μ as its single variable parameter.

B.4 Optimization of Sampling Rate Conversion

Although sampling rate conversion by a noninteger factor is often used in TDPs and RDPs, it is rarely employed for D&R. Therefore, only the optimization of integer-factor conversion is outlined below. It comprises optimization of the converter structure and optimization of its filtering. The discussion on the latter is focused on FIR filtering. The optimization criterion is the minimum number of multiplications for given conversion factor and quality.

A single-stage conversion structure is efficient when the conversion factor is a prime number. When it can be factorized, cascade structures with a prime conversion factor at each stage allow reducing the amount of calculations. The largest reduction for downsampling is achieved when the stages' conversion factors follow in nondescending order. In this case, the decimating filter of the first stage, operating at the highest sampling rate f_{s1}, has the lowest order due to the smallest ratio f_{s1}/B_t where B_t is the filter transition band. Indeed, according to simplified Kaiser formula, the length of an equiripple FIR filter can be approximated as

$$N_{FIR} \approx \frac{-2\log_{10}(10\delta_p\delta_s)}{3(B_t/f_s)} + 1 \qquad (B.1)$$

where δ_p and δ_s are ripple in the filter passband and stopband, respectively. For the same reasons, the sequence of the stages' conversion factors in upsampling structures should be opposite.

Nyquist filters are important for sampling rate conversion. They originate from Nyquist's work on ISI. To avoid ISI, a symbol of length τ_{sym} should be shaped by a filter with impulse response $h(t)$ that meets conditions:

$$\begin{cases} h(0) = 1 \\ h(\pm k\tau_{sym}) = 0 \end{cases} \quad (B.2)$$

where $k = 1, 2, \ldots$. Filters with such $h(t)$ are called Nyquist filters. Their $h(t)$ have regularly spaced zeros at $k\tau_{sym}$. Well-known examples of these filters are raised-cosine filters (see Chapters 3 and 4). A condition equivalent to (B.2) for the Nyquist filter transfer function $H(f)$ is:

$$\frac{1}{\tau_{sym}} \sum_{k=-\infty}^{\infty} H\left(f - \frac{k}{\tau_{sym}}\right) = 1 \quad (B.3)$$

As follows from (B.3), the transfer function of a Nyquist filter has the partition of unity property.

A digital Nyquist FIR filter has $h(t)$ zeros spaced at N sampling periods T_s. Its coefficients (counted from the center) meet conditions

$$\begin{cases} h_0 = \dfrac{1}{N} \\ h_{\pm kN} = 0 \end{cases} \quad (B.4)$$

where $k = 1, 2, \ldots$. In (B.4), h_0 is scaled for unit gain. The equivalent frequency-domain condition is:

$$\sum_{k=-\infty}^{\infty} H\left(f - \frac{k}{NT_s}\right) = 1 \quad (B.5)$$

A digital LPF meeting (B.4) and (B.5) is also called Nth-band filter. An ideal rectangular Nth-band LPF has one-sided bandwidth $B = (f_s/2)/N$. In practice, $B < (f_s/2)/N$. As follows from (B.4), every Nth coefficient (counting from the middle) of an Nth-band FIR LPF is zero. This reduces the number of multiplications required for the LPF realization. At first glance, this property makes such LPFs perfect for sampling rate conversion by factor N. Actually, they are not always advantageous because Nth-band LPFs provide much smaller δ_p than practically required. Indeed, their δ_p and δ_s relate as

$$\delta_p \approx (N - 1)\delta_s \quad (B.6)$$

Formula (B.6) is exact for $N = 2$. It is fairly accurate for $N \leq 5$ and $\delta_s \geq 10^{-4}$. When $N > 5$ and $\delta_s < 10^{-4}$, (B.6) gives the upper bound of δ_p. Excessively small δ_p increases the filter length N_{FIR}, according to (B.1). Therefore, Nth-band LPFs are longer than other FIR LPFs, and the presence of zero coefficients may not compensate the length increase. Thus, the use of Nth-band LPFs for sampling rate conversion requires careful substantiation in each particular case if $N > 2$. Nth-band LPFs with $N = 2$, called half-band filters (HBFs), minimize the number of multiplications required for sampling rate conversions with factor two in virtually all practical cases due to the maximum number of zero coefficients. Indeed, every other HBF coefficient is zero except the center coefficient that is equal to 0.5 in a unit-gain HBF. The one-sided AFR $H(f)$ of an HBF is symmetric about $H(0.25f_s)$, that is,

$$H(f) = 1 - H(0.5f_s - f) \quad \text{for} \quad 0 < f < 0.5f_s \qquad (B.7)$$

Figure B.4 shows the triangular-shaped amplitude spectra $|S_{q1}(f)|$ of signals intended for upsampling by three different factors M and the AFRs $|H_{i,f}(f)|$ of two types of equiripple interpolating FIR LPFs: Nth-band LPFs (with $N = M$) not utilizing "don't care" bands (depicted with dotted lines) and conventional LPFs utilizing "don't care" bands (depicted with dash-dotted lines). Note that Nth-band LPFs also can utilize "don't care" bands, and the types of the filters in Figure B.4 were selected exclusively for illustrative purpose. In every case, presented in Figure B.4, $f_{s1}/B = 3$ and $\delta_s = 60$ dB, whereas δ_p are limited by ±0.25 dB for the conventional LPFs and determined by (B.6) for Nth-band LPFs. Since HBFs are optimal for sampling rate conversions with $N = 2$, only the HBF AFR is shown in Figure B.4(a). This filter requires, on average, three multiplications per output sample. The average number of multiplications per output sample in the third-band and conventional LPFs with the AFRs shown in Figure B.4(b) are seven and six, respectively. The fifth-band LPF and conventional LPF with the AFRs in Figure B.4(c) require, on average, 8.2 and 6.4 multiplications per output sample, respectively. For different f_{s1}/B, δ_s, and δ_p, the filter design results would be different. The general rule is that the advantage of

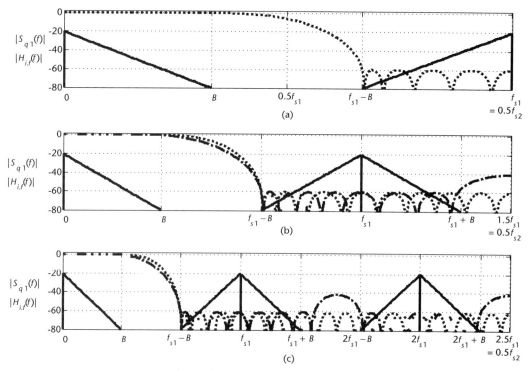

Figure B.4 Amplitude spectra $|S_{q1}(f)|$ of signals and AFRs $|H_{i,f}(f)|$ of interpolating FIR LPFs for upsampling by different factors M: (a) $M = 2$, (b) $M = 3$, and (c) $M = 5$.

Nth-band LPFs diminishes with increasing N and decreasing δ_s. At the same time, the importance of utilizing "don't care" bands increases as f_{s1}/B and conversion factor grow and δ_s decreases.

B.5 Generalization

The sampling rate conversion described above for baseband real-valued signals can be generalized for baseband complex-valued and bandpass real-valued signals. For downsampling and upsampling of a baseband complex-valued signal $Z_q(nT_s) = I_q(nT_s) + jQ_q(nT_s)$, it is sufficient to perform the required processing separately with its I and Q components. If sampling rate conversion in D&R circuits is combined with linear predistortion or postdistortion to compensate the distortions in mixed-signal and/or analog circuits, this combining should be performed at the stage with the lowest sampling rate. The coefficients of the combined filters at that stage can become complex-valued, and four real-valued filters, instead of two, are required in this case.

The major difference between sampling rate conversions of bandpass and baseband signals is that it may be accompanied by signal spectrum inversion for bandpass signals, depending on the conversion factors and positions of the signal spectral replicas within Nyquist zones. Downsampling without and with spectrum inversions is illustrated by the spectral diagrams in Figure B.5 for the conversion factor equal to two. The spectral diagrams in Figures B.5(a, c) show the amplitude spectra

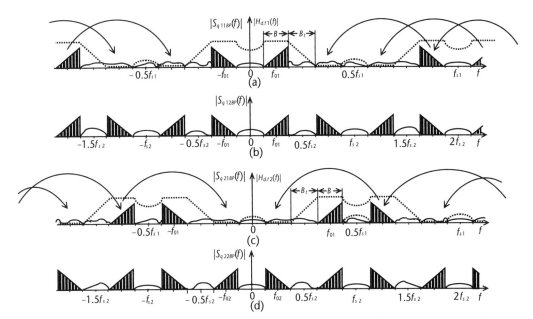

Figure B.5 Downsampling of bandpass signals $u_{q11BP}(nT_{s1})$ and $u_{q21BP}(nT_{s1})$ by factor $L = f_{s1}/f_{s2} = 2$: (a) amplitude spectrum $|S_{q11BP}(f)|$ of $u_{q11BP}(nT_{s1})$ and AFR $|H_{d.f1}(f)|$ of the first decimating filter (dotted line), (b) amplitude spectrum $|S_{q12BP}(f)|$ of $u_{q12BP}(lT_{s2})$, (c) amplitude spectrum $|S_{q21BP}(f)|$ of $u_{q21BP}(nT_{s1})$ and AFR $|H_{d.f2}(f)|$ of the second decimating filter (dotted line), and (d) amplitude spectrum $|S_{q22BP}(f)|$ of $u_{q22BP}(lT_{s2})$.

$|S_{q11BP}(f)|$ and $|S_{q21BP}(f)|$ of signals $u_{q11BP}(nT_{s1})$ and $u_{q21BP}(nT_{s1})$, respectively, prior to downsampling. These diagrams also depict the AFRs $|H_{d.f1}(f)|$ and $|H_{d.f2}(f)|$ of the corresponding decimating filters. The spectral diagrams in Figures B.5(b, d) show the amplitude spectra $|S_{q12BP}(f)|$ and $|S_{q22BP}(f)|$ of signals $u_{q12BP}(lT_{s2})$ and $u_{q22BP}(lT_{s2})$, respectively, obtained after downsampling. The diagrams in Figure B.5 demonstrate that when the conversion factor is equal to two, the position of the lowest-frequency spectral replica of the signal between 0 and $0.25f_{s1}$ prevents the spectrum inversion (see Figures B.5(a, b)), whereas its position between $0.25f_{s1}$ and $0.5f_{s1}$ causes this inversion (see Figures B.5(c, d)).

APPENDIX C
On the Use of Central Limit Theorem

The central limit theorem is one of the most widely used theorems of probability theory. Its multiple versions for random variables and functions have been proven under strictly specified conditions. In applied science and engineering, however, the theorem is typically used in its most general and inexact forms. For stochastic (random) processes, it is often formulated as follows: the probability distribution of a sum of statistically independent arbitrarily distributed stationary (or locally stationary) stochastic processes with comparable statistical characteristics tends toward Gaussian (i.e., normal) as the number of the processes grows. The constraints imposed on the partial processes can differ. For instance, the condition of stationarity may not be introduced, while identicalness or at least similarity of their probability distributions may be required. The possibility of their distributions being non-Gaussian is always implied. A certain neglect of mathematical rigor is unavoidable in applications because there is no real physical object that precisely corresponds to its theoretical model. Thus, there is nothing strange or bad in vague formulation of the central limit theorem when it is used for practical purposes. Problems arise if not all aspects of the neglected constraints are properly understood.

Prior to discussing problems of this nature, it is necessary to concisely explain the motivations for the use of the central limit theorem in electrical engineering and communications. First, linear transformations of Gaussian processes produce processes that are also Gaussian. Therefore, a Gaussian process after passing through a linear circuit remains Gaussian, and only its correlation function (or PSD) must be calculated to fully characterize it. Second, for this and some other reasons, the problems of optimal filtering and demodulation in presence of Gaussian noise have closed-form solutions that are well substantiated theoretically and well validated by their long practical use. Third, many signals and physical phenomena in electrical engineering and communications are sums of multiple non-Gaussian partial signals or physical phenomena, respectively. While it is difficult or impossible to find a closed-form solution for each of them, such a solution can be easily obtained for their sum if it has Gaussian distribution. To effectively use this approach, it is necessary to know the limits and conditions of its applicability. The following paradox helps to understand them.

C.1 Paradox Statement

Let us assume that M narrowband non-Gaussian stochastic signals $X_m(t)$ with comparable powers and nonoverlapping spectra pass through a Rx preselector. Let us further assume that M is so large that the sum

$$Y(t) = \sum_{m=1}^{M} X_m(t) \tag{C.1}$$

can be considereed Gaussian based on common sense. It is known that linear transformations of Gaussian signals produce signals that are also Gaussian. Therefore, if $Y(t)$ is sent through an ideal linear channel filter that selects only one signal, for example, $X_{m=a}(t)$, without any distortion and rejects all others (see Figure C.1), signal $X_{m=a}(t)$ at the filter output should be Gaussian. However, as stated above, all the signals $X_m(t)$, including $X_{m=a}(t)$, are non-Gaussian.

C.2 Paradox Resolution

To resolve this paradox, recall that a stochastic process is Gaussian if and only if its n-dimensional (i.e., n-variate) probability distribution (represented, for instance, by an n-dimensional PDF) with $n \to \infty$ is Gaussian (see Section 1.2.2). As the number M of processes $X_m(t)$ in the sum $Y(t)$ increases, its one-dimensional PDF approaches a Gaussian PDF first. Then, one by one, the PDFs with higher dimensionalities follow it. However, as the PDF dimensionality increases, M required for approaching the Gaussian distribution grows so fast that, in most practical cases, only two-dimensional PDFs at best can become close to Gaussian with a sufficient degree of confidence. Thus, strictly speaking, $Y(t)$ is not a Gaussian process even when M is very large but finite. The n-dimensional PDF of $Y(t)$, with $n \to \infty$, is Gaussian only in the asymptotic case when $M \gg n \to \infty$ within a finite preselector bandwidth. In

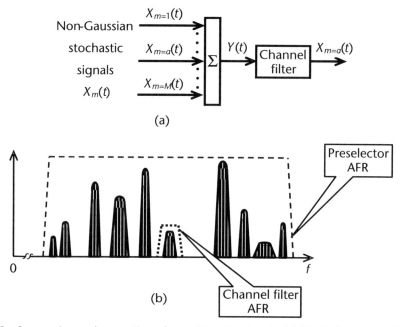

Figure C.1 Summation and separation of non-Gaussian signals: (a) block diagram and (b) spectral diagram.

this case, the widths of all $X_m(t)$ spectra and distances among them tend to zero. As a result, the number of $X_m(t)$ passing through any realizable channel filter tends to infinity, and, consequently, the process at the filter output remains Gaussian.

C.3 Discussion

Thus, only those stochastic processes that are sums of a virtually infinite number of partial stochastic processes (e.g., thermal or shot noise) can be considered strictly Gaussian. In most cases, however, a limited number M of partial signals or physical phenomena make only one-dimensional or two-dimensional distribution of their sum close to Gaussian. Two well-known examples are: (1) Rayleigh and Rician models of fading channels are based on summing the signals arriving over relatively small numbers of varying propagation paths, and (2) instantaneous values of the sum of ISs within a preselector passband are usually considered Gaussian when their number exceeds 5.

For sums of stationary stochastic processes, their accurate two-dimensional Gaussian approximation is sufficient for solving virtually all practical problems.

Increasing M guarantees better approximation of the sums of partial signals or physical phenomena by Gaussian processes, which is especially noticeable at the tails of their probability distributions. Symmetric distributions of partial processes ensure faster convergence of their sums to Gaussian processes for a given M. Relative simplicity of obtaining closed-form solutions using Gaussian distributions makes them attractive as a first approximation, even if it is not perfectly accurate. The central limit theorem is also used for the substantiation of log-Gaussian (log-normal) distributions of products of many comparable random factors.

The material above shows that, although applying the central limit theorem to the sums of a finite number of partial stochastic processes allows solving many important problems, these sums are not Gaussian processes, but processes whose one-dimensional or two-dimensional distributions are close to Gaussian ones. Therefore, the extent, to which the properties of Gaussian processes are relevant to specific sums, should be determined on a case-by-case basis.

APPENDIX D
Sampling Theorem for Bandlimited Signals

Chapter 5 contains statements of the sampling theorem but does not provide their proofs. These proofs are presented below in a way that clarifies their physical substance. Note two inconsistencies related to this theorem. First, real-world signals undergoing sampling have finite duration, contradicting the finite-bandwidth assumption, and strict bandlimiting cannot be provided by any physically realizable filter. Second, sampling functions used in all discussed versions of the sampling theorem have infinite duration and cannot exactly interpolate finite-duration signals. Thus, practical S&I are always accompanied by aliasing errors, caused by the impossibility of perfect bandlimiting, and time-domain truncation errors, caused by representing finite-duration signals by infinite-duration functions. There are also other sources of errors in S&I circuits (e.g., jitter errors caused by deviation of actual sampling instants from the expected ones, linear and nonlinear distortions). Hence, real-world S&I are always nonideal.

D.1 Sampling Theorem for Baseband Signals

D.1.1 Theorem

An analog baseband real-valued square-integrable signal $u(t)$ with one-sided bandwidth B can be represented by its instantaneous values $u(nT_s)$, taken uniformly with period $T_s = 1/(2B)$, and reconstructed from them according to

$$u(t) = \sum_{n=-\infty}^{\infty} u(nT_s)\varphi_{nBB}(t) = \sum_{n=-\infty}^{\infty} u(nT_s)\varphi_{0BB}(t - nT_s) \quad (D.1)$$

where $\varphi_{nBB}(t)$ are the baseband sampling functions:

$$\varphi_{nBB}(t) = \operatorname{sinc}\left[2\pi B(t - nT_s)\right] = \frac{\sin\left[2\pi B(t - nT_s)\right]}{2\pi B(t - nT_s)} \quad (D.2)$$

Note that (D.1) and (D.2) correspond, respectively, to (5.3) and (5.4).

D.1.2 Proof

Set of sampling functions $\{\varphi_{nBB}(t)\}$ forms an orthogonal basis in the function space of bandlimited square-integrable functions. According to (1.37), the coefficients of the generalized Fourier series for $u(t)$ with respect to orthogonal basis $\{\varphi_{nBB}(t)\}$ are

$$c_n = \frac{1}{\|\varphi_{nBB}(t)\|^2} \int_{-\infty}^{\infty} u(t) \varphi_{nBB}^*(t) \, dt \tag{D.3}$$

Using the Fourier transform property called Parseval's formula (do not confuse it with Parseval's identity (1.38))

$$\int_{-\infty}^{\infty} g(t) h^*(t) \, dt = \int_{-\infty}^{\infty} S_g(f) S_h^*(f) \, df = \int_{-B}^{B} S_g(f) S_h^*(f) \, df, \tag{D.4}$$

(5.6), (5.7), and (1.48), (D.3) can be rewritten as

$$c_n = \frac{1}{T_s} \int_{-B}^{B} S_u(f) S_{\varphi nBB}^*(f) \, df = \int_{-B}^{B} S_u(f) \exp(j2\pi f n T_s) df = u(nT_s) \tag{D.5}$$

where $S_u(f)$ and $S_{\varphi nBB}^*(f)$ are, respectively, spectra of $u(t)$ and $\varphi_{nBB}^*(t)$. Thus, the optimal coefficients c_n are the signal values $u(nT_s)$. Since $u(t)$ is bandlimited and square-integrable, this series converges to $u(t)$ for any t. This proves (D.1).

D.1.3 Discussion

It can also be proven that $u(t)$ can be reconstructed from $u(nT_s)$ if $T_s < 1/(2B)$. The fact that reducing T_s does not prevent such reconstruction is intuitively clear and can be illustrated by the timing diagram in Figure 5.2. Still, timing diagrams do not show why $T_s = 1/(2B)$ is a critical point, while spectral diagrams make it obvious. Indeed, according to (5.10), sampling causes proliferation of the spectrum $S(f)$ of the sampler's input signal $u(t)$ that is shown, for instance, in Figure 5.3. When the sampling rate $f_s = 2B$, that is, $T_s = 1/(2B)$, the neighboring replicas of $S(f)$ in $S_d(f)$ border each other. When $f_s > 2B$, that is, $T_s < 1/(2B)$, there are gaps between all neighboring replicas of $S(f)$ in $S_d(f)$. Thus, overlapping of these replicas is avoided in both cases, and $u(t)$ can be accurately reconstructed from its samples $u(nT_s)$. When $f_s < 2B$, that is, $T_s > 1/(2B)$, the neighboring replicas of $S(f)$ in $S_d(f)$ inevitably overlap, and $u(t)$ cannot be accurately reconstructed from its samples $u(nT_s)$. Thus, $f_s = 2B$ is the minimum acceptable sampling rate.

D.2 Sampling Theorem for Bandpass Signals

Baseband S&I of a bandpass signal $u(t)$ require representing it by the I and Q components $I(t)$ and $Q(t)$ or by the envelope $U(t)$ and phase $\theta(t)$ of its baseband complex-valued equivalent $Z(t)$ according to (1.86) and (1.87). The proof of the sampling theorem for bandpass signals represented by $I(t)$ and $Q(t)$ is provided in Section D.2.1, and for those represented by $U(t)$ and $\theta(t)$ it is provided in Section D.2.2. Bandpass S&I of bandpass signals require representing them by the instantaneous values, and the sampling theorem for that case is proven in Section D.2.3.

D.2.1 Sampling of Bandpass Signals Represented by $I(t)$ and $Q(t)$

D.2.1.1 Theorem

An analog bandpass real-valued square-integrable signal $u(t)$ with center frequency f_0 and bandwidth B can be represented by the samples $I(nT_s)$ and $Q(nT_s)$ of $I(t)$ and $Q(t)$, and reconstructed from them according to

$$\begin{aligned} u(t) &= \left[\sum_{n=-\infty}^{\infty} I(nT_s)\varphi_{nBBE}(t)\right]\cos(2\pi f_0 t) - \left[\sum_{n=-\infty}^{\infty} Q(nT_s)\varphi_{nBBE}(t)\right]\sin(2\pi f_0 t) \\ &= \left[\sum_{n=-\infty}^{\infty} I(nT_s)\varphi_{0BBE}(t - nT_s)\right]\cos(2\pi f_0 t) \\ &\quad - \left[\sum_{n=-\infty}^{\infty} Q(nT_s)\varphi_{0BBE}(t - nT_s)\right]\sin(2\pi f_0 t) \end{aligned} \quad \text{(D.6)}$$

where $T_s = 1/B$ and $\varphi_{nBBE}(t)$ are the sampling functions of the baseband equivalent $Z(t)$:

$$\varphi_{nBBE}(t) = \operatorname{sinc}\left[\pi B(t - nT_s)\right] = \frac{\sin\left[\pi B(t - nT_s)\right]}{\pi B(t - nT_s)} \quad \text{(D.7)}$$

D.2.1.2 Proof

The one-sided bandwidth B of $u(t)$ is two times wider than the one-sided bandwidth B_Z of $Z(t)$, that is, $B_Z = 0.5B$ (see, for instance, Figure 5.17). The one-sided bandwidths of $I(t)$ and $Q(t)$ are also equal to $B_Z = 0.5B$. Since $I(t)$ and $Q(t)$ are baseband real-valued square-integrable signals, they can be represented by their samples $I(nT_s)$ and $Q(nT_s)$ taken with period $T_s = 1/(2B_Z) = 1/B$ according to (D.1):

$$\begin{aligned} I(t) &= \sum_{n=-\infty}^{\infty} I(nT_s)\varphi_{nBBE}(t) = \sum_{n=-\infty}^{\infty} I(nT_s)\varphi_{0BBE}(t - nT_s) \\ Q(t) &= \sum_{n=-\infty}^{\infty} Q(nT_s)\varphi_{nBBE}(t) = \sum_{n=-\infty}^{\infty} Q(nT_s)\varphi_{0BBE}(t - nT_s) \end{aligned} \quad \text{(D.8)}$$

For a given f_0, $I(t)$ and $Q(t)$ completely determine $u(t)$ (see (1.85)). Substituting (D.8) in (1.85) yields (D.6).

D.2.2 Sampling of Bandpass Signals Represented by $U(t)$ and $\theta(t)$

D.2.2.1 Theorem

An analog bandpass real-valued square-integrable signal $u(t)$ with center frequency f_0 and bandwidth B can be represented by the samples $U(nT_s)$ and $\theta(nT_s)$ of envelope $U(t)$ and phase $\theta(t)$, and reconstructed from them according to

$$\begin{aligned}
u(t) &= \sum_{n=-\infty}^{\infty} U(nT_s)\varphi_{nBBE}(t)\cos\left[2\pi f_0 t + \theta(nT_s)\right] \\
&= \sum_{n=-\infty}^{\infty} U(nT_s)\varphi_{0BBE}(t - nT_s)\cos\left[2\pi f_0 t + \theta(nT_s)\right]
\end{aligned} \quad \text{(D.9)}$$

where $T_s = 1/B$ and functions $\varphi_{nBBE}(t)$ are defined by (D.7).

D.2.2.2 Proof

According to (D.6), $u(t)$ is completely determined by $I(nT_s)$ and $Q(nT_s)$ for a given f_0. Simultaneously, as follows from (1.88),

$$I(nT_s) = U(nT_s)\cos\left[\theta(nT_s)\right] \text{ and } Q(nT_s) = U(nT_s)\sin\left[\theta(nT_s)\right] \quad \text{(D.10)}$$

Substituting (D.10) into (D.6) yields

$$\begin{aligned}
u(t) = &\left[\sum_{n=-\infty}^{\infty} U(nT_s)\cos\left[\theta(nT_s)\right]\varphi_{nBBE}(t)\right]\cos(2\pi f_0 t) \\
&- \left[\sum_{n=-\infty}^{\infty} U(nT_s)\sin\left[\theta(nT_s)\right]\varphi_{nBBE}(t)\right]\sin(2\pi f_0 t)
\end{aligned} \quad \text{(D.11)}$$

Applying identity $\cos(\alpha)\cos(\beta) - \sin(\alpha)\sin(\beta) = \cos(\alpha + \beta)$ to (D.11), we obtain

$$u(t) = \sum_{n=-\infty}^{\infty} U(nT_s)\varphi_{nBBE}(t)\cos\left[2\pi f_0 t + \theta(nT_s)\right] \quad \text{(D.12)}$$

which proves (D.9).

D.2.2.3 Discussion

It is apparent that representation of bandpass signals by the pairs of samples $I(nT_s)$ and $Q(nT_s)$ or $U(nT_s)$ and $\theta(nT_s)$ requires a two-channel structure with the minimum acceptable sampling rate $f_s = B$ in each channel. Consequently, the minimum total sampling rate is equal to $2B$ (i.e., it is identical for baseband and bandpass signals with the same B).

D.2.3 Sampling of Bandpass Signals' Instantaneous Values

D.2.3.1 Theorem

An analog bandpass real-valued square-integrable signal $u(t)$ with center frequency f_0 and bandwidth B can be represented by its instantaneous values $u(nT_s)$, taken uniformly with period $T_s = 1/(2B)$, and reconstructed from them according to

$$u(t) = \sum_{n=-\infty}^{\infty} u(nT_s)\varphi_{nBP}(t) = \sum_{n=-\infty}^{\infty} u(nT_s)\varphi_{0BP}(t - nT_s) \tag{D.13}$$

where $\varphi_{nBP}(t)$ are the bandpass sampling functions

$$\varphi_{nBP}(t) = \operatorname{sinc}\left[\pi B(t - nT_s)\right]\cos\left[2\pi f_0(t - nT_s)\right] = \varphi_{nBBE}(t)\cos\left[2\pi f_0(t - nT_s)\right] \tag{D.14}$$

if and only if

$$f_0 = |k \pm 0.5|B \tag{D.15}$$

where k is an integer.

Note that (D.13), (D.14), and (D.15) correspond, respectively, to (5.27), (5.28), and (5.26).

D.2.3.2 Proof

When $T_s = 1/(2B)$, sampling functions $\varphi_{nBP}(t)$ form an orthogonal basis if and only if (D.15) is true. In this case, coefficients c_n of the generalized Fourier series for $u(t)$ with respect to the orthogonal basis $\{\varphi_{nBP}(t)\}$ are

$$c_n = \frac{1}{\|\varphi_{nBP}(t)\|^2} \int_{-\infty}^{\infty} u(t)\varphi_{nBP}^*(t)\,dt \tag{D.16}$$

Using (D.4), (5.30), and (5.31), (D.16) can be rewritten as

$$\begin{aligned}
c_n &= \frac{1}{T_s}\int_{-\infty}^{\infty} S_u(f)S_{\varphi nBP}^*(f)\,df \\
&= \int_{-(f_0+0.5B)}^{-(f_0-0.5B)} S_u(f)\exp(j2\pi fnT_s)\,df \\
&\quad + \int_{f_0-0.5B}^{f_0+0.5B} S_u(f)\exp(j2\pi fnT_s)\,df = u(nT_s)
\end{aligned} \tag{D.17}$$

where $S_u(f)$ and $S_{\varphi nBP}^*(f)$ are, respectively, spectra of $u(t)$ and $\varphi_{nBP}^*(t)$. Thus, the optimal coefficients c_n are the signal values $u(nT_s)$. Since $u(t)$ is bandlimited and square-integrable, this series converges to $u(t)$ for any t. This proves (D.13).

D.2.3.3 Discussion

Sampling of bandpass signals' instantaneous values requires a single-channel structure with the minimum acceptable sampling rate $f_s = 2B$. Thus, all the versions of the sampling theorem presented above confirm the fact that signals with bandwidth B and duration T_s require $2BT_s$ samples for their discrete-time representation if no other constraints are imposed on them, that is, the signal dimensionality (or its number of degrees of freedom) is $2BT_s$ in this case.

Additional constraints imposed on a signal allow reducing its dimensionality. For instance, an amplitude-modulated sinewave with a known initial phase can be represented just by the samples of its amplitude. Similarly, a phase-modulated sinewave with a known amplitude can be represented just by the samples of its phase. Other signal properties can also be used for the dimensionality reduction.

List of Acronyms

ac	Alternating current
A/D	Analog-to-digital conversion or converter (depending on the context)
AE	Antenna element
AFR	Amplitude-frequency response
AGC	Automatic gain control
AJ	Antijam
ALC	Automatic level control
AM	Amplitude modulation or analog memory (depending on the context)
AMB	Analog and mixed-signal back-end (of a transmitter)
AMF	Analog and mixed-signal front-end (of a receiver)
AQ-DBPSK	Alternating quadratures DBPSK
AQr	Auxiliary quantizer
ASIC	Application-specific integrated circuit
ASK	Amplitude-shift keying
AtC	Attenuator control circuit
AWGN	Additive white Gaussian noise
BA	Buffer amplifier
BAW	Bulk acoustic wave
BPF	Bandpass filter
BPSK	Binary phase-shift keying
CDF	Cumulative distribution function
CDM	Code-division multiplexing
CDMA	Code-division multiple access
CDP	Central digital processor
CI	Controlled inverter
CNF	Conjunctive normal form (in Boolean algebra)
COFDM	Coded orthogonal frequency-division multiplexing
CR	Cognitive radio
CT	Code transformer

D/A	Digital-to-analog conversion or converter (depending on the context)
D&R	Digitization and reconstruction
DBPSK	Differential BPSK
dc	Direct current
DCA	Digitally controlled amplifier
DDS	Direct digital synthesis or synthesizer (depending on the context)
DF	Direction finding
DFC	Digital functional conversion or converter (depending on the context)
DFI	Digital filter-interpolator
DIP	Digital image processor
DMS	Digital multipole switch
Dmx	Demultiplexer
DNF	Disjunctive normal form (in Boolean algebra)
DPCM	Differential pulse-code modulation
DPD	Digital part of digitization circuit
DPR	Digital part of reconstruction circuit
DQPSK	Differential QPSK
DS	Direct sequence
DSB-FC	Double-sideband full-carrier
DSB-RC	Double-sideband reduced-carrier
DSB-SC	Double-sideband suppressed-carrier
DSP	Digital signal processing or processor (depending on the context)
DWFG	Digital weight function generation or generator (depending on the context)
ECS	Electronic cyclic switch
EHF	Extremely high frequency
ELF	Extremely low frequency
ENOB	Effective number of bits
ESD	Energy spectral density
EVM	Error vector magnitude
EW	Electronic warfare
FC	Format converter
FDM	Frequency-division multiplexing
FDMA	Frequency-division multiple access
FIR	Finite impulse response

FM	Frequency modulation
FPGA	Field programmable gate array
FPIC	Field programmable integrated circuit
FQr	Fast quantizer
FSK	Frequency-shift keying
GAt	Guard attenuator
GMSK	Gaussian minimum shift keying
GNSS	Global navigation satellite system
GPP	General-purpose processor
GPS	Global positioning system
GPU	Graphics processing unit
GS	Grayscale
HBF	Half-band filter
HF	High frequency
IC	Integrated circuit
IF	Intermediate frequency
IIR	Infinite impulse response
IMP	Intermodulation product
IP	Intercept point
IS	Interfering signal
ISI	Intersymbol interference
ITU	International Telecommunication Union
JPQ	Joint-processing quantizer
ksps	Kilosamples per second
LAN	Local area network
LF	Low frequency
LNA	Low-noise amplifier
LO	Local oscillator
LPF	Lowpass filter
LS	Least squares
LSB	Lower sideband or least significant bit (depending on the context)
LTI	Linear time-invariant (system)
MBQr	Multibit quantizer
MB D/A	Multibit D/A
MD/A	Multiplying D/A
MDS	Minimum detectable signal
MEMS	Microelectromechanical systems

MF	Medium frequency
MFS	Master frequency standard
MIMO	Multiple-input multiple-output
MR	Memory register
MSB	Most significant bit
MSK	Minimum shift keying
Msps	Megasamples per second
MTC	Modulo-two counter
Mx	Multiplexer
NDC	Novel digitization circuit
NF	Noise figure
NRC	Novel reconstruction circuit
NUS	Nonuniform sampling
OFD	Original-frequency distortion
OFDM	Orthogonal frequency-division multiplexing
OFDMA	Orthogonal frequency-division multiple access
OOK	On-off keying
OQPSK	Offset QPSK
PA	Power amplifier
PAM	Pulse-amplitude modulation
PAT	Phase and amplitude tuner
PCM	Pulse-code modulation
PDF	Probability density function
PFR	Phase-frequency response
PM	Phase modulation
PMF	Probability mass function
PN	Pseudonoise (or pseudorandom)
PPM	Pulse-position modulation
PS	Pulse shaping or shaper (depending on the context)
PSD	Power spectral density
PSK	Phase-shift keying
PWM	Pulse-width modulation
QAM	Quadrature amplitude modulation
QPSK	Quadrature phase-shift keying
RC	Reconstruction circuit
RDP	Receiver digital portion
RF	Radio frequency

RFID	RF identification
rms	Root mean square
ROM	Read-only memory
Rx	Receiver
S&I	Sampling and interpolation
SAW	Surface acoustic wave
SDR	Software-defined radio
SHA	Sample-and-hold amplifier (integrating)
SHAWI	Sample-and-hold amplifier with weighted integration
SHF	Super high frequency
SLF	Super low frequency
SNR	Signal-to-noise ratio
sps	Samples per second
SPU	Specialized processing unit
SS	Spread spectrum
SSB	Single sideband
TDM	Time-division multiplexing
TDMA	Time-division multiple access
TDP	Transmitter digital portion
THA	Track-and-hold amplifier
THF	Tremendously high frequency
TV	Television
Tx	Transmitter
UCA	Uniform circular array
UHF	Ultrahigh frequency
ULA	Uniform linear array
ULF	Ultralow frequency
USB	Upper sideband
VAM	Virtual antenna motion
VCA	Voltage-controlled amplifier
VCO	Voltage-controlled oscillator
VGA	Variable-gain amplifier
VHF	Very high frequency
VLF	Very low frequency
VOR	Very high frequency omnidirectional radio range (navigation system)
VSB	Vestigial sideband (modulation)

WFG	Weight function generation or generator (depending on the context)
WKS	Whittakers-Kotelnikov-Shannon
WPS	Weighted pulse shaping or shaper (depending on the context)

About the Authors

Yefim S. Poberezhskiy received his M.S. and Ph.D. degrees in electrical engineering, respectively, from Kharkiv Polytechnic Institute, Ukraine, and Moscow Radio Communications Research Institute, Russia. His professional interests include communication systems and their units such as receivers, transmitters, transceivers, and antenna amplifiers; algorithms and hardware for modulation/demodulation, encoding/decoding, synchronization, and control in radio systems; as well as digitization and reconstruction of signals and images in various applications. He is an author of over 200 publications and over 35 inventions (USSR invention certificates and U.S. patents). A book, *Digital Radio Receivers* (in Russian; Radio & Communications: Moscow, 1987), is among his publications. He has held positions in both industry and academia. From 1976 to 1995, he was with Omsk State Technical University, Russia, initially as an associate professor of applied mathematics and later as a professor of electrical engineering. He also was the head of the Digital Signal Processing Laboratory at that university. His latest positions were with Rockwell Scientific, Thousand Oaks, California, and SAIC, San Diego, California. At present, he is a consultant in signal processing and communication systems.

Gennady Y. Poberezhskiy received his M.S. and Ph.D. degrees in electrical engineering, respectively, from Moscow Aviation Institute, Russia, and the University of Southern California, Los Angeles. His professional interests include communication and navigation systems, digital and mixed-signal processing, signal detection and tracking, channel equalization, adaptive arrays, and direction finding. He is an author of more than 30 publications and patents. Currently, he is a senior principal systems engineer at Raytheon Space and Airborne Systems, El Segundo, California.

Index

Ac. *See* Alternating current
Adder, 236, 262
Additive white Gaussian noise (AWGN), 45, 54, 100, 103, 148, 151, 156
See also Noise Gaussian white
AE. *See* Antenna, element
AFR. *See* Amplitude-frequency response
AGC. *See* Automatic gain control
AJ. *See* Anti-jam
ALC. *See* Automatic level control
Aliasing, 169, 174–176, 240, 299
Alternating current, 13, 20, 23
AM. *See* Modulation and demodulation, amplitude; Analog memory
AMB. *See* Analog and mixed-signal back-end
AMF. *See* Analog and mixed-signal front-end
Amplitude-frequency response, 28, 35–38, 87–95, 20, 136–144, 175–192, 203–228, 288–294
Analog and mixed-signal back-end, 30, 53, 78–79, 108, 215, 220–221
Analog and mixed-signal front-end, 30, 53, 127–147, 158–159, 212, 234–240
Analog decoding, 4, 39, 163, 209
Analog memory, 257–263
Analog-to-digital converter (A/D), 4–5, 114–120, 126–128, 140–147, 199, 235, 249–252
   generalized successive-approximation, 250–252
   multi-iteration, 251
   multistage, 251–252
   pipelined, 114, 251–252
   sigma-delta, 114, 117, 224, 230, 252–254
   subranging, 251–252
   successive-approximation, 250–251

A/D. *See* Analog-to-digital converter
   superconductor, 126,
   See also Quantizer; Quantization
Antenna
   array, 3, 49–50, 62–68, 76–77, 236–238
   beam, 61–66, 243
   community, 59
   coupling, 78–78, 108, 122, 145, 158, 244
   direction finding (DF), 65–67, 238
   directional, 49, 55, 61–68
   element (AE), 66–67, 236–244
   loop, 61, 66
   null, 61, 65–66
   pattern, 65–66, 237–238
   rotating, 61–62, 65–67
   satellite TV, 60
   small, 60–62, 66, 77
   subarray, 236–238
   transceiver, 125–126, 158
   transmitter (Tx), 3, 29–30, 44–59, 77–78, 122–125, 234–245,
   velocity vector, 66–67, 239–241
   virtual, 66–67, 238–244
   virtual antenna motion (VAM), 238–244
   whip, 66
Antialiasing filter. *See* Filter, antialiasing
Anti-jam, 58, 71, 124, 234
Application-specific integrated circuit, 50, 77, 114, 124–125
Approximation theory, 165
AQ-DBPSK. *See* Modulation and demodulation, alternating quadratures DBPSK
AQr. *See* Quantizer, auxiliary
ASIC. *See* Application-specific integrated circuit
ASK. *See* Amplitude-shift keying
AtC. *See* Attenuator control circuit

Attenuator control circuit (AtC), 236–237
Automatic gain control (AGC), 116, 120–123, 126, 145, 158, 287
Automatic level control, 78
AWGN. *See* Additive white Gaussian noise

BA. *See* Buffer amplifier
Bandpass filter. *See* Filter, bandpass
Bandwidth-efficient modulation. *See* Modulation and demodulation, bandwidth-efficient
Bandwidth-efficient signal. *See* Signal, bandwidth-efficient
Basis, 16–17, 19, 23, 40, 228, 281, 300, 303
BAW. *See* Filter, bulk acoustic wave
Beamforming, 236, 238, 243–244
BER. *See* Bit error rate
Bit error rate, 130
BPF. *See* Filter, bandpass
BFSK. *See* Modulation and demodulation, binary FSK
BPSK. *See* Modulation and demodulation, binary PSK
Broadcast systems. *See* Radio systems, broadcast
Buffer amplifier, 87, 89, 92, 200–201, 204, 207, 251, 258, 265

Cable, 45–46, 59–61
CDF. *See* Cumulative distribution function
CDM. *See* Code-division multiplexing
CDMA. *See* Code-division multiple access
CDP. *See* Central digital processor
Central digital processor, 239, 242
Central limit theorem, 8, 137, 295–297
Channel decoding, 30, 43, 53–54, 57–58, 124
Channel encoding, 30, 43, 52–53, 55, 57, 78–79, 100, 267
Channel mismatch
  compensation, 224–226
  estimation, 225–226
  mitigation, 222–224
Chayka, 62, 67
CI. *See* Controlled inverter
CNF. *See* Conjunctive normal form
Code-division multiple access (CDMA), 51, 53, 58, 64, 79

Code-division multiplexing (CDM), 58
Coded OFDM (COFDM), 60
Code transformer (CT), 221
COFDM. *See* Coded OFDM
Communication channel, 37, 45, 48, 50, 52, 57–59, 71, 124, 131, 164
Communication systems. *See* Radio systems, communication
Companding, 52, 255
Complex envelope, 30, 33, 53
Complex-valued equivalent, 53–54, 89–96, 140–142, 145–148, 187–189, 301
Compression factor, 273–274
Compressive quantization, 264–275
Compressive sampling, 166, 275
Conjunctive normal form, 82, 84–86
Controlled inverter, 259, 262–263
Convolution, 28, 172, 181–182, 223, 229, 280–281, 285, 288
Convolution property, 23, 28, 172
Correlation (autocorrelation) function, 12, 25–29, 255, 295
Correlation coefficient, 10–11
Correlation interval, 37, 35, 37, 137, 271
Covariance, 10, 238
Covariance (autocovariance) function, 11–12, 25, 27
CR. *See* Radio, cognitive
Crest factor, 55, 57–58, 77–79, 101–103, 107–109, 148–151, 157
Cross-correlation, 55, 150
Cross-correlation function, 12–13, 27–28, 131
CT. *See* Code transformer
Cumulative distribution function (CDF), 7–9, 11

D/A. *See* Digital-to-analog converter
D&R. *See* Digitization and reconstruction
DBPSK. *See* Modulation and demodulation, differential BPSK
Dc. *See* Direct current
DCA. *See* Digitally controlled amplifier
DDS. *See* Direct digital synthesizer
Decryption, 30, 52
Delta function, 4, 21, 171, 177–179, 191, 222, 279, 281, 282–284
Delta modulation. *See* Modulation and demodulation, delta

Demodulation
   coherent, 23, 54, 56, 102, 148–151, 157
   frequency-invariant, 103, 151–157
   hard-decision, 149, 152
   noncoherent, 54, 101–103, 148–152, 156–157
   soft-decision, 57
Demultiplexer (Dmx), 144, 214, 219
DF. *See* Radio systems, direction finding
DFC. *See* Digital functional converter
DFI. *See* Digital filter-interpolator
Differential decoder, 149–152, 154
Differential approach, 66
Differential encoder, 102–107, 152
Differential modulation. *See* Modulation and demodulation
Differential quantization. *See* Quantization; Quantizer
Digital filter-interpolator (DFI), 240, 242–243
Digital functional converter (DFC), 80–81, 85, 109
Digital image processor (DIP), 266–269
Digital multipole switch (DMS), 266–269
Digital part of digitization circuit (DPD), 54, 124, 147
Digital part of reconstruction circuit (DPR), 53, 79, 89–91, 94, 96, 99–100
Digital signal processing (DSP), 7, 59, 66–70, 75–77, 86–89, 114–117, 157, 183–185, 228
Digital signal processor (DSP), 4–5, 49–51, 77, 114, 257–261, 262
Digital-to-analog converter (D/A), 4–5, 80–82, 89–92, 96–100, 108, 181–189, 205–206, 213–221, 250–254, 262–268
   fast D/A, 267–268
   multibit (MB D/A), 267–269
   multiplying (MD/A), 218, 220, 233
   sigma-delta, 114
   superconductor D/A, 126,
Digital weight function generator (DWFG), 218
Digitally controlled amplifier (DCA), 218, 220–221, 233
Digitization and reconstruction
   general, 1, 5
   in radio systems, 43–72, 163, 167, 181
   in Rxs, 51–52, 55, 61, 114–117, 145, 290, 293
   in Txs, 51–52, 55, 61, 75–79, 86–100, 290, 293
Digitization
   baseband signals, 86–89
   bandpass, 1, 34–36, 114–117, 142–147
   baseband of bandpass signals, 1, 61, 116, 140–142, 140–142, 145–146, 188
DIP. *See* Digital image processor
Direct current (dc), 13, 19–24, 27, 29, 95–99, 133–134, 145–147, 193, 205, 280–283
Direct digital synthesizer (DDS), 75, 79–85, 109, 233
Direct sequence (DS), 54, 57–58, 63, 100–102, 129–131, 148–150
Discontinuity, 20, 38, 249, 264–268, 270–273, 285
Disjunctive normal form (DNF), 82–86
Dithering, 256
DMS. *See* Digital multipole switch
Dmx. *See* Demultiplexer
DNF. *See* Disjunctive normal form.
Doppler effect, 15, 23, 66–67, 157, 238–242
Double-sideband suppressed-carrier amplitude modulation (DSB-SC AM), 23, 29
Double-sideband full-carrier amplitude modulation (DSB-FC AM), 23, 29–30
Double-sideband reduced-carrier (DSB-RC AM), 29
Downlink, 59
Downsampling, 87, 89, 109, 115, 117, 140, 142, 185, 188, 252, 287–294
DPD circuit. *See* Digital part of digitization circuit
DPR circuit. *See* Digital part of reconstruction circuit
DQPSK. *See* Modulation and demodulation, differential QPSK
DS. *See* Direct sequence
DSP. *See* Digital signal processing; Digital signal processor
DWFG. *See* Digital weight function generator

Dynamic range,
  in-band, 126
  out-of-band, 126,
  receiver (Rx), 113–117, 119–124, 126–140, 147, 204–205, 212, 219, 233–238
  single-tone, 127
  two-tone, 116, 127–128, 136–140
  transmitter (Tx), 76, 96–99, 221

ECS. *See* Electronic cyclic switch
Effective number of bits (ENOB), 235, 250
Electronic cyclic switch, 66, 239–240, 242–243
Encryption, 30, 49, 52, 77–78, 108
Energy-efficient modulation. *See* Modulation and demodulation, energy-efficient
Energy-efficient signals. *See* Signals, energy-efficient
Energy signal. *See* Signal, energy
Energy spectral density, 25–27
Enhanced Loran (eLoran), 62
ENOB. *See* Effective number of bits
Environment, 47–48, 53, 61, 68, 70, 76, 115, 133, 136
Error vector magnitude (EVM), 78
ESD. *See* Spectral density, energy
EVM. *See* Error vector magnitude
EW Rx. *See* Receiver (Rx), electronic warfare
EW system. *See* Radio systems, electronic warfare
Extremely high frequency, 46, 69
Extremely low frequency, 46, 48

FDM. *See* Frequency-division multiplexing
FDMA. *See* Frequency-division multiple access
FC. *See* Format converter
Field programmable gate array (FPGA), 50, 77, 114, 125
Filter
  analog, 38, 88, 97, 105, 120, 142, 192, 194, 234
  analog interpolating, 4–5, 37–39, 77–100, 178–195, 205–208, 213–215, 219–221, 227–228
  See also Interpolation, analog; analog bandpass; analog baseband
  antialiasing, 4–5, 37–39, 54, 87–92, 117–132, 140–147, 175–189, 200–209, 216–219, 223, 227–228
  bandpass (BPF), 37–38, 92–100, 142–147, 189–193, 204–214, 221, 244
  bulk acoustic wave (BAW), 96, 99, 145, 147, 193, 204
  ceramic, 96, 99, 145, 147, 193
  complex-valued, 89, 140–143, 193
  crystal, 96, 99, 116–117, 145, 147, 193, 204
  digital, 102, 114, 120, 184, 224, 226
  digital decimating, 87, 89, 121, 140–144, 185, 287–294,
  digital interpolating, 79, 89–93, 102, 185–187, 288–294
  See also Interpolation, digital
  electromechanical, 96, 99, 145, 147, 193
  finite impulse response (FIR), 87, 115–117, 132, 140–144, 186, 208–209, 227–233, 287–292
  half-band (HBF), 87–92, 140–144, 186, 291–292
  Gaussian, 102,
  highpass (HPF), 37
  infinite impulse response (IIR), 87, 131, 287
  Kalman, 63
  linear, 54, 130–132
  lowpass (LPF), 34–38, 80–82, 87–96, 140–146, 178–189, 221, 287–293
  matched, 54, 57, 118, 124, 130–131, 152, 231–233
  Nyquist, 60, 290–293
  optimal, 69, 180, 195, 235, 295
  passband, 37, 87–92, 123–128, 132–147, 179–193, 208, 218, 227–230, 234, 290
  real-valued, 89, 140–143, 193
  raised cosine, 102, 157, 191
  root raised cosine, 102, 105, 149, 152, 156–157
  surface acoustic wave (SAW), 96, 99, 145, 147, 193, 204
  symbol-shaping, 78–79, 102–107, 156–157

Index   317

stopband, 37, 123, 142–147, 179–193, 208–209, 218–220, 227–234, 285, 290
transition band, 37, 87–99, 140–147, 179–193, 227, 290
FM. *See* Modulation and demodulation, frequency
Format converter, 102–103, 105
Fourier series
　generalized, 1, 17–18, 172, 300, 303
　trigonometric and complex exponential, 1, 18–21, 25, 132, 166, 284
Fourier transform, 1, 21–28, 33, 37, 170, 172, 209, 281–285, 300
FPGA. *See* Field programmable gate array
Frame (in data transmission), 54, 124
Frame (in function space), 228, 285
Frame (in image processing), 264–265, 269–274, 276
Frequency-division multiple access (FDMA), 51, 58, 64, 79
Frequency-division multiplexing (FDM), 29, 58, 165
Frequency-hopping, 54
Frequency-shifting, 23
Frequency synthesizer, 53, 78–79, 122, 124–125
FSK. *See* Modulation and demodulation, frequency-shift keying

Gain
　AMB, 98, 215, 221
　AMF, 115–119, 123–131, 147, 158–159, 201–202, 212, 235
　amplifier, 97–98, 115, 220–221, 251–252, 258
　antenna, 44–45, 65, 237–238
　channel, 222–225
　coding, 57, 71
　control. *See* Automatic gain control (AGC)
　modulation, 56, 58
　processing, 57–58, 71, 102, 131
Generalized successive-approximation A/D. *See* Analog-to-digital converter, generalized successive-approximation
Generalized successive-approximation quantizer. *See* Quantizer, generalized successive-approximation

General-purpose processor (GPP), 77, 124–125
Global navigation satellite systems (GNSSs), 46, 62–65, 67
　See also Global positioning system; Radio systems, navigation, positioning, and geolocation
Global positioning system (GPS), 62–65
　See also Global navigation satellite systems
Gaussian (normal) distribution, 8–12, 295–297
GMSK. *See* Modulation and demodulation, Gaussian MSK
GNSS. *See* Global navigation satellite system
GPP. *See* General-purpose processor
GPS. *See* Global positioning system
GPU. *See* Graphics processing unit
Graphics processing unit (GPU), 77

HBF filter. *See* Filter, half-band
HPF filter. *See* Filter, highpass

Image compression, 264, 266
IMP. *See* Intermodulation product
Impulse response, 28, 35, 37, 87, 178–179, 182–183, 191, 200
Intercept point (IP), 127, 140
Interchannel interference, 38
Interfering signal, 87–88, 122–145, 158, 184–185, 188, 222, 234–241, 244–245
Intermodulation product, 93–99, 113–116, 127–139, 145–147, 193, 202–207, 226, 235
Interpolation
　analog, 4–5, 37–39, 227–228
　analog bandpass, 90–96, 99–100, 192–193, 205–207, 212–215, 219–221
　analog baseband, 89–90, 96, 185–187
　See also Filter, analog interpolating
　digital, 79, 89–93, 102, 185–187, 288–294
　See also Filter, digital interpolating
　theory, 165–166
Intersymbol interference, 38, 102, 149, 152, 290
IS. *See* Interfering signal

ISI. *See* Intersymbol interference

Joint-processing quantizer (JPQ), 256–264
JPQ, *See* Joint-processing quantizer

Least significant bit (LSB), 256–259, 262
Linear time-invariant (LTI), 1, 18–19, 28, 164
Log-Gaussian distribution, 137
Long-distance link, 180–181
Loran-C, 62, 67
Lower sideband (LSB), 23–24
LPF. *See* Filter, lowpass
LSB. *See* Lower sideband; Least significant bit
LTI. *See* Linear time-invariant

Master frequency standard (MFS), 53, 78–79, 121, 124–125
Mathematical (theoretical) model
  IS statistical, 136–138
    Rx signal path, 132, 136–137
    signals and circuits, 2–4, 16,
    stochastic, 6–13, 295–297
    THA, 200–202
    thermal noise, 118
MB D/A. *See* Digital-to-analog converter, multibit
MD/A. *See* Digital-to-analog converter, multiplying
MDS. *See* Minimum detectable signal
Memory register (MR), 265, 267–268
MEMS. *See* Microelectromechanical system
MFS. *See* Master frequency standard
Microelectromechanical system (MEMS), 50, 53, 79
MIMO. *See* Multiple-input multiple-output
Minimum detectable signal (MDS), 127–128, 138–140
Mobile phones, 46, 51, 65
Mode
  active, 265
  duplex, 48, 125, 158
  extreme, 238
  full duplex, 125–126
  half-duplex, 48, 125, 158
  idle, 265
  saturation, 57, 100, 108
  simplex, 48–49,
  standard mode, 237, 238
Modulation and demodulation
  alternating quadratures DBPSK (AQ-DBPSK), 100, 102–107, 151–157
  amplitude (AM), 23–24, 29–30, 46, 59
  analog, 29–30
  angle, 29–30
  bandwidth-efficient, 55–57, 107–108
  See also Signal, bandwidth-efficient
  binary FSK (BFSK), 56, 100–101
  binary PSK (BPSK), 55–56, 58, 100–102, 107, 148, 151, 157
  delta, 252, 255
  differential BPSK (DBPSK), 100–103, 148–152
  differential PCM (DPCM), 252, 255, 266–267, 270–271, 273
  differential QPSK (DQPSK), 101, 148, 152
  digital, 31, 56
  energy-efficient, 55–57, 100–103
  See also Signal, energy efficient
  frequency (FM), 29, 46, 59–60, 62
  frequency-shift keying (FSK), 30–31, 55–56
  Gaussian MSK (GMSK), 51, 151
  generalized, 52–55, 57, 71, 124
  minimum shift keying (MSK), 101, 151
  phase (PM), 29, 102–103, 107
  phase-shift keying (PSK), 30–31, 51, 55–58, 78
  pulse-amplitude (PAM), 56, 58, 173, 181
  pulse-code (PCM), 49, 167, 181, 183, 249, 252–254, 264–267, 270
  pulse-position (PPM), 55, 181
  pulse-width (PWM)
  quadrature amplitude (QAM), 30, 51, 56–58, 60, 78–79
  quadrature PSK (QPSK), 51, 56, 60, 101–102, 148, 151
  sigma-delta, 108, 114, 252
  single-sideband (SSB), 29
  vestigial sideband (VSB), 29, 60
Modulo-two counter (MTC), 103, 105

Moment, 9–10, 12, 37, 39
Moment function, 11–12, 39
Most significant bit (MSB), 80–82, 251, 257
MR. *See* Memory register
MSB. *See* Most significant bit
MSK. *See* Modulation and demodulation, minimum shift keying
MTC. *See* Modulo-two counter
Multi-iteration A/D. *See* Analog-to-digital converter, multi-iteration
Multi-iteration quantizer. *See* Quantizer, multi-iteration
Multipath, 45, 53, 55, 57–58, 60, 114
Multiple-input multiple-output (MIMO), 50, 68
Multiplexer (Mx), 209, 215, 226, 265, 269
Multistage A/D. *See* Analog-to-digital converter, multistage
Multistage quantizer. *See* Quantizer, multistage
Mx. *See* Multiplexer

Narrowband interference or interfering signals (ISs), 53, 69, 128–131
Navigation systems. *See* Radio systems, navigation
NDC. *See* Novel digitization circuit
NF. *See* Noise figure
Noise factor, 117–119, 158
Noise figure (NF), 117–119
Noise Gaussian white, 45, 118, 156
See also Additive white Gaussian noise (AWGN)
Noise Gaussian nonwhite, 54, 130–131
Noise non-Gaussian, 54, 119
Nonlinear distortion, 120–124, 179, 185–189, 193, 195, 200, 202, 234, 299
Nonlinearity, 97, 116–117, 124–140, 158, 201
Nonlinear product, 65, 132–133
Novel digitization circuit (NDC), 215–216, 218–220, 222–224, 226–233, 235
Novel reconstruction circuit (NRC), 215, 219–222, 226–227, 229, 233
NRC. *See* Novel reconstruction circuit
Nulling, 127, 238, 243–244,
Nyquist zone, 94–96, 206, 234, 287, 293

OFDM. *See* Orthogonal frequency-division multiplexing
OFDMA. *See* Orthogonal frequency-division multiple access
Optimal demodulation, 49, 54, 103, 129–131, 295
Optimal filtering, 69, 180, 234, 295
Optimal sampling and interpolation, 167, 172–173, 177, 180, 194–195
Optimal sampling rate, 93–94, 114, 117, 127, 135, 137, 144, 193,
Orthogonal expansions, 16–18,
Orthogonal frequency-division multiple access (OFDMA), 51, 58
Orthogonal frequency-division multiplexing (OFDM), 49, 60

PA. *See* Power amplifier
PAM. *See* Modulation and demodulation, pulse-amplitude modulation
Parseval's formula, 300
Parseval's identity, 17, 25, 300
PCM. *See* Modulation and demodulation, pulse-code modulation
PFR. *See* Phase-frequency response
Phase accumulator, 80–81
Phase-frequency response, 28, 87, 89, 140, 143, 175, 179, 209, 287
Phase-locked loop (PLL), 79
PLL. *See* Phase-locked loop,
PM. *See* Modulation and demodulation, phase
PN. *See* Pseudonoise or pseudorandom
Positioning and geolocation. *See* Radio systems, positioning and geolocation
Power amplifier (PA), 75–79, 97–98, 107–108, 221
PPM. *See* Modulation and demodulation, pulse-position
Power signal. *See* Signal, power
Preselector, 114, 121–123, 126–128, 295–297
Probability, 6–13, 56, 115, 137–138
Probability distribution, 6–11, 39, 295–297
PS. *See* Pulse shaper
PSD. *See* Spectral density, power
Pseudonoise or pseudorandom (PN), 54, 102, 148–150

PSK. *See* Modulation and demodulation, phase-shift keying
Pulse
  Gaussian, 39, 281
  triangular, 280–281, 285
  rectangular, 21–22, 92, 206, 279–281, 285
  gating, 89–90, 92, 179, 186, 206
  short, 173, 177, 179–181
Pulse shaper (PS), 89–92, 96, 99, 186, 206
PWM. *See* Modulation and demodulation, pulse-width

QPSK. *See* Modulation and demodulation, quadrature PSK
Quantization
  compressive, 264–275
  differential, 252, 266–268
  logarithmic, 255
  nonuniform, 52, 255
  predictive, 52, 255–256, 264–269
  uniform, 119, 249
  See also Analog-to-digital converter (A/D); Quantizer
Quantization noise, 119, 152–159, 252–259
Quantization step, 86, 119, 128, 249, 255, 257–265
Quantizer, 5, 87–88, 199, 209, 215–218, 235, 249–259, 262–273
  auxiliary (AQr), 259, 261–262
  composite, 249–252
  compressive, 264–275
  differential, 252, 266–268
  fast (FQr), 265–272
  multibit (MBQr), 265–272
  multi-iteration, 251
  multistage, 251–252
  pipelined, 114, 251–252
  predictive, 52, 255–256, 264–269
  sigma-delta, 252–254
  subranging, 251–252
  successive-approximation, 250–251
  See also Analog-to-digital converter (A/D); Quantization

Radar. *See* Radio systems, radar
Radio
  bands, 45–48
  broadcast, 43–44, 46–49, 59–63
  See also Radio systems, broadcast
  channel, 1, 29, 48, 53
  communications, 44, 46, 49
  See also Radio systems, communication
  cognitive (CR), 75–77, 108, 117, 124, 128, 136, 235
  digital, 50–53, 75–79, 113–117, 145, 163, 167–169, 183–184, 193–195, 199–205, 287
  high frequency (HF), 45, 48
  link, 48–49
  multipurpose and multistandard, 31, 51, 75–77, 108
  software defined (SDR), 75–77, 108, 117, 124, 136
  transmission, 44
  ultrawideband, 44, 168
  waves, 43–44, 47–48, 65, 67–68
Radio frequency (RF) spectrum, 45–48
Radio systems, 43–72,
  broadcast, 43–44, 46–49, 59–63
  See also Radio, broadcast; Satellite broadcast communications, 44–60, 125
  See also Radio, communications; Satellite communications
  direction finding (DF), 65–68
  electronic warfare (EW), 44, 49, 65, 68–70, 236
  navigation, positioning, and geolocation, 43–47, 49–50, 61–68
  See also Global navigation satellite systems
  radar, 37–38, 43–49, 61, 65–70, 116,
Random events, 6–7
  certain, 6–7
  dependent, 7
  impossible, 6–7
  independent, 7
  mutually exclusive, 6
  mutually nonexclusive, 7
Random processes, 6, 11–13, 295
See Stochastic processes
Random variables, 6–12
  discrete (digital), 7–8, 10
  continuous (analog), 7–10
  independent, 9, 11
  multidimensional, 9–10

Reconstruction
  general,
  bandpass,
  baseband,
Reciprocal mixing, 117, 120–121, 126
RF spectrum. *See* Radio
    frequency spectrum
RC. *See* Reconstruction circuit
RDP. *See* Receiver digital portion
Read-only memory (ROM), 81
Receiver (Rx)
  analog, 121
  broadcast, 43–44, 59–61, 116
  communication, 43–44, 51–55, 61, 68,
      113–159
  digital, 30, 34, 45, 51–55, 60–61,
      113–159
  electronic warfare (EW), 44, 70
  navigation, 44, 62, 67–68, 236
  radar, 68–70
Receiver digital portion (RDP), 30–34,
    53–54, 70, 76–7, 113–116, 120–157,
    185–189, 234–242
Reconstruction circuit (RC), 31, 52, 54, 124
ROM. *See* Read-only memory

S&I. *See* Sampling and interpolation
Sample-and-hold amplifier (SHA),
    114–116, 199–204, 207, 214, 234,
    267–268
Sampler, 4–5, 87, 120, 128, 174, 199–200,
    250, 257
Sampling
  bandpass, 115, 190–193, 204–209, 213,
      222–224, 244
  baseband, 132, 184, 187–190, 224, 290
  nonuniform, 168
  uniform, 86–87, 281
Sampling and interpolation (S&I), 37, 163–
    194, 199–244, 299–304
Sampling theorem
  constructive nature, 163, 166, 169, 172–
      173, 183
  direct interpretation, 178–180, 183–184,
      191
  for bandpass signals, 163, 167, 187–194,
      300–304
  for baseband signals, 163, 167, 169–187,
      299–300

  function, 18, 165, 169–172, 179–182,
      299–303
  hybrid interpretation, 179, 181–185, 187,
      189, 193–194
  indirect interpretation, 178–180, 183–
      189, 191–193
  nonuniform, 167–168
  uniform, 163, 166, 169, 187
  Whittakers-Kotelnikov-Shannon (WKS),
      166, 169, 179, 228
Sampling theory, 165–169
Satellite broadcast, 46, 60
See also Radio system, broadcast
Satellite communications, 45–47, 60, 125
See also Radio systems, communication
Satellite navigation systems, 46, 62–65,
    67
See also Radio systems, navigation,
    positioning, and geolocation
SAW filter. *See* Filter, surface acoustic wave
Selectivity, 117, 120–123, 126, 136
Sensitivity
  quantizer, 240, 250, 253–254, 256
  Rx, 63, 117–121, 126–128, 238
SDR. *See* Radio, software defined radio
SHA. *See* Sample-and-hold amplifier
SHAWI. *See* SHA with
    weighted integration
SHA with weighted integration (SHAWI),
    207–215, 234
Sigma-delta A/D. *See* Analog-to-digital
    converter, sigma-delta
Sigma-delta D/A. *See* Digital-to-analog
    converter, Sigma-delta modulation,
    See Modulation and demodulation,
    sigma-delta
Sigma-delta quantizer. *See* Quantizer,
    sigma-delta
Signals
  analog, 1–5, 37, 54, 78, 97, 123–124,
      173–174
  antipodal, 55, 100
  bandpass, 29–38, 89–99, 115, 141–145,
      167–170, 187–193, 199–204, 293,
      300–304
  bandwidth-efficient, 56–57, 77–78,
      107–108
  See also Modulation and demodulation,
      bandwidth-efficient

Signals *(Continued)*
  baseband, 29–39, 86–99, 141–145, 169–187, 199–205, 287–293, 299–303
  biorthogonal, 55–56, 101
  complex-valued, 18, 25–34, 53–54, 89–96, 105–109, 140–155, 187, 293, 301
  constellation, 56–58, 78–79, 103–107
  deterministic, 1, 5–6, 25, 27–28, 39, 226
  dimensionality, 58–59, 255, 296, 304
  digital, 1–5, 53–54, 76, 115–116, 185, 287–288
  discrete-time (sampled), 1–4, 29, 87–99, 140–145, 167–176, 187, 205, 222–223
  energy, 3–4, 25, 27, 35, 37
  energy-efficient, 77–78, 100–102, 107–108, 148–155
  See also Modulation and demodulation, energy-efficient
  orthogonal, 55–56, 101
  power, 3–4, 25–27
  random (stochastic), 1, 5–7, 25–28, 137, 165
  real-valued, 20–35, 53, 89–96, 140–143, 191, 299–303
  simplex, 55–56
Signal energy equation, 3, 25
Signal power equation, 3, 25–27, 119
Signal spectral sidelobes, 57, 100, 101, 105, 148
Source coding, 51–52, 54, 71, 254–255, 276
Sparsity, 168, 249, 275
Specialized processing unit (SPU), 77, 126
Spectral density
  cross-, 27
  energy (ESD), 25–27
  power (PSD), 26–28, 54, 118–123, 128–131, 137, 158–159, 253, 295
  sampling function, 170, 190
  signal, 21–23, 39, 280–285
Spectral sidelobes, 27, 101, 105, 148
SPU. *See* Specialized processing unit
Spurious outputs, 117, 120–121, 126
Spurious responses, 99, 116, 120–122, 147
Spurious signals, 121–122

Stochastic function, 6, 137, 165
Stochastic(random) processes or signals, 6, 11–13, 295–297
  ergodic, 11–13, 27–28
  Gaussian, 12, 39, 295–297
  log-Gaussian, 137
  locally stationary 13, 295
  nonergodic, 13
  stationary, 11–13
  strictly stationary, 11–13, 27–28
  wide-sense stationary, 12
  stationary, 11–13, 295, 297
Subranging A/D. *See* Analog-to-digital converter, subranging
Subranging quantizer. *See* Quantizer, subranging
Successive-approximation A/D. *See* Analog-to-digital converter, successive-approximation
Surface acoustic wave, 96, 99, 145, 147, 193, 204
Synchronization, 54–55, 124, 156–157

TDM. *See* Time-division multiplexing
TDMA. *See* Time-division multiple access
TDP, *See* Transmitter digital portion
Television (TV), 48, 59–61, 68
Terrestrial over-the-air broadcast, 60
THA. *See* Track-and-hold amplifier
Time-division multiple access (TDMA), 51, 55, 79
Time-division multiplexing (TDM), 29, 164–165, 167, 169, 180–181
Time-frequency scaling, 22–23, 37, 240
Time-interleaved SHAWIs, 208–212, 215
Time shifting, 14–15, 22–23
Time-shift invariant, 11
Track-and-hold amplifier (THA), 4, 87, 115, 135, 200–204, 207, 251–252
Transceiver, 125–126
Transmitter (Tx)
  analog, 30
  broadcast, 43–44, 59–61
  communication, 43, 51–55, 72–109
  digital, 13, 40, 45, 51–55, 72–109
Transmitter digital portion (TDP), 30–33, 53, 75–79, 86–89, 100–101, 107–109, 114–157, 184–189

Transmitter drive (exciter), 77–78, 98
Trellis coding, 57, 78
TV. *See* Television
Tx. *See* Transmitter

UCA. *See* Uniform circular array
ULA. *See* Uniform linear array
Uniform circular array (UCA), 66
Uniform linear array (ULA), 236–237, 239–241
Uplink, 59
Upper sideband (USB), 23–24
Upsampling, 79, 87, 89–92, 109, 185–187, 240, 287–293
USB. *See* Upper sideband

VAM. *See* Antenna, virtual antenna motion
Variable-gain amplifier (VGA), 97
Variance, 10, 12, 134, 260
VCA. *See* Voltage-controlled amplifier
VCO. *See* Voltage-controlled oscillator
Very high frequency omnidirectional radio range (VOR), 61–62
VGA. *See* Variable-gain amplifier
Voltage-controlled amplifier (VCA), 218

Voltage-controlled oscillator (VCO), 79, 98–99
VOR. *See* Very high frequency omnidirectional radio range
VSB. *See* Modulation and demodulation, vestigial sideband

Walsh spectral coefficients, 256, 258–259, 262, 276
Walsh (or Walsh-Hadamard) functions, 55, 101
Waveform, 1–22, 63, 82
Weight function generator, 85, 207, 209, 214, 231, 233–234
Weight function spectral sidelobes, 212, 217, 229
Weighted integration, 115–116, 207, 285
Weighted pulse shaper (WPS), 212–215, 219
WFG. *See* Weight function generator
Window
  hopping, 256–258, 262–262
  sliding, 256–257, 261–264
WKS. *See* Sampling theorem, Whittakers-Kotelnikov-Shannon
WPS. *See* Weighted pulse shaper

Recent Titles in the Artech House Signal Processing Library

Complex and Hypercomplex Analytic Signals: Theory and Applications, Stefan L. Hahn and Kajetana M. Snopek

Computer Speech Technology, Robert D. Rodman

Digital Signal Processing and Statistical Classification, George J. Miao and Mark A. Clements

Handbook of Neural Networks for Speech Processing, Shigeru Katagiri, editor

Hilbert Transforms in Signal Processing, Stefan L. Hahn

Introduction to Direction-of-Arrival Estimation, Zhizhang Chen, Gopal Gokeda, and Yi-qiang Yu, Editors

Phase and Phase-Difference Modulation in Digital Communications, Yuri Okunev

Signal Digitization and Reconstruction in Digital Radios, Yefim S. Poberezhskiy and Gennady Y. Poberezhskiy

Signal Processing in Digital Communications, George J. Miao

Signal Processing Fundamentals and Applications for Communications and Sensing Systems, John Minkoff

Signals, Oscillations, and Waves: A Modern Approach, David Vakman

Statistical Signal Characterization, Herbert L. Hirsch

Statistical Signal Characterization Algorithms and Analysis Programs, Herbert L. Hirsch

Voice Recognition, Richard L. Klevans and Robert D. Rodman

For further information on these and other Artech House titles, including previously considered out-of-print books now available through our In-Print-Forever® (IPF®) program, contact:

Artech House
685 Canton Street
Norwood, MA 02062
Phone: 781-769-9750
Fax: 781-769-6334
e-mail: artech@artechhouse.com

Artech House
16 Sussex Street
London SW1V 4RW UK
Phone: +44 (0)20 7596-8750
Fax: +44 (0)20 7630-0166
e-mail: artech-uk@artechhouse.com

Find us on the World Wide Web at: www.artechhouse.com